METHODS OF
SATELLITE OCEANOGRAPHY

Published under the auspices of the

SCRIPPS INSTITUTION OF OCEANOGRAPHY

University of California, San Diego

SCRIPPS STUDIES IN EARTH AND OCEAN SCIENCES

General Editors

Gustaf Arrhenius
Charles S. Cox
Richard H. Rosenblatt

METHODS OF

SATELLITE OCEANOGRAPHY

Robert H. Stewart

Scripps Institution of Oceanography
University of California, San Diego
and
Jet Propulsion Laboratory
California Institute of Technology

UNIVERSITY OF CALIFORNIA PRESS

Berkeley   Los Angeles   London

University of California Press
Berkeley and Los Angeles, California

University of California Press, Ltd.
London, England

Library of Congress Cataloging in Publication Data
Stewart, Robert H.
    Methods of satellite oceanography.

    Bibliography: p.
    Includes index.
    1. Astronautics in oceanography. I. Title.
GC10.4.A8S73   1984   551.46'0028   83-18017
ISBN 0–520–04226–3

Printed in the United States of America

1 2 3 4 5 6 7 8 9

CONTENTS

PREFACE

Within the past decade, many electromagnetic techniques for the study of the earth and planets have been applied to the study of the ocean. These methods, described in a varied and scattered body of literature, are mostly diffusing into oceanography from the outside and are being put to use with increasing rapidity to study the seas.

In this book I wish to (a) elucidate the principles of the interaction of electromagnetic radiation with the atmosphere and ocean, over the range of frequencies from radio through light; (b) describe the techniques for exploiting these interactions to study the ocean; (c) mention the types of instruments and satellite systems now being used; and (d) examine the accuracy and usefulness of the measurements. In doing so, I have paid particular attention to experiments that compare surface measurements with remote observations, rather than to theoretical studies. While satellite techniques are emphasized, I have not excluded shore and aircraft based measurements, as they are often identical to those from space, differing only in being more convenient. Neither instruments nor the processing of signals from them have been described in depth. Rather, I have only outlined these subjects in sufficient detail to allow the user to understand those aspects that influence the usefulness or resolution of the data.

Nor have I described in depth particular contributions to oceanography, although I have mentioned instances in which satellite data have benefited ocean studies. Satellites are only platforms, and it makes no more sense to write about satellite oceanography than it does to discuss shipboard oceanography. Both platforms carry useful instruments; both contribute to oceanography, but advances in our knowledge of the oceans stem from a blending of theory with observations from a variety of sources, and are rightfully the subject of oceanographic texts.

The exposition will assume some familiarity with physics, calculus, and the theory of Fourier transforms, but will provide more of a physical description of processes than a rigorous mathematical statement of the techniques. As such, the book is for scientists with varied backgrounds interested in applying satellite techniques to their work.

The notation used in each chapter is mostly that of common usage, but this notation can cause confusion. The material comes from many fields, including physical oceanography, meteorology, planetology, astrophysics, optics, and radio science. As a result, the same or similar words, terms, and notation sometimes refer to very different concepts. To minimize the problem, I have defined terms where appropriate, and have given units and a few numerical values when possible.

In referring to the literature, I have preferred to cite easily available material, especially that which clearly explains the ideas being considered. Hence I have avoided the grey literature of contractor reports, conference proceedings, and technical memoranda, because most are nearly impossible to find. Nor have I traced ideas back to their original sources, although I hope I have not slighted those who first developed the techniques described here.

Many colleagues have helped in the preparation of this book, but special thanks must go to Walter Munk and William Nierenberg; without their support and encouragement this book would not have been possible. It is also my pleasure to thank those who have made helpful comments. Among them are W. Alpers, K. Baker, R. Bernstein, F. Carsey, M. Chahine, B. Douglas, B. Farmer, K. Kelly, J. Marsh, E. Njoku, C. Paulson, R. Stevenson, Y. Yasuda, and C. Wunsch. I am grateful to Ruth Zdvorak for drawing the figures and Elaine Blackmore for typing, correcting, and typesetting the manuscript. This work was generously supported by the Office of Naval Research, by the Graduate Department of Scripps Institution of Oceanography, and by NASA through a contract to the Jet Propulsion Laboratory of the California Institute of Technology.

R. H. S.

1

INTRODUCTION

For centuries, oceanography has depended on sparse measurements made to great depths from a few ships plodding their way through the sea. Now satellites promise nearly total coverage of the world's oceans using only a few days of observations. But how relevant are these surface observations to today's oceanographic problems? How well matched to the oceanic time and space scales are the satellite samples? What oceanic processes can be observed, and in what way? How accurate are the observations? And finally, what instruments and systems exist to provide oceanic data from space?

Answers to these questions are scattered through a diverse body of literature. By and large, the development and deployment of instruments suitable for observing the oceans from space has been done outside the traditional academic oceanographic community. Radio astronomers, seeking to study extraterrestrial phenomena, first needed to understand the propagation of electromagnetic radiation through the atmosphere. Planetologists, scattering radio waves from the Moon, Mars, Venus, and Mercury using transmitters on Earth and on interplanetary spacecraft, demonstrated the practical ways in which such radiation can probe rough surfaces. Radio scientists, studying the propagation of radio signals, often turned their techniques around and studied the phenomena that influenced the propagation, thus converting communications noise into geophysical signal. And meteorologists, carefully analyzing the infrared signals observed by their satellites, showed they could map the thermal features at the sea surface. All have noticed the varied ways electromagnetic radiation interacts with matter, and all have exploited this interaction to their benefit. In some cases, their studies have been of the general properties of scatter and emission from rough surfaces, and they have catalogued a surprising number of ways that the sea surface might be observed. In other cases, their studies have led to the design and construction of specific oceanographic instruments. To date, these have been shown to be able to measure surface winds, waves, temperature, currents, phytoplankton, fish schools, salinity, internal waves, sea ice, and perhaps even bathymetry.

[1]

Because many of these new, ocean-observing techniques are often special examples of more general techniques developed to observe other related phenomena, it is natural that they were first described in publications outside the oceanographic literature. This fact, however, makes more difficult an assessment of the state of development of a particular technique—say, that of microwave radiometry—and oceanographers tend to be best informed of particular applications of interest to them, such as microwave measurements of sea-surface temperature, rather than of the general theory of radiometry through absorbing and scattering atmospheres. Thus, new users of a technique may encounter problems if they are not aware of important assumptions leading to the interpretation of data from a particular instrument. These difficulties are even more acute in satellite oceanography, where the rapid development of the field has ensured that many techniques are either poorly documented or else not yet described outside of technical reports.

The rapid development of the field is due to a number of influences, but certainly the design and launch of several new satellites for observing the sea—notably Seasat, Geos-3, Nimbus-7, and Tiros-13—are paramount, as is the development of new techniques for extracting oceanographic information from sensors on other satellites. A notable example of the latter is the processing of infrared observations made by meteorologic satellites in order to show thermal patterns at the sea surface. The application of these new satellite data to the study of oceanic problems has now begun, and is expected to become more important in the future.

1.1 Why remote-sensing of the oceans?

Before going into the details of satellite oceanography, let us first answer the question, why would anyone wish to study the sea from space, or even from aircraft? Three reasons are immediately apparent:
(a) to obtain a global picture of the oceans in order to study basin-wide phenomena;
(b) to observe regions not easily studied by ship, for example, the Southern Ocean around Antarctica in winter;
(c) to make measurements that are either impossible or difficult by ordinary means, such as observing oceanic rainfall or the distribution of small waves on the sea surface.

Global or basin-wide observations are particularly appropriate to two newly emerging areas of ocean studies: a) climate and the interaction of the ocean and the atmosphere over periods of months to years, and b) the study of ocean dynamics and the role of planetary or Rossby waves and mesoscale eddies.

The climate studies seek estimates of the quantities of heat stored in the surface layers of the ocean and the rate at which the heat is exchanged with the atmosphere or advected to other regions by currents. Measurements of surface wind, sea temperature, and the distribution of clouds are particularly important. For example, Namais (1978) has shown that the distribution of anomalous sea surface temperature in the North Pacific influences the position of oceanic

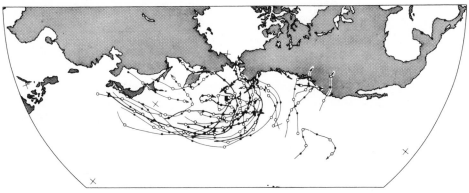

Storm Tracks – January 1977

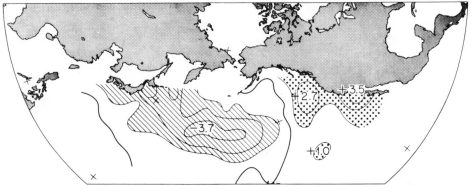

Sea Surface Temperature – Winter 1977

Figure 1.1 The influence of sea surface temperature on storms. Upper: tracks of the centers of major sea-level cyclones over the North Pacific (from Mariner's Weather Log, 1977). Lower: departure of sea surface temperature from seasonal mean for December, January, and February in °F. Shaded areas are in excess of 1°F (from Namais, 1978).

storms (fig. 1.1), which, in turn, influence the weather over North America; and that a very unusual distribution of sea surface temperature in the Pacific could have contributed to the abnormally cold winter of 1976–77 on the east coast of North America. To investigate these phenomena quantitatively, meteorologists are calculating the general circulation of the atmosphere using numerical models that attempt to include the influence of the ocean. But to be successful, these computations require, as boundary conditions, timely observations of such variables as sea surface temperature from entire ocean basins.

Studies of ocean dynamics require even more detailed observations. Recently, physical oceanographers have begun to realize the importance of fluctuating currents in ocean dynamics, and have shifted their attention away from a description of mean ocean motion, a description that was the center of their interest for over a century. This change in emphasis results from a realization, developed in the last ten years, that fluctuating velocities contain far

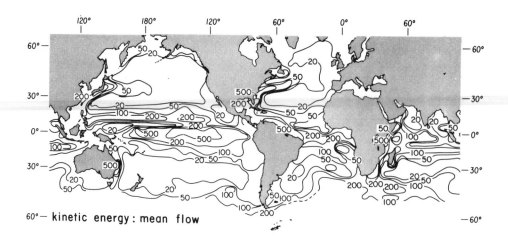

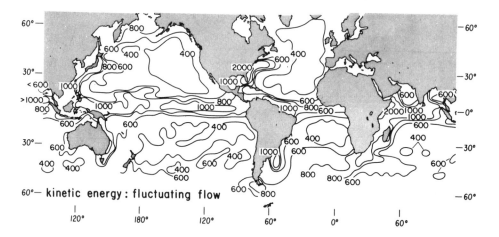

Figure 1.2 Kinetic energy per unit mass (cm²/sec²) of sea surface currents averaged over 5° squares. Upper: kinetic energy of the mean flow. Lower: kinetic energy of fluctuating currents (from Wyrtki, Magaard, and Hager, 1976).

more kinetic energy than does the mean motion (fig. 1.2) and that variability cannot be ignored. Now many are trying to classify the types of wavelike and turbulent motions, to assess their role in ocean dynamics, to estimate the importance of the various instabilities of the mean motion, and to calculate the interactions among wave fields. Included in these motions are both baroclinic and barotropic Rossby waves (planetary waves), tides, and mesoscale eddies. Again, understanding and parameterizing these motions requires repeated observations of large areas of the ocean under many different conditions, with spatial resolution on order of tens of kilometers (Duing, 1978).

Despite their known importance, global studies of climate and ocean dynamics and of most other oceanic phenomena are hindered by lack of data. The studies rely on a few observations, which traditionally have come from routine observations made by ships of commerce and by warships, because the world's

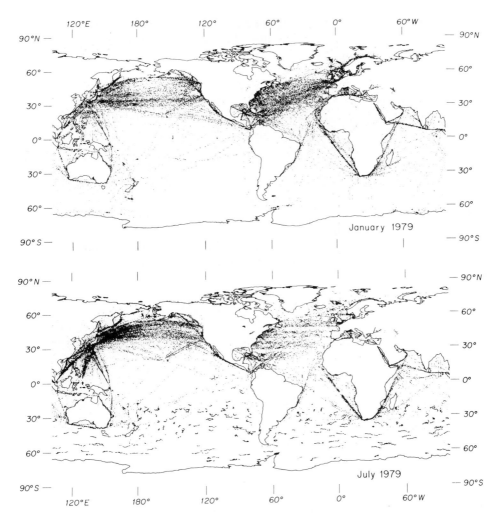

Figure 1.3   Monthly distribution of surface observations made by ships and buoys and received by the NOAA Pacific Marine Environmental Group in Monterey, California (plot by Douglas McClain). Note the increased observations in the Southern Ocean in July 1979 produced by drifting buoys deployed for the Global Weather Experiment, and the incorrect data that appear over land.

oceanographic fleet is much too small to engage in global studies. This limitation leads to large gaps in coverage, particularly in the southern hemisphere, but also in the west-central Pacific (fig. 1.3). The paucity of data is immediately apparent when we note (NASA, 1981) that the present network for collecting oceanographical information is comparable to a meteorological network manned by university scientists who make observations for a few days at a time over a small area, occasionally supplemented by information from a few ambitious groups that drive across the country, stopping every few hundred miles to measure barometric pressure and make a radiosonde observation.

Clearly, reliable systematic and global observations of even the most rudimentary kinds, made by unskilled observers, are not yet available. Quantitative, accurate measurements of ocean-surface currents, plankton, wave height, or air-sea temperature difference are even rarer and more difficult to obtain. Provided the time and space scales are appropriate, satellite observations promise to fill these gaps.

Finally, while the global studies attract attention, they frequently require parametric representation of the smallest scales of the sea surface, for these scales govern the interchange of momentum, heat, and water between the ocean and the atmosphere. What is the nature of the transition between these two large reservoirs of fluid? What is the role of small wavelets and surface films? The former roughen the boundary and perturb the airflow; the second damp out wavelets and inhibit the exchange of mass. Studies designed to help answer these questions are difficult to conduct from ships, particularly in storms; yet exchanges are largest when the wind is strongest. For example, the exchange of energy varies as wind speed cubed, and a typhoon wind blowing at 80 meters per second ($m.s^{-1}$) for a day transfers as much energy to the sea as does an 8 $m.s^{-1}$ trade wind blowing for nearly three years.

Fortunately, the scatter of centimeter wavelength radiowaves, and to some extent the emission of radio noise from the surface, depend on the small scale roughness at the sea surface, and can be used conveniently to study these phenomena. In fact, most of what is known about ocean wavelets, their relationship to the wind, their distribution as a function of phase of long waves, and their response to currents results from observations of microwaves scattered from the sea. Even much longer waves, those with dekameter wavelength, can be studied fruitfully. In this instance, LORAN-A radio signals of 160 m wavelength have provided the most accurate measurements of the directional distribution of 80 m, 7-s waves in the deep ocean, and of their response to the wind.

1.2 General works on the subject

No monograph treats the particular subject of satellite oceanography, although a number of books on remote sensing contain chapters or sections on the observation of the ocean, and several oceanographic journals have published special issues dedicated to particular aspects of the subject. Included here is a list of more recent popular books and special journal volumes on the subject, together with a few comments on their general usefulness. Of course, many particular subjects relating to remote sensing of the ocean are covered in various textbooks, and these will be referred to in the pertinent chapters.

Barrett, E. C. 1976. *Introduction to environmental remote sensing.* London: Curtis, 336 pp.

  A general introduction to the topic for undergraduates, but it contains no detailed information on particular techniques or instruments.

———— 1974. *Climatology from satellites.* London: Methuen, 418 pp.

  This is a book on climatology emphasizing data obtained from satellites, and it provides a fine description of the uses of data from meteorological satellites. It tends to accept satellite data as correct, and does not delve deeply into analyses of accuracy of data.

Derr, V. E., ed. 1972. *Remote sensing of the troposphere.* Washington: U.S. Government Printing Office.

This is another fine book, including chapters on remote sensing of the ocean; its only fault is being slightly out of date.

Ewing, G. C., ed. 1965. *Oceanography from space.* Woods Hole: Woods Hole Oceanographic Institution.

This pioneering collection of papers, most of which look into the future, defines what is or could be possible as inferred from aircraft studies of the sea. It is remarkable that almost all present techniques were discussed in this book.

Gower, J. F. R., ed. 1981. *Oceanography from space.* New York: Plenum Press.

This wide ranging collection of papers, marking the 16th anniversary of the first meeting on this topic, provides a good summary of the field as of mid-1980.

Pouquet, Jean. 1974. *Earth sciences in the age of satellites.* Holland: D. Reidel, 169 pp.

A slim book with interesting figures and tables and many references, mostly devoted to examples of observing land, but not neglecting the atmosphere and the ocean.

Reeves, Robert G., ed. 1975. *Manual of remote sensing.* Falls Church VA: American Society of Photogrammetry, 2123 pp.

An excellent and comprehensive manual, including much detailed information on remote sensing of the ocean.

Schanda, Erwin, ed. 1976. *Remote sensing for environmental sciences.* Berlin: Springer-Verlag, 367 pp.

Another fine book, including two excellent chapters on radio observations of the sea, but the optical chapters are less useful.

Slater, P. N. 1980. *Remote sensing, optics and optical systems.* Reading: Addison-Wesley, 575 pp.

This book provides a good description of optical instruments used in remote sensing, and together with the volumes by Ulaby, Moore, and Fung, provides a firm foundation for the technology of satellite remote sensing.

Swain, P. H., and Davis, S. M., eds. 1978. *Remote sensing: The quantitative approach.* New York: McGraw-Hill, 396 pp.

Detailed but simple descriptions of remote-sensing techniques, using visible and infrared light with emphasis on observations of land and crops, are contained in this useful book.

Ulaby, F. T., Moore, R. K., and Fung, A. K. *Microwave remote sensing: Active and passive.*
Vol. 1. 1981. *Microwave remote sensing fundamentals and radiometry.* 456 pp.
Vol. 2. 1982. *Radar remote sensing and surface scattering and emission theory.* 608 pp.
Vol. 3. *Volume scattering and emission theory, advanced systems, and applications.*
Reading: Addison-Wesley. Forthcoming

These books provide a good description of radio frequency and foundations of radio remote sensing.

The journals that have issued special numbers or volumes devoted to various aspects of satellite oceanography include:

IEEE Journal of Oceanic Engineering, 1977, **OE2(1)**, 1–159, (C. T. Swift, ed.), Radio oceanography.

Boundary-Layer Meteorology, 1978, **13**, 1–435, (R. H. Stewart, ed.), Radio oceanography.

Journal of Geophysical Research, 1979, **84(B8)**, 3779–4079, Scientific results of the Geos-3 mission.

Science, 1979, **204**(4400), 1405–1424, Initial evaluation of the performance of Seasat.

Boundary-Layer Meteorology, 1980, **18(1-3)**, 1–358 (J. F. R. Gower, ed.), Passive radiometry.

IEEE Journal of Oceanic Engineering, 1980, **OE5(2)**, 71–180, (D. E. Weissman, ed.), Seasat sensors.

Journal of the Astronautical Sciences, 1980, **28**(4), 313–428, Analysis of the Seasat ephemeris.

Journal of Geophysical Research, 1982, **87**(C5), 3173–3438, (R. L. Bernstein, ed.), Seasat geophysical evaluation.

Journal of Geophysical Research, 1983, **88**(C3), 1529–1952, (A. D. Kirwan, T. J. Ahrens, and G. H. Born, eds.), Scientific results of Seasat.

2

THE NATURE OF THE SEA SURFACE AND THE AIR ABOVE

An understanding of the region about the sea surface, the atmospheric and surface boundary layers, ocean waves, and surface currents is important to our discussions for many reasons. Satellites observe only the sea surface and not deeper layers; waves strongly influence the propagation of radiation at the surface; and the oceanic variables of interest, such as surface current velocity or wind speed, must be explicitly defined. But most importantly, the region is interesting in itself. It is dynamically very active, it influences weather and climate, it is the domain of commerce and fisheries, and, more and more it is the workshop and playground for millions who live near the sea.

Wind blowing over the sea at first ruffles the surface and then begins to form small waves. Stronger, more persistent winds blowing over greater distances generate larger waves; and strong winds blowing for days over hundreds of kilometers generate large waves typical of storms. The extraction of energy and momentum from the air not only builds waves; it also slows the wind immediately above the sea surface, creating the atmospheric boundary layer, and drags along the surface layers of the sea, producing surface currents (fig. 2.1). With the passage of time, days, and seasons, the working of the wind, the heating by the sun, evaporation, and cooling all combine to produce the basin-wide flow of ocean currents. But the process doesn't end there; the currents redistribute Earth's heat, influence the atmosphere, and contribute to the generation of winds that eventually stir the oceans.

The next few pages provide only the barest outline of some of these processes plus a few examples useful for discussing the techniques of satellite oceanography. A more complete treatment of the subject can be found in the standard textbooks, a few of which are referred to in the appropriate places.

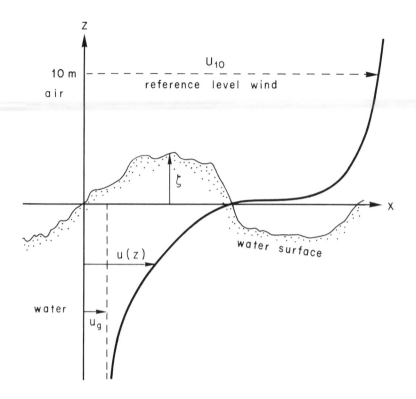

Figure 2.1 Sketch of the atmospheric and surface boundary layers, showing the variation of the velocity $u(z)$ of both wind and current in the vicinity of the wavy sea surface. At distances of a few hundred meters away from the surface the flow blends into the geostrophic flow u_g of the atmosphere and the ocean. Note that this sketch really only applies to the zone within a few meters of the sea surface. Above and below this zone the direction of the flow rotates in direction, the Ekman turning.

2.1 Airflow over the sea

Airflow over the sea is turbulent, and molecular processes dominate only in the millimeter or so above and below the surface. The first few meters above the sea are characterized by loss of momentum from the air and a slow increase in velocity with height. At higher levels, the flow is influenced more by changes in density induced by differences in temperature between the air and the water, and by Earth's rotation. Ultimately, at greater heights, the influence of the surface is weakened, and the flow is governed by the general circulation of the atmosphere. For our discussion, only the lowest layers are important, those within 20 m of the surface.

Our knowledge of the atmospheric and surface boundary layers is based largely upon empirical correlations whose mathematical formulation was guided by judicious use of dimensional analysis. The process began years ago with attempts to describe the airflow over flat and curved surfaces: the study of aerodynamics, useful for design of wings and airplanes. Wind-tunnel studies of airflow showed that some flows, such as those over a flat plate, could be described by a few simple "laws" or correlations, and a perusal of such textbooks as Hinze (1975) and Schlichting (1968) provides a foundation for under-

standing part of what follows. Later, the studies spilled out into fields and bays, where the same correlations were observed to be useful. This work culminated in a number of very careful measurements from towers and the stable Floating Instrument Platform (FLIP), producing the description of the airflow over the sea that is sketched out here. More information, and a firmer theoretical basis for the discussions, can be found in textbooks such as those by Kraus (1972) or Kitaigorodskii (1973), and practical implementations are discussed in Brown and Liu (1982) and Kondo (1975).

In the region between 1 cm and 10 m above the surface, the atmosphere is dominated by the influence of the surface, and is described by three variables of fundamental importance: the fluxes of horizontal momentum, sensible heat and latent heat across the sea surface (see Leavitt and Paulson, 1975). Assuming that the wind is constant and horizontally homogeneous, the fluxes are defined by the correlations between the vertical component of the turbulent wind and the horizontal wind, temperature, and water vapor, respectively, all measured within a few meters of the surface:

$$\tau = \overline{\rho u w} \approx \rho \; \overline{u w} = \rho u_*^2$$

$$H = \overline{c_p \rho \; w t} \approx \rho \; c_p \; \overline{w t}$$

$$E = L_E \; \overline{w q}$$

where the overbar indicates a time-averaged mean quantity, usually made over an hour's time, and where the notation is that noted in table 2.1. These correlations are all very difficult to measure, and they are often expressed by empiri-

Table 2.1   Notation for Describing Airflow Over the Sea

C_D	Drag coefficient
C_E	Moisture transfer coefficient
C_H	Sensible heat transfer coefficient
c_p	Specific heat capacity of air [1030 J.kg^{-1}.K^{-1}]
E	Flux of latent heat [W.m^{-2}]
g	Acceleration due to gravity [9.8 m.s^{-2}]
H	Flux of sensible heat [W.m^{-2}]
L	Monin-Obukhov scaling length [m]
L_E	Specific heat of evaporation [2.4 × 10^6 J.kg^{-1}]
q	Water vapor density [kg.m^{-3}]
Q_a	Mean water-vapor density at reference height [kg.m^{-3}]
Q_s	Mean water-vapor density close to the sea surface [kg.m^{-3}]
t	Air temperature [K]
T_a	Mean air temperature at the reference level [K]
T_s	Mean sea-surface temperature [K]
u_*	Friction velocity [m.s^{-1}]
u , w	Horizontal and vertical components of the wind [m.s^{-1}]
$U(z)$	Mean horizontal wind [m.s^{-1}]
U_{10}	Mean wind at a reference height of 10 m above the mean sea level [m.s^{-1}]
z	Height above mean sea level [m]
z_0	Roughness height [m]
κ	Karmen's constant [0.4]
ρ	Air density [1.15 kg.m^{-3}]
τ	Flux of momentum through the sea surface (wind stress) [Pa]

cal functions relating them to more easily measured variables such as wind speed and air-sea temperature differences. These "bulk coefficient" relations are:

$$\tau = \rho \ C_D \ U_{10}^2$$

$$H = \rho \ c_p \ C_H \ U_{10} \ (T_s - T_a)$$

$$E = C_E \ L_E \ U_{10} \ (Q_s - Q_a)$$

Note that although the fluxes are nearly independent of height in the region near the surface, the bulk relations depend on wind speed, a quantity that depends on elevation. Thus measurements are usually related to a standard height, almost always 10 m. In some specialized satellite literature this height is taken to be 19.5 m, but this figure results from a historical accident: a few early weather ships had anemometers at 19.5 m, and rather than convert their observations to wind at a 10 or even 20 m reference level, the data were simply tabulated and the level noted. In the following discussion, we will be particularly interested in the drag coefficient C_D and the variation in wind speed with height. Other relationships among C_D, C_H, and L_E and their relationship with wind speed are discusssed by Kondo (1975).

The wind speed close to the surface varies strongly with height, and the shape of the velocity profile depends on both wind stress and the stability of the air column. Cold air blowing over warm water is heated from below and tends to convect. The air column becomes unstable, and the transfer of heat and momentum no longer depends solely on the turbulence generated by wind shear, but it is augmented with the turbulence generated by the temperature instability. In contrast, turbulence is reduced when warm air blows over cold water.

Buoyancy tends to dominate at some considerable height above the water, and shear always dominates close to the surface. The level at which the two are roughly comparable is given by the Monin-Obukhov scaling length:

$$L = - \frac{u_*^3 \ T_a \ \rho \ c_p}{\kappa \ g \ H}$$

$$L \approx - \frac{(C_D)^{1/2} \ T_a \ U_{10}^2}{\kappa \ g \ (T_s - T_a)}$$

The second equation follows from the first through the use of the bulk coefficient equations assuming $C_D = C_H$; and both assume that the influence of temperature dominates that of humidity. This is generally true at midlatitudes, but the influence of humidity may need to be considered in tropical regions and during times of rapid evaporation from the surface. A more general form of the scaling length that includes the influence of humidity on stability is given by Pond et al. (1971). The sign convection is such that

$$z/L < 0 \qquad \text{unstable air}$$

$$z/L > 0 \qquad \text{stable air}$$

$$z/L \approx 0 \qquad \text{neutral stability}$$

To obtain a rough estimate of the size of L, consider a very stable (or unstable) boundary layer with:

$$U_{10} = 10 \text{ m/s}$$

$$C_D = 1.1 \times 10^{-3}$$

$$(T_s - T_a) = 10\text{K}$$

$$T_a = 300\text{K}$$

which yield:

$$L = 25 \text{ m}$$

Because the air-sea temperature difference rarely exceeds 10K over the ocean, and because it is almost always less than 2K, we expect L to be greater than 30 m unless the winds are very light. Thus the influence of stability is usually small at those levels near the sea that are of interest to our discussions.

If stability can be neglected, the velocity profile above the sea surface is logarithmic with height; and when stability is considered, the dimensionless gradient of the velocity profile is given by

$$\frac{\kappa z}{u_*} \frac{dU}{dz} = \phi_M$$

$$\phi_M = \text{function } (z/L)$$

Integration of the gradient then gives the profile. For the particular case when ϕ_M is equal to unity, this yields the logarithmic profile. The functional form of ϕ_M is known only empirically, primarily from a series of measurements made over wheat fields in Kansas and reported by Businger et al. (1971). They found (fig. 2.2):

$$\phi_M = (1 - 15 \, z/L)^{-1/4} \qquad z/L < 0$$

$$\phi_M = (1 + 4.7 \, z/L) \qquad z/L > 0$$

but other forms, based on the same data, have been reported (Busch, 1973).

If the influence of stability is weak, ϕ_M can be expanded in a power series to yield:

$$\frac{\kappa z}{u_*} \frac{dU}{dz} = 1 + \beta \, z/L$$

$$\beta = 4.7 \qquad z/L > 0$$

$$\beta = 3.8 \qquad z/L < 0$$

which has the solution

$$U = \frac{u_*}{\kappa} \left[\ln (z/z_0) + \beta \left(\frac{z - z_0}{L} \right) \right]$$

This form of the velocity profile implies that the wind vanishes at the reference height z_0; however, in reality this does not happen. The quantity z_0 is only a

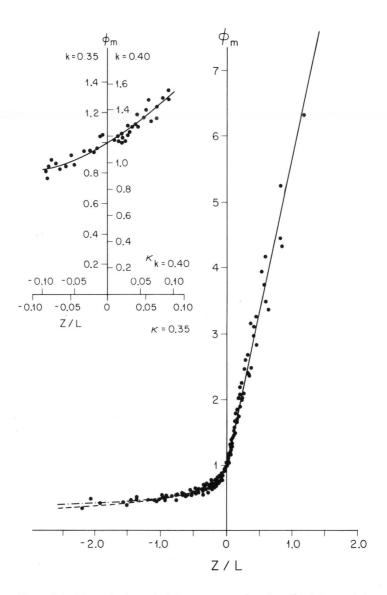

Figure 2.2 Dimensionless wind shear ϕ_m as a function of height z relative to the Monin-Obukhov length L. The inset shows the best fitting value of Karman's constant is 0.35 (from Businger et al., 1971).

mathematical fiction to express the data conveniently. Viscosity, which has been ignored in this discussion, dominates within a centimeter or so of the surface, and the equation is valid only from a few centimeters above the surface up to a level where $z/L \approx 1$.

From the above we see that the velocity profile is determined by three empirical coefficients: β, u_*, and z_0. The first is nearly constant; and the second is fixed by the drag coefficient. The third is less well known and, contrary to

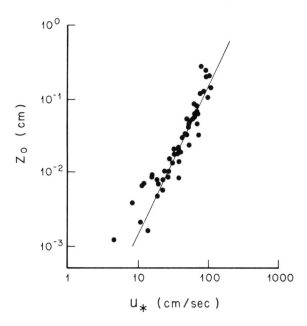

Figure 2.3 Laboratory and field measurements of the roughness height z_0 as a function of friction velocity u_* (from Wu, 1980). The solid line is Charnock's relation $z_0 = 0.0156 \, u_*^2/g$.

boundary layers over solid surfaces, it is not determined by the height of the roughness (wave height). Rather, the boundary layer over waves is influenced by the momentum transfer to the waves (which changes the structure of the turbulence in the boundary layer), and by the existence of a thin surface current. A number of studies of $U(z)$ have been used to determine z_0 empirically. The data have considerable scatter when plotted as a function of wind speed, but Charnock's (1955) relation (fig. 2.3):

$$z_0 = a \, u_*^2 / g$$

$$a = 0.0156$$

not only provides a good fit to the observations (Wu, 1980) but, by relating z_0 to u_*, it also fixes the form of the drag coefficient. To see this, recall that $C_D = (u_* / U_{10})^2$ and use z_0 in the logarithmic profile to obtain:

$$C_D = [\kappa / \ln(gz / a \, u_*^2)]^2$$

Using $z = 10$ m, $a = 0.0156$, and $\kappa = 0.4$, this can be rewritten to yield C_D as a nonlinear function of U_{10} and C_D, which can then be solved numerically to obtain $C_D (U_{10})$.

The drag coefficient and the heat fluxes have been measured many times over the ocean, using many different techniques. Earlier work used indirect methods because of the experimental difficulties of measuring correlations

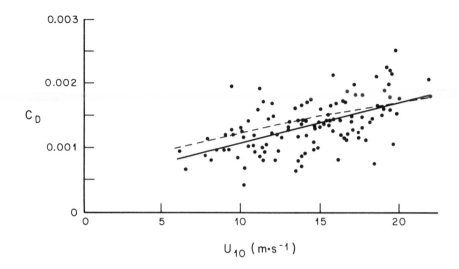

Figure 2.4 Measurements of the drag coefficient C_D as a function of wind speed U_{10} at 10 m above the sea surface. The solid line is $C_D = (0.44 + 0.063\ U_{10}) \times 10^{-3}$, and the dashed line follows from Charnock's equation for z_0 (from Smith, 1980).

between turbulent quantities; and published values showed considerable scatter. Since 1965, however, accurate correlations have been measured over the sea, resulting in improved determinations of C_D. Smith (1980), in particular, has carefully observed wind stress and heat fluxes over a wide range of wind speed and finds that (fig. 2.4)

$$C_D = (0.44 + 0.063\ U_{10}) \times 10^{-3}$$

not only provides a good fit to the measurements of the drag coefficient but also agrees closely with the values obtained from Charnock's equation using $a = 0.01$. Furthermore, his studies indicate that drag is only weakly dependent on wave height and stability.

The principal influence of stability is to shift the velocity profile away from the logarithmic relation, thus requiring that either the velocity measurement or the drag coefficient be corrected in order to use the coefficient to estimate drag. The percentage difference between the logarithmic profile $(\beta = 0)$ and the correct profile, assuming $z \gg z_0$ and $z/L < 1$, is

$$\frac{\Delta U}{U} \approx 100\ \frac{\beta z}{L\ \ln(z/z_0)}\ \%$$

Alternatively, the effect of stability on the drag coefficient can be estimated from Kraus (1972:174) using

$$C_D' = \left(\frac{u_*}{U}\right)^2 = \frac{\kappa^2}{[\ln(z/z_0) + \beta z/L]^2}$$

$$C_D' \approx C_D\ [1 - 2\beta z/[L\ \ln(z/z_0)]]$$

for

$$z/L \ll 1$$

where the prime denotes the value of the drag coefficient corrected for the influence of stability.

In principle, we can now relate a measurement of wind speed made near the surface to wind at nearby levels and to drag on the sea surface. Knowing U at some height z, and the air-sea temperature differences, we determine L, the Monin-Obukhov length, and estimate the size of z/L. If it is negligibly small, the value of U at 10 m can be calculated from the logarithmic velocity profile using Charnock's value of z_0. If z/L is small but not negligible, U is corrected to the value it would have at that level if z/L were zero; then the logarithmic profile is used to find U at 10 m. If z/L is not small the velocity profile can be integrated numerically to estimate U_{10}; but it is preferable to measure U at a lower level where z/L is small, because extrapolations of the velocity profile for large z/L are not reliable.

2.2 The spectrum of ocean waves

The ocean-surface waves generated by the wind range in length from millimeters to hundreds of meters. Those with wavelength less than 2 cm are dominated by surface tension and are called capillary waves. The longer wavelengths, ranging up to the giant storm waves, are dominated by gravity, and are called gravity waves. Much longer waves, with lengths of hundreds of kilometers but with amplitudes of only a few tens of centimeters, are produced by currents, tidal forces, and earthquakes.

The statistical properties of the wind-generated waves vary only slowly in time and space; and the sea surface can be locally described by the three-dimensional Fourier transform $A(\vec{k})$:

$$A[k_x, k_y, f] = \int_x \int_y \int_t \zeta(x,y,t) \exp[2\pi i (k_x x + k_y y - 2\pi f t)] \, dx \, dy \, dt$$

where ζ is surface elevation, k_x and k_y are the horizontal components of the wavenumber vector, t is time, and x, y are horizontal distances (see table 2.2).

The spectrum of the gravity waves is then calculated from:

$$X(\vec{k}, f) \equiv AA^*$$

where (*) is the complex conjugate and the normalization is such that, when the spectrum is integrated over all frequencies and wavenumbers, it yields the variance of sea-surface elevation:

$$\overline{\zeta^2} = \int_k \int_f X(\vec{k}, f) \, d\vec{k} \, df$$

Through using this spectral form, the sea surface is represented by a superposition of waves of all possible wavelengths, $L = 2\pi/k$, and periods, $T = 1/f$, travelling in all possible directions, $\theta = \arctan(k_y/k_x)$. Although this produces

Table 2.2   Notation for Describing Ocean Waves

c	Phase velocity of a wave [m.s^{-1}]
f	Wave frequency [Hz]
f_m	Frequency at the peak in the spectrum [Hz]
$\tilde{f}_m$	Dimensionless frequency at the peak in the spectrum
G	Peak-enhancement parameter
g	Acceleration of gravity [9.8 m.s^{-2}]
$H_{1/3}$	Significant wave height [m]
$\vec{k}$	Wavenumber vector [m^{-1}]
k_x , k_y	Horizontal components of $\vec{k}$ [m^{-1}]
L	Wavelength [m]
t	Time [s]
T	Wave period [s]
U_{10}	Mean wind at a height of 10 m above mean sea level [m.s^{-1}]
$\tilde{x}$	Dimensionless fetch
x , y , z	Downwind (fetch), cross-wind, and vertical coordinate [m]
α	Dimensionless high-frequency parameter
γ	Ratio of surface tension to water density [7.2×10^{-5} m^3.s^{-2}]
$\zeta (x , y , t)$	Sea surface elevation [m]
θ	Direction toward which a wave propagates [deg]
ν	Kinematic viscosity of water [10^{-6} m^2.s^{-1}]
σ	Peak-width parameter
$\Phi (f)$	Frequency spectrum of surface elevation [m^2.s]
$X(\vec{k} , f)$	Wavenumber-frequency spectrum of surface elevation [m^4.s]
$\Psi (\vec{k})$	Wavenumber spectrum of surface elevation [m^4]
$\Psi_0 (f , \theta)$	Directional frequency spectrum [m^2.s^2]

a complete description of the sea surface, the spectrum is not easily measured and we seek simpler representations of the surface.

The first step toward simplification is to assume that the longer waves ($L > 1$ m) obey the dispersion relation applicable to infinitesimal amplitude gravity waves:

$$4\pi^2 f_0^2 = gk$$
$$k = (k_x^2 + k_y^2)^{1/2}$$

This reduces the dimension of the spectrum by one to yield the directional spectrum

$$\Psi (\vec{k})\delta (f - f_0) = X (\vec{k} , f)$$
$$\Psi (k , \theta) \, k \, dk \, d\theta = \Psi_0 (f , \theta) \, f \, df \, d\theta$$

In this form, the sea is represented by a superposition of plane waves of all possible wavelengths or frequencies travelling in all possible directions.

The directional spectrum can be simplified still further by the integration of Ψ_0 over all directions (angles) to yield the frequency spectrum

$$\Phi (f) = f \int_0^{2\pi} \Psi_0 (f , \theta) d\theta$$

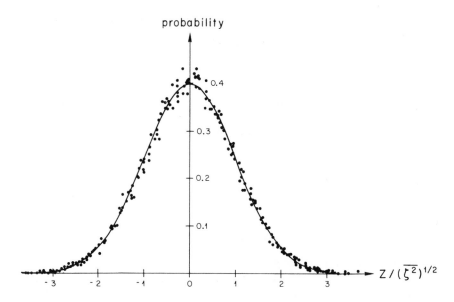

Figure 2.5 Probability distribution of sea surface elevation normalized by the standard deviation $(\overline{\zeta^2})^{1/2}$ of the observations, for significant wave heights of 1.3 to 3.7 m. The observations show little influence of nonlinear processes and are well fit by the Gaussian curve (from Carlson, Richter, and Walden, 1967).

Alternatively, this spectrum can be (and usually is) obtained directly from the Fourier transform of sea-surface elevation measured as a function of time at a point.

The various spectra describing the sea surface are poorly known, the simpler spectra being much better known than the more complete. In fact, $X(\vec{k}, f)$ has rarely been measured. Therefore, in discussing what is known about the waves, it is convenient to consider the simplest statistics first, beginning with the distribution of wave heights, continuing with an example of the spectrum of gravity waves, and then proceeding to a description of the spectral form appropriate for wavelets and capillary waves. A fuller description of the various statistics of the ocean surface, and of the spectra described here, can be found in Phillips (1977).

The distribution of sea-surface elevation relative to mean sea level is fundamental for describing the sea surface. If we assume that the sea surface is represented by a large number of independent sinusoids, then the distribution should be a Gaussian (fig. 2.5); however, the assumption is not precisely true. The various components of the wave spectrum interact weakly, and the non-linearity leads to small deviations from the Gaussian distribution (Longuet-Higgins, 1963), one measure of the deviation being the skewness of the distribution. This is zero for a Gaussian, but for the sea surface, it is on order of the mean-square surface slope. Thus the deviations from a Gaussian curve should be greatest for short, steep waves at the high frequency tail of the spectrum. These ideas are supported by measurements. The surface-elevation distribution of 1 to 3 m high waves is well fit by the Gaussian curve (fig. 2.5), while the distribution of smaller waves reported by Kinsman (1965:344) and the distribution of wave slopes reported by Cox and Munk (1954) are better fit by a slightly modified Gaussian curve, the Gram-Charlier distribution.

Another important statistical parameter describing the sea surface is ocean wave height. Wave height, the distance from wave trough to crest, is easily estimated by anyone looking out to sea, but the concept is not easily described mathematically. Frequently, the significant wave height, $H_{1/3}$, is used to denote the mean of the highest one-third of waves observed. Although this is not very precise, it does seem to correspond to the heights reported by conscientious observers. A more detailed mathematical description of a random sea (Barber and Tucker, 1962; Hoffman and Karst, 1975) makes the concept more precise and leads to a relation between wave height and the standard deviation of surface elevation:

$$H_{1/3} = 4 \ (\overline{\zeta^2})^{1/2}$$

This quantity is both accurately defined and easily measured by observers and by instruments on buoys, ships, and satellites. Because it is easily measured, wave height is the most commonly reported wave statistic, with thousands of observations being reported worldwide each day. These have been thoroughly studied and summarized in various places (see U.S. Navy Marine Climatic Atlas of the World), with the result that wave heights are fairly well known and predictable except in remote regions and in the most severe storms.

Of the various spectra, the frequency spectrum is best known. It is commonly measured with wave staffs or accelerometers mounted on buoys or ships, and its relations to wind speed, duration, and fetch (distance over which the wind blows) are reasonably well known for wavelengths greater than 1 to 2 m. For example, in the idealized case of constant wind blowing away from a lee shore, the frequency spectrum has the somewhat forbidding form (Hasselmann, et al., 1973):

$$\Phi(f:x) = \alpha(x)g^2(2\pi)^{-4}f^{-5} \exp\left[5/4 \ (f_m/f)^4\right] G^{\exp\left[\frac{-(f-f_m)^2}{2\sigma^2 f_m^2}\right]}$$

where the nondimensional variables

$$\tilde{f}_m = 3.5\tilde{x}^{-0.33}$$

$$\alpha = 0.076\tilde{x}^{-0.22}$$

$$G = 3.3$$

$$\sigma = 0.08$$

are based on many observations of waves blowing away from lee shores. The dimensionless fetch $\tilde{x}$ and frequency at the peak in the spectrum $\tilde{f}_m$ are defined by

$$\tilde{f}_m \equiv f_m U_{10}/g$$

$$\tilde{x} \equiv xg/U_{10}^2$$

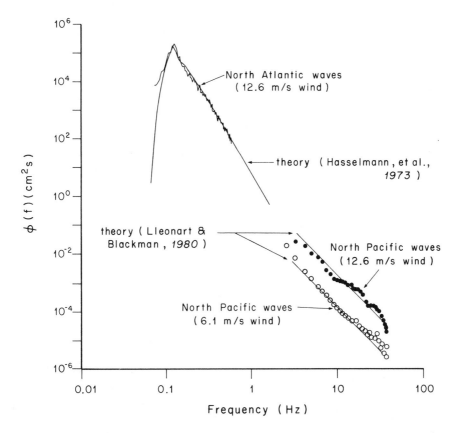

Figure 2.6 Observations of the frequency spectrum of sea surface elevation. Broken line: the spectrum of waves produced by 12.6 m.s⁻¹ winds in the North Atlantic. Points: the high-frequency tail of the spectrum of waves produced by 12.6 and 6.1 m.s⁻¹ winds near Japan (from Kondo, Fujinawa, and Naito, 1973). Upper smooth curve: from the Hasselmann et al. (1973) form of the wave spectrum. Lower smooth curves: from the Lleonart and Blackman (1980) form of the high-frequency tail of the spectrum.

where U_{10} is the mean wind speed 10 m above the sea surface, x is fetch, and g is gravity.

Although this spectrum is strictly valid only for waves going away from a lee shore, it also adequately represents any wave field generated by a steady wind blowing for time t, when x is replaced by

$$x = gt/(4\pi f)$$

The information contained in this form of the spectrum can be made explicit by considering its various terms (fig. 2.6). The variables α and $\tilde{f}_m$ describe the evolution of the spectrum with fetch $\tilde{x}$, and all are made nondimensional by scaling with g and U_{10} according to the ideas of Kitiagordskii. In dimensional form, the dependence on fetch of α, f_m, and the variance of surface elevation is given by:

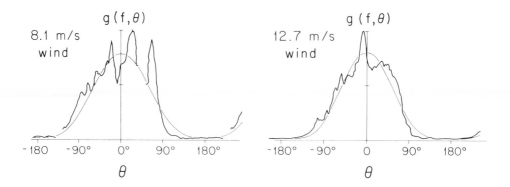

Figure 2.7   Directional distribution of 0.14 Hz waves measured near Wake Island by Tyler et al. (1974) as a function of angle θ relative to the mean wind. The measurements (solid line) have a resolution of 3–5°, and the light curve is the best-fitting function of the form $[\cos^s(\theta/2) + \text{constant}]$.

$$\alpha = 0.076 \, [U^2/(xg)]^{0.22}$$

$$f_m = 3.5[g^2/(Ux)]^{1/3}$$

$$\overline{\zeta^2} = 1.6 \times 10^{-7} \, xU^2/g$$

Thus, for a constant wind, the variance of wave height increases with fetch while the frequency at the peak decreases. For frequencies considerably higher than the peak frequency, the spectrum decreases asymptotically according to:

$$\Phi(f) = \alpha g^2 (2\pi)^{-4} f^{-5}$$

And at low frequencies, the exponential term sharply cuts off the spectrum at a frequency just below the peak frequency. The G and σ terms enhance the peak, and make it narrower and higher than it might otherwise be, but both have minor influences on the shape of the spectrum.

The directional spectrum $\Psi_0(f, \theta)$ is less well known than the frequency spectrum. It has been measured a few times in a few selected places, for $f < 1$ Hz, but its dependences on wind speed, duration, and fetch are yet to be described adequately. A number of measurements by pitch-roll wave buoys and by dekameter radars (Longuet-Higgins, Cartwright, and Smith, 1963; Tyler et al., 1974) suggest it is of the form (fig. 2.7):

$$\Psi_0(f, \theta) = \Phi(f) \, g(\theta, f)$$

$$\int_0^{2\pi} g(\theta, f) d\theta = 1$$

$$g(\theta, f) = \cos^s(\theta/2) \, H(s)$$

$$H(s) = 2\pi^{1/2} \, \Gamma(s/2 + 1/2)/\Gamma(s/2 + 1)$$

where $H(s)$ is a normalizing factor necessary to satisfy the integral of $g(\theta, f)$ over all angles, and $g(\theta, f)$ is the distribution of wave energy as a function of angle.

This form of the spectrum has considerable wave energy going at large angles to the wind, and allows a few waves to go upwind, the amount depending on the exponent s. The exponent is not well known, but does tend to be large at $f = f_m$ and small for $f \gg f_m$. An approximate relation for s that fits the observations reported by Hasselmann, Dunckel, and Ewing (1980) is:

$$s = a \, (f/f_m)^b$$

$$a = 19.5 \qquad\qquad b = -2.3 - 1.45(U_{10}/c - 1.17) \qquad f > 1.05 \, f_m$$

$$a = 14.0 \qquad\qquad b = 4.1 \qquad\qquad\qquad\qquad\qquad\qquad f < 1.05 \, f_m$$

The asymptotic form of the wavenumber spectrum at short wavelengths can be estimated from Hasselmann's equation using the dispersion relation and the Jacobian relating the frequency spectrum to the wavenumber spectrum (Phillips, 1977:106, 144):

$$\Psi_0(f, \theta) = \alpha g^2 \, (2\pi)^{-4} \, f^{-5} \, g(\theta) \, H(s)$$

$$\Psi_0(f, \theta) \, f \, df \, d\theta = \Psi(k, \theta) \, k \, dk \, d\theta$$

$$\Psi(k, \theta) = \alpha \, g(\theta) \, H(s)/(2 \, k^4)$$

These relationships apply only for wavelengths greater than about 1 m. For very short waves or high frequencies, neither the dispersion relation nor the frequency spectrum used in this derivation is correct.

In general, the spectrum of wind-generated gravity waves is distinguished by: (a) a narrow peak; (b) little energy at frequencies lower than the peak; (c) a high-frequency asymptote of f^{-5} or k^{-4}; (d) symmetry about the wind; (e) a narrow beamwidth for waves travelling as fast as the wind; and (f) a wider beamwidth for both faster and slower waves. The model adequately describes waves generated by steady, homogeneous winds; but variations in the wind field, such as the rotating winds in storms, produce a more complex wave field. This can be described by integrating the equations for the transport of wave energy density, including source terms, as described in Hasselmann et al. (1973) and Hasselmann, et al. (1976).

The higher-frequency, shorter-wavelength part of the spectrum is not nearly as well known as the larger, longer-wave part. The dispersion relation of short waves depends on currents and the orbital velocity of long waves, so that frequency and wavenumber spectra are not easily related; and direct measurements of the frequency or wavenumber spectra are difficult using conventional techniques. In fact, only a few measurements have been reported for frequencies above 1 Hz, and the function $\Phi(k)$ is almost unknown at equivalent wavelengths (waves shorter than 1 m). As a result, radar methods described in later chapters are providing the new information on these short waves and wavelets, their relation to wind and surface tension, and their variability as a function of the phase of larger waves.

Cox and Munk (1954) and Hughes, Grant, and Chappell (1977) measured the mean-square slope of ocean waves, which are dominated by the shortest waves, and found:

$$\overline{\zeta_y^2} = 1.9 \times 10^3 \ U_{10}$$
$$\overline{\zeta_x^2} = 3.16 \times 10^{-3} \ U_{10}$$
$$\sigma = \zeta_x^2 + \zeta_y^2 = 5.12 \times 10^{-3} \ U_{10}$$

Thus the slopes in the direction of the wind are larger than cross-wind slopes, and both depend on wind speed.

The high-frequency tail of the wave spectra has been measured a few times in the laboratory and at sea. The various measurements have been summarized by Lleonart and Blackman (1980), who found that the observations are well fit by (fig. 2.6):

$$\Phi(f) = D \ u_*^2 \ (u_* \nu / \gamma)^{1/2} \ f^{-3}$$

with

$$D = 4.12 \times 10^{-3}$$

over the range

$$15 \ \text{Hz} < f < 100 \ \text{Hz}$$
$$0.2 \ \text{cm} < L < 2 \ \text{cm}$$

where u_* is a nearly linear function of wind speed, ν is viscosity, and γ is surface tension. Thus the high-frequency tail of the ocean wave spectrum must depend on the local wind speed, a dependence that has important consequences, as described in chapter 12.

2.3 Surface currents

The momentum and energy transferred to the sea from the atmosphere ultimately produce surface currents and turbulence in the upper layers of the ocean. But the process is not entirely direct. A fraction of the momentum and energy first goes into surface waves, and later is released into currents and turbulence when the waves break. Initially these currents are closely coupled to the wind, but with time they respond to the influence of Earth's rotation and merge into the general circulation of the sea.

The surface boundary layer in the ocean is thought to be a mirror image of the surface boundary layer in the air, with three notable differences. First, water is 900 times denser than air, and u_*, which scales as the square root of the density ratio, is 30 times smaller in the water than in the air, assuming that stress is continuous across the boundary. Second, ocean waves carry momentum, reducing the amount available to produce currents; however, the wave momentum is generally small and transient. Most waves break, and only a small fraction of wave momentum is ultimately carried away from the windy area where the waves are generated. Third, the ocean is very stratified, with cold, dense water lying just below the surface. The change in density inhibits vertical motions, and stirring by the wind produces a thin mixed layer, approximately 100 m thick, superposed on a region of strong stability, the thermocline.

The currents within a few meters of the ocean surface are difficult to measure, and our understanding of the surface motions is based more on analogy with laboratory and atmospheric flows than on experience. Usually it is assumed that stress, friction, and the Earth's rotation influence the flow in a manner first described by Ekman (1905), and so the wind-driven surface currents are often called Ekman flows. Buoyancy appears to be more important than friction, however, and the flow tends not to be that predicted by Ekman theory. About all that can be said is that the surface layers tend to move as a slab in response to the wind, that a thin boundary layer may exist within the upper 5 to 10 m of the surface, but that the dynamics of this thin layer have not been well measured.

Once free of the direct forcing by the wind, the currents tend to move slowly and frictionlessly under the influence of pressure gradients and the Coriolis force; and to a good approximation, we can assume that the two forces are in balance. Such a flow is said to be geostrophic. At the sea surface, the pressure gradient is manifest as a surface slope $(\partial \zeta / \partial x; \partial \zeta / \partial y)$, and the two horizontal components of the geostrophic velocity are

$$u_s = - \frac{g}{f_0} \frac{\partial \zeta}{\partial y}$$

$$v_s = \frac{g}{f_0} \frac{\partial \zeta}{\partial x}$$

where ζ is the sea-surface elevation, x and y are horizontal coordinates, g is gravity, and the Coriolis parameter f_0 is (table 2.3):

$$f_0 = 2\Omega \sin(\theta)$$

where Ω is the rotation rate of Earth and θ is latitude.

Table 2.3 Notation for Describing Geostrophic Currents

D	Vertical scale height of density [m]
f_0	Coriolis parameter [s^{-1}]
g	Gravity [m.s^{-2}]
L	Length scale of horizontal motion [m]
N	Buoyancy frequency [s^{-1}]
R	Radius of Earth [6.4×10^6 m]
T	Time scale of horizontal motion [s]
u_s, v_s	Horizontal components of the surface geostrophic current [m.s^{-1}]
x, y, z	Local Cartesian coordinates [m]
β	Latitudinal variation of Coriolis parameter [s^{-1}.m^{-1}]
$\zeta(x,y)$	Mean sea-surface height relative to a level surface [m]
θ	Latitude [deg]
ρ	Water density [kg.m^{-3}]
Ω	Earth's rotation rate [7.3×10^{-5} rad.s^{-1}]

Two features of the geostrophic flow are worth noting. First, the slopes associated with surface flows are very small. For example, the Gulf Stream has a surface velocity of around 2 m.s^{-1} at 30°N, and this produces a slope of only 74 cm in 100 km. Second, the flow cannot be in strict geostrophic balance, because if it were, the flow could not evolve but would be fixed for all time. Nevertheless, the deviations from geostrophic balance tend to be small, and oceanic flows are usually quasigeostrophic.

The spatial and temporal variability of the quasigeostrophic flow in both the atmosphere and the ocean are important to any system for measuring the flow, and the typical scale of the variability can be coarsely estimated in simple ways (Flatté et al., 1979:3). We assume that the fluid contains waves governed by velocity and planetary rotation, and that these waves are close to instability, that is, the fluid particles move nearly as fast as the wave. Then the length scale is approximately the Rossby radius of deformation (Pedlosky, 1979:336):

$$L = DN_0/f_0$$

where D is a characteristic vertical scale of the buoyancy frequency

$$N = [-g/\rho \; d\rho/dz]^{1/2} \approx N_0 \exp [z/D],$$

and where N_0 is a characteristic buoyancy frequency and ρ is water density. The buoyancy frequency (often called the Brunt-Väisälä frequency) is a measure of the fluid's stability to vertical motion, and is, in an idealized way, the frequency of oscillation of a parcel of fluid displaced vertically from its equilibrium position.

The length scale can be estimated using values typical of the oceanic and atmospheric stability, together with $f = 7.3 \times 10^{-5}$ rad/s, the value of the Coriolis parameter at 30° latitude. For the oceans,

$$N_0 \approx 3 \text{ cycles/hour} = 5 \times 10^{-3} \text{ rad/s}$$

$$D \approx 1\text{km} = 10^3 \text{ m}$$

$$L_{\text{OCEANS}} = 70 \text{ km}$$

For the atmosphere,

$$N_0 \approx 1/60 \text{ rad/s (Eckart, 1960)} = 1.7 \times 10^{-2} \text{ rad/s}$$

$$D \approx 8 \text{ km} = 8 \times 10^3 \text{ m}$$

$$L_{\text{ATMOSPHERE}} = 1900 \text{ km}$$

The time scale is obtained from the approximate period of Rossby waves:

$$T = 2\pi/(\beta L)$$

where β is defined as the local variation of Coriolis frequency f with latitude:

$$f = f_0 + \beta y$$

$$\beta = 2\Omega \; (\cos \theta)/R$$

where y is the distance in the northward direction and R is the radius of Earth. Using L defined above,

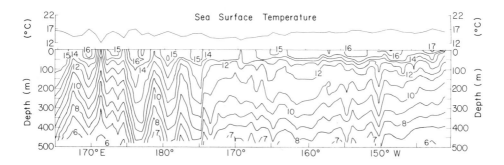

Figure 2.8 Temperature of the surface layer of the Pacific measured along 38°N in 1975 by a freighter of the American President Line, showing the spatial scale of oceanic variability (from Bernstein and White, 1977).

$$T = 2\pi\,R\,\tan\theta\,/\,(N_0 D)$$

Again, using typical values, we obtain

$$T_{\text{OCEAN}} \approx 60 \text{ days}$$

$$T_{\text{ATMOSPHERE}} \approx 2 \text{ days}$$

Thus a satellite measuring every 50 km every few days barely samples adequately in space, but oversamples in time if it is observing the ocean.

A dramatic example of this space scale, measured by observers on a freighter of the American President Line along 38°N using expendable bathythermographs, devices to measure temperature as a function of depth while the ship steams along (fig. 2.8), shows many eddies extending from the surface to deeper than 500 m with typical scales of 200–400 km in the western portion of the section and shorter scales in the eastern portion (Bernstein and White, 1977). The time scales of oceanic motion have been observed by the Mid-Ocean Dynamics Experiment and by satellite. For example, infrared images from the very high resolution radiometer on the NOAA's satellite have shown the slow motion of an eddy moving into California coastal waters with a time scale of about a month (Bernstein et al., 1977). Similar variability has been observed in the northwest Atlantic, where correlation lengths of 100 km and correlation times of 60 days have been reported (Flatté et al., 1979:41).

It is important to remember that these estimates of time and space scales are only crude approximations meant to show that the atmosphere and ocean are very different, and that systems optimized to observe one may not provide useful measurements of the other. The true variability depends on a number of factors, and can differ substantially from the estimates (see Warren and Wunsch, 1981:342 for a much more complete discussion of the variability of oceanic currents). In particular, the oceans have variability with higher frequency and shorter scales in regions of strong flow, such as the Gulf Stream or the Drake Passage, and near the equator where the Coriolis frequency vanishes. Even the flow shown in figure 2.8 has scales in the east that differ substantially from those in the west.

3

ELECTROMAGNETIC RADIATION

Electromagnetic radiation used for remote sensing spans a range of frequencies from a few cycles per hour, used to study the conductivity of the upper lithosphere, to 10^{20} cycles per second, the frequency of gamma rays used to search for radioactive ore bodies or to study the composition of the moon. The total range is a phenomenal 24 orders of magnitude. We will be content to use a more restricted range of nine orders of magnitude, from 10^6 cycles per second to 10^{15} cycles per second, spanning the spectrum from radio waves to ultraviolet light. Very low radio frequencies, X-rays, and gamma rays play little role in satellite observations of the sea.

3.1 The nature of electromagnetic radiation

Electromagnetic radiation consists of alternating electric and magnetic fields, and the radiation is classified by the frequency of the fluctuation, the standard unit of measurement being the number of cycles per second. In early textbooks this was abbreviated cps, but more recently the unit has been renamed the hertz (Hz), after Heinrich Hertz, who proved by direct experiment Maxwell's conjecture that light was electromagnetic radiation.

Thus, regardless of its frequency, all radiation obeys the same set of equations. It can be refracted, diffracted, or polarized, and it is shifted in frequency on reflection from a moving object (Doppler effect). The only difference is the distance or scale over which phenomena operate. If all dimensions are scaled by the wavelength of the radiation, then light cannot be distinguished from radio waves, and textbooks on optics can be used to predict the propagation of radio signals. Of course, at the very shortest wavelengths, quantum effects increase in importance, and short waves interact with matter in different ways than do longer waves.

The dynamics of electric and magnetic fields in the presence of matter is described by Maxwell's equations. Of particular importance to our discussions is the solution for a propagating wave, which may be written (see Born and Wolf, 1970:10ff., 613ff.; Kraus, 1966:134ff.)

[28]

$$E_y = E_0 \, exp \, [-j \, (kx - \omega t)]$$

$$k^2 = \omega^2 \mu \epsilon - j\omega\mu\rho$$

$$\omega = 2\pi f$$

$$\mu = \mu_0 \mu_r$$

$$\epsilon = \epsilon_0 \, \epsilon_r = \epsilon_0 (\epsilon' - j\epsilon'')$$

where the notation used in this section is given in table 3.1. For most materials of interest to remote sensing, the relative permeability is unity ($\mu_r = 1$), and the complex wavenumber k can be written

$$k^2 = \omega^2 \mu_0 \epsilon_0 \, [\epsilon' - j\sigma/(\epsilon_0 \omega)]$$

where the quantity in brackets is the complex dielectric constant

$$\epsilon_r = [\epsilon' - j\sigma/(\epsilon_0 \omega)]$$

In free space, $\epsilon_r = 1$, $\sigma = 0$ and the wave propagates at a velocity

$$c^2 = 1/(\mu_0 \epsilon_0)$$

the usual expression for the velocity of light in a vacuum. In matter, the wave velocity is

$$v = \frac{\omega}{k} = \frac{c}{\sqrt{\epsilon_r}}$$

Table 3.1 Notation for Describing Electromagnetic Radiation

c	velocity of light in free space [3.0×10^8 m.s^{-1}]
d	skin depth [m]
E_x, E_y	components of the electric-field intensity [V.m^{-1}]
f	frequency [Hz]
j	$\sqrt{-1}$
k	complex wavenumber [m^{-1}]
n	complex index of refraction
t	time [s]
u	velocity [m.s^{-1}]
v	velocity of light in the medium [m.s^{-1}]
x, y, z	components of a Cartesian coordinate system [m]
$\Delta\omega$	Doppler shift [Hz]
ϵ	permittivity of the medium [F.m^{-1}]
ϵ', ϵ''	components of the dielectric constant (relative permittivity)
ϵ_0	permittivity of free space [8.85×10^{-12} F.m^{-1}]
ϵ_r	complex dielectric constant
η, χ	components of the index of refraction
κ	attenuation coefficient [m^{-1}]
μ	permeability of the medium [H.m^{-1}]
μ_0	permeability of free space [$4\pi \times 10^{-7}$ H.m^{-1}]
μ_r	relative permeability [≈ 1 for most material]
σ	conductivity of the medium [S.m^{-1}]
ω	frequency [rad.s^{-1}]

and the ratio of this velocity divided into the velocity of light in a vacuum is the index of refraction

$$n = c/v$$

$$n = \sqrt{\epsilon_r}$$

Thus the complex index of refraction is the square root of the dielectric constant, a result of particular importance for the discussion of reflection and emission from surfaces described in chapter 6.

Because the index of refraction is complex, radiation propagating through matter is not only slowed but also attenuated. The attenuation is due to the imaginary component of n, and can be evaluated by writing

$$n = \eta + j\chi$$

and noting that

$$k = \omega n/c$$

$$k = \omega\eta/c + j\omega\chi/c$$

Thus the electric-field intensity can be written

$$E_y = E_0 \exp[-j(\omega\eta\chi/c - \omega t)] \exp[\omega\chi x/c]$$

$$E_y = E_0 \exp[\kappa x] \exp[-j(\omega nx/c - \omega t)]$$

$$\kappa = \omega x/c .$$

Here, κ is an attenuation coefficient and the electric field is reduced to $1/e$ of its original value in a distance d, the skin depth,

$$d = c/(\omega\chi)$$

$$d = c/(2\pi f\chi)$$

If the material is a reasonably good conductor such that $\epsilon' \ll \epsilon''$, then the skin depth is that of a metal:

$$d = 1/\sqrt{2\pi f\mu_0\sigma} \qquad \epsilon' \ll \sigma/(\epsilon_0\omega)$$

Finally, the index of refraction and the skin depth can be expressed in terms of the material properties by taking the square root of the dielectric constant to obtain

$$\eta^2 = \frac{|\epsilon_r| + \epsilon'}{2}$$

$$\chi^2 = \frac{|\epsilon_r| - \epsilon'}{2}$$

$$|\epsilon_r| = [\epsilon'^2 + \sigma^2/(\epsilon_0\omega)^2]^{1/2}$$

To put these equations into context, consider a simple example. At 10 gigahertz (GHz) and 20°C sea water has a dielectric constant of approximately

$$\epsilon_r = 52 - 37\,j$$
$$n = \sqrt{\epsilon_r} = 7.10 - 3.66\,j$$
$$\chi = -3.66$$
$$d = 1.3 \text{ mm}$$

Thus sea water is almost completely opaque to this frequency and acts very much like a perfect conductor.

With this as an introduction to the theory of the propagation of electromagnetic radiation in matter, it is useful to define a few additional terms used to describe the radiation. Two beams of radiation are said to be *coherent* if there exists a systematic relationship between the phases of their electric-field vectors. If this relationship is random, the radiation is *incoherent*. Polarization is defined by the orientation of the electric-field vector. If the vector is confined to a plane, the radiation is *plane polarized*; but in general it is not. For the more general case, consider a wave travelling in the z direction in a Cartesian coordinate system, with electric field components

$$E_x = E_1 \sin(\omega t - kz)$$
$$E_y = E_2 \sin(\omega t - kz + a)$$

where E_1 and E_2 are constants and a is a phase difference, also constant. If $E_1 = E_2$ and $a = \pm 90°$, the wave is *circularly polarized*, and the electric field vector rotates. If $a = -90°$, the wave is right circularly polarized; and if one could view the electric vector of a receding wave, the rotation would be clockwise, the same direction as that of a right-hand screw, hence the term. There exist two conflicting conventions for the direction of polarization, however. The one just described is that accepted by the Institute of Radio Engineers (IRE, 1942). The other, and opposite, definition is that of classical physics; thus descriptions of polarization must be read carefully to determine what convention has been used. Further discussion on the subject can be found in Kraus (1966).

The polarization may also be defined relative to the surface upon which radiation impinges. If linearly polarized radiation has an electric field parallel to the sea surface, it is *horizontally polarized*; if it is perpendicular to the surface it is *vertically polarized*. Of course, this distinction disappears if the *angle of incidence*, the angle between the direction of propagation of the radiation and the perpendicular to the surface, is 0°. In this case, the electric field is always parallel to the surface.

If a beam of radiation strikes the edge of an obstacle, part of the energy will will change direction and *diffract* into the shadow zone behind the obstacle. If the beam enters a region of different velocity its direction can also change, but this process is called *refraction*, the change in angle being predictable from the change in velocity using Snell's Law. If the beam is reflected by a moving target, its frequency is Doppler shifted by an amount

$$\Delta\omega = 2\,\omega u/c$$

where ω is the original frequency, $\Delta\omega$ is the change in frequency, and u is the velocity of the target. The equation assumes that $u \ll c$ and that u is parallel to the direction of propagation of electromagnetic radiation. If u is in the same direction, $\Delta\omega$ is negative; if it is in the opposite direction, $\Delta\omega$ is positive, and the reflected wave has higher frequency than the incident wave.

3.2 Names of radiation bands

The various regions of the electromagnetic spectrum have been explored at different times, by different disciplines, using markedly different apparatus. Naturally this diversity has resulted in a wide range of terms to describe the radiation. Broadly speaking, and as a result of historical development, radiation can be divided into two classes handled in completely different ways. The lowest frequencies, the radio bands, are characterized by the use of coherent sources of radiation, with modulation rates that are a significant fraction of the frequency, received by equipment that observes the phase of the radiation. For example, FM radio signals, with frequency near 100 MHz, are modulated at a frequency of 20 KHz and the phase of the fluctuation is detected by the receiver to reproduce the original information encoded into the signals, the radio broadcast program. The highest frequencies, light beams, are characterized by incoherent sources, or by coherent sources with relatively low rates of modulation, received by sensors that detect the square of the amplitude of the signal, the transmitted power, rather than its phase. The border between these two classes of radiation is fuzzy. It continues to evolve toward higher frequencies, and it is now at frequencies around a terahertz (10^{12} Hz) or greater (Baird, 1983).

The wide spectrum of radio signals is systematically classified into well defined bands, each a factor of 10 wide, with boundaries of 3×10^N Hz, which corresponds to bands of wavelengths with boundaries of 10^{8-N} m. This division, and the adjectives and acronyms associated with the bands, are now widely accepted and understood around the world (table 3.2). For example, the

Table 3.2 Nomenclature of the Radio Frequency Bands

Band	Frequency Range	Metric Subdivision	Adjectival Designation
2	30 to 300 Hz	Megametric waves	ELF Extremely low frequency
3	300 to 300 Hz		VF Voice frequency
4	3 to 30 kHz	Myriametric waves	VLF Very low frequency
5	30 to 300 kHz	Kilometric waves	LF Low frequency
6	300 to 3000 kHz	Hectometric waves	MF Medium frequency
7	3 to 30 MHz	Dekametric waves	HF High frequency
8	30 to 300 MHz	Metric waves	VHF Very high frequency
9	300 to 3000 MHz	Decimetric waves	UHF Ultra high frequency
10	3 to 30 GHz	Centimetric waves	SHF Super high frequency
11	30 to 300 GHz	Millimetric waves	EHF Extremely high frequency
12	300 to 3000 GHz	Decimillimetric waves	— Far Infrared

The bands extend from 0.3×10^N to 3×10^N Hz. The upper limit is included in each band; the lower limit is excluded (from Reference Data for Radio Engineers).

frequencies between 0.3 and 3.0 GHz are known everywhere as Ultra High Frequencies (UHF). This designation is much more precise than the older, catch-all term "microwaves." But despite this widespread usage of the new terms, older ones are slow to die away. One older set of terms in particular causes much confusion: that employed during the Second World War to maintain secrecy in microwave radar research (table 3.3). One still finds frequent reference to X- and L-band radars and to S-band transponders, although the subband designations are less frequently used, except for K_a band.

Table 3.3 Alternate (Letter) Nomenclature of Radio Bands

Band	Frequency Range [Gigahertz]	Band	Frequency Range [Gigahertz]
P	0.225 − 0.390	K	10.90 − 36.00
L	0.390 − 1.550	Q	36.0 − 46.0
S	1.55 − 5.20	V	46.0 − 56.0
C	3.90 − 6.20	W	56.0 − 100.0
X	5.20 − 10.90		

This use of letter designations is discouraged, and is included here only as a key to the earlier literature. Subdivisions of these bands are designated by a lower-case letter subscript, i.e. K_a-band. Further description of the bands is given in Reference Data for Radio Engineers.

The use of the radio bands is governed by treaties among nations subscribing to the Radio Regulations of the International Telecommunications Union. The treaties have arisen from recommendations made at regular meetings among nations, the World Administrative Radio Conferences, to encourage the harmonious use of radio bands and to minimize interference among users. At present, the frequencies are allocated up to 40.0 GHz for general use and to 275 GHz for space uses, the latter being allocated at the Extraordinary Administrative Radio Conference of 1963, the World Administrative Radio Conference for Space Telecommunications of 1971, and the last regular meeting of the World Administrative Radio Conference, held in 1979.

These international agreements have important consequences for remote sensing. Use of particular radio frequencies most useful for studying one or another natural phenomenon may be prohibited, or other users of the frequency may preclude its use for remote sensing. At first these may seem like minor problems, but three points are worth noting. First, it takes only one strong transmission at a frequency observed by a radiometer to destroy its ability to make measurements while the transmitter is visible, even though it may not be in the direction viewed by the radiometer. Secondly, UHF/VHF transmissions from Earth are so strong that Earth should be observable using radio telescopes at distances of ten to twenty light years. Finally, these conflicts, while merely annoying at present, threaten to become more serious in the future as ever higher frequencies are exploited for communication, radars, radio navigation, and hundreds of other uses (fig. 3.1), frequencies close to or identical to those used by spaceborne instruments.

As a result of the 1979 meeting of the World Administrative Radio Conference, many bands were allocated for remote sensing of the Earth, either as pri-

Figure 3.1 Point to point surface microwave transmissions in the New York area (from Bowers and Frey, 1972). Each beam is projected toward the horizon and can be seen from space by satellites in that direction. Similar crowded transmissions at many other places around the world threaten to interfere with satellite-based observations of the sea.

mary or secondary allocations (tables 3.4 and 3.5). Because secondary users of a band may not interfere with primary users, nor claim protection from interference by a primary user, the allocations take into consideration the compatibility of the uses. In the bands listed, interference is expected to be tolerable for observations of the sea. Other bands have also been allocated for space use, but they are either too narrow or insufficiently protected to be useful for satellite oceanography. Neither these bands nor those above 100 GHz are listed. Complete listings, together with an assessment of possible interference, are contained in Litman and Nicholas (1982).

Table 3.4 Recommended Bands for Spaceborne Radiometers

Frequency (GHz)	Bandwidth (MHz)	Allocation
1.400 − 1.427	27	primary, no emissions permitted
4.200 − 4.400	200	secondary
6.425 − 7.075	650	secondary
10.60 − 10.70	100	mostly secondary
18.60 − 18.80	200	secondary
21.20 − 21.40	200	primary
22.21 − 22.50	290	primary
23.60 − 24.00	400	primary, no emissions permitted
31.30 − 31.50	200	primary, no emissions permitted
36.00 − 37.00	1000	primary
50.20 − 50.40	200	primary
51.40 − 54.25	2850	primary, no emissions permitted
54.25 − 58.20	3950	primary
58.20 − 59.00	800	primary, no emissions permitted
64.00 − 65.00	1000	primary, no emissions permitted
86.00 − 92.00	6000	primary, no emissions permitted

Allocated by the 1979 World Administrative Radio Conference (1–100 GHz band; compiled from information in Litman and Nicholas, 1982).

Table 3.5 Recommended Bands for the Spaceborne Radars

Frequency (GHz)	Bandwidth (MHz)	Allocation
1.215 − 1.300	85	secondary
3.10 − 3.30	200	secondary
5.25 − 5.35	100	secondary
8.50 − 8.75	250	secondary
9.50 − 9.80	300	secondary
13.40 − 14.00	600	secondary
17.20 − 17.30	100	secondary
24.05 − 24.25	200	secondary
35.20 − 36.00	800	primary
78.00 − 79.00	1000	primary

Allocated by the 1979 World Administrative Radio Conference (compiled from information in Litman and Nicholas, 1982).

The designation of light bands (table 3.6) is somewhat less precise and systematic than are designations for radio bands, and reflects the much older and more diverse history of exploration of these frequencies, as well as the vagaries of nature that led to the particular sensitivity of the human eye. As wideband instruments for observing this radiation are developed, and as ever-wider bands are used for remote sensing and communications, these older designations will probably fall away, to be replaced by more precise terms. Meanwhile, we must live with them.

Table 3.6 Nomenclature of Light Bands

Frequency Range [Hz]	Adjectival Designation	Wavelengths [μm]
$0.3 \times 10^{12} - 3.0 \times 10^{12}$	Far Infrared*	$1000 - 100$
$3.0 \times 10^{12} - 1.0 \times 10^{14}$	Middle Infrared	$100 - 3$
$2.0 \times 10^{13} - 1.0 \times 10^{14}$	Thermal Infrared	$14 - 3$
$1.0 \times 10^{14} - 4.3 \times 10^{14}$	Near Infrared	$3 - 0.7$
$4.3 \times 10^{14} - 7.5 \times 10^{14}$	Visible Light	$0.7 - 0.4$
$7.5 \times 10^{14} - 3.0 \times 10^{16}$	Ultraviolet Light	$0.4 - 0.01$

*Also known as decimillimetric waves.

3.3 Radiometric units and terms

The terms used to describe the flux of radiation are many and varied, with different conventions used for different purposes. In what follows, I have attempted to be consistent with the standard terms adopted by the American National Standards Institute (ANSI Z7.1-1967) and described by Driscoll and Vaughan (1978, sec. 1) and Reeves (1975:58). These are generally consistent with the terminology proposed for optical oceanography by the International Association of Physical Sciences of the Ocean (Morel and Smith, 1982). Of course, no one set of terms can be consistent with the diverse subjects encompassed by satellite oceanography. Thus radio science invariably uses the term brightness as a synonym for radiance, and ϵ is widely used for both emissivity and dielectric constant.

Electromagnetic radiation is a form of energy. It can do work, it can be used to heat a body absorbing the radiation, and it can produce a response in an instrument designed to detect radiation. *Radiant energy* (Q) is the amount of energy carried by the radiation, and is measured in joules (J). But a fixed quantity of radiation is not as important as is the rate at which energy is transported, the *radiant flux* ($\Phi = dQ/dt$), measured in joules per second (watts, W). The flux is the rate at which the radiation can do work, and it is the power available for producing a response in an instrument. Of even greater interest is the rate at which energy is collected by a surface such as a lens, an antenna, or a photographic plate. This unit, the radiant flux per unit area, is the *flux density*, and it is known by two names. If the radiation is incident upon the surface, the incoming flux density is the *irradiance* ($E = d\Phi/dA$), where dA is the infinitesimal element of area. If it is emitted from the surface, it is called the *exitance* or sometimes the *emittance* ($M = d\Phi/dA$).

It should be apparent that the flux of sensible or latent heat described in chapter 2 had units of watts per square meter, and that this corresponds to the flux density of radiation just defined. Thus it is worth emphasizing once again that the diverse heritage of remote sensing has led to different concepts having similar names, and that units, notation, and concepts must be read with care to avoid confusion.

All the above quantities are defined independently of the angular distribution of the beam of radiation; the next level of complexity accounts for this distribution. But first, we must define the concept of the *solid angle*. To do this, consider a source of small extent at the center of a sphere of radius R beaming

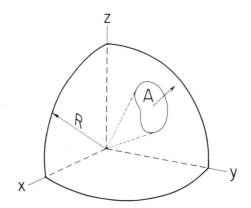

Figure 3.2   Definition of the unit of solid angle $\Omega = A/R^2$, the steradian.

radiation outward in a cone of directions (fig. 3.2) that strikes the surface of the sphere. The boundaries of the cone outline an area A on the sphere, and the *solid angle* (Ω), measured in steradians (sr), of the cone of radiation is the area divided by the square of the radius ($\Omega = A/R^2$). Because the total area of a sphere is $4\pi R^2$, there are 4π steradians of solid angle over a sphere.

In describing the angular distribution of radiation from a body, an important distinction is made between point sources and extended sources. If the body is a point source, then the radiant flux per solid angle is the *radiant intensity* ($I = d\Phi/d\Omega$). If the body is extensive, then the radiation per unit area of the body is important, and this leads to the concept of brightness or radiance—a concept to which we will return time and again in discussing the remote sensing of the sea.

Brightness and *radiance* are both defined to be the radiant flux per solid angle per area projected in the direction of the radiation (fig. 3.3), with the direction

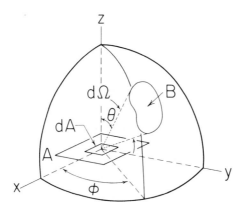

Figure 3.3   Brightness *B* is the incident radiant flux per solid angle $d\Omega$ per area projected onto the surface ($dA \cos\theta$).

of radiation differentiating the two terms. Incoming radiation is brightness; outgoing radiation is radiance. If we consider a small surface area dA oriented as shown in the figure, then the projected area is $dA \cos\theta$, and brightness B and radiance L are defined by

$$B = d\Phi/(dA \; d\Omega \; \cos\theta)$$

$$L = -B$$

Integrating these expressions over all angles and over the area of the surface, we obtain the total power received by a collector having an area A :

$$\Phi = \int_{\Omega}\int_{A} B(\phi,\theta) \; \cos\theta \; dA d\Omega$$

If the radiance emitted by a surface does not change as a function of angle, the surface appears equally bright regardless of orientation, and the surface is said to be a Lambert surface. For the simple case of a plane Lambertian radiator of area A

$$\Phi = AB \int_{0}^{2\pi}\int_{0}^{\pi/2} \cos\theta \; \sin\theta \; d\theta d\phi$$

$$\Phi = \pi AB$$

where Ω has been explicitly written out in spherical coordinates with $d\Omega = \sin\theta \; d\theta d\phi$, and then integrated over the half space defined by the plane A .

Finally, one further level of subdivision is useful for describing radiation. The radiance used in remote sensing is usually observed over narrow bands of frequencies, and the frequency and wavenumber spectra of the radiation are important. By dividing the quantities defined above by the bandwidth of the observation, the corresponding spectral forms of the expressions are obtained. For example, the spectral radiance L_λ is obtained by dividing the radiance L by the bandwidth (in units of wavelength) of the instrument making the observation:

$$L_\lambda = L/d\lambda$$

To keep the various spectral quantities distinct, the subscript λ is appended when the bandwidth is measured in units of wavelength; ν is appended when the units are frequency; and $\tilde{\nu}$ is appended when the units are wavenumber, where

$$c = \nu\lambda \quad \text{and} \quad \tilde{\nu} = 1/\lambda$$

and where c is the velocity of light, ν is the frequency of the radiation, λ the wavelength, and $\tilde{\nu}$ the wavenumber. Thus the spectral radiance L_λ used in optics has units of $W.m^{-3}.sr^{-1}$, but this is usually written as $W.m^{-2}.sr^{-1}.\mu m^{-1}$ to emphasize that the bandwidth is measured in units of wavelength in micrometers (μm). The equivalent units for $L_{\tilde{\nu}}$ are $W.m^{-2}.sr^{-1}.cm$ because the wavenumber is usually in units of cm^{-1}. Note that wavenumber $\tilde{\nu}$ in optics is λ^{-1}, while wavenumber k in oceanography is $2\pi\lambda^{-1}$.

Just as the frequency and wavenumber spectra of ocean waves could be related through the dispersion relation, so too can the spectra of electromagnetic radiation be related using the expression for the complex wavenumber, a relationship made explicit in the next section where we consider the frequency and wavelength spectra of radiation emitted by a blackbody.

As a historical footnote, it is worth noting that the spectrum of light was usually observed using gratings that produce the distribution of radiant energy as a function of wavelength, while radio signals were usually observed using receivers that produce frequency spectra, practices that tended to accentuate the differences between the two bands. As coherent radiation can now be produced at light frequencies using lasers, however, the distinction between the two bands has begun to fade.

The peculiarities of the response of the human eye to radiation has led to yet another set of quantities for measuring radiation, the luminous quantities, which are essentially the radiant quantities weighted by the response of the human eye. These quantities are based on the subjective response of our eyes and will not be used in our discussions. They are mentioned only because luminous quantities are frequently discussed in the photographic literature; and it is important to know they are distinct from the radiometric quantities defined above. A further treatment of the terms can be found in Reeves (1975:64–65).

The above extensive description of the various terms used to describe radiation can be summarized readily by writing the most precise term, B_λ, in terms of the successively less precise terms to obtain:

$$dB_\lambda = \frac{dB}{d\lambda} = \frac{dI}{d\lambda\,dA\,\cos\theta} = \frac{d\Phi}{d\Omega\,d\lambda\,dA\,\cos\theta} = \frac{dQ}{dt\,d\Omega\,d\lambda\,dA\,\cos\theta}$$

With an ability to describe the fluxes of radiation, it is now possible to begin to consider the interaction of radiation with matter. Three more terms, reflectance, transmittance, and absorptance, are important. A fourth, emissivity, is described in the next section. An object or surface viewed by an instrument is usually illuminated by a source such as the sun or a radar transmitter. The incident flux of energy, the irradiance, is either reflected, absorbed, or transmitted through the surface, and it is the reflected portion of energy that allows the surface to be seen. To make these ideas precise, the *hemispherical reflectance* (ρ) is defined as the ratio of the exitance reflected from a plane of matter to the irradiance incident on that plane. The *hemispherical transmittance* (τ) is the ratio of the exitance leaving the opposite side of the plane to the irradiance incident on the plane. Because energy is conserved, the *hemispherical absorptance* (α) is

$$\alpha = 1 - \tau - \rho$$

Usually these quantities vary with frequency, and the spectral reflectance, transmittance and absorptance are the proper generalizations of the terms. Farther along, we will find that these quantities also depend on angle, and the terms can be further generalized. In much of what follows, however, the subscripts, and sometimes the word spectral, will be dropped for simplicity in discussing radiation. In almost all applications of remote sensing, the radiation is band limited and spectral terms apply, so the simplification should cause no confusion.

These concepts and their symbols and units are summarized in table 3.7. They play a central role in describing the radiation used in remote sensing; they allow physical phenomena to be described precisely, thus allowing measurements by various instruments to be compared.

Table 3.7 Summary of Radiometric Concepts

Name, [units]	Symbol	Concept
Radiant energy [J]	Q	Capacity of radiation within a specified spectral band to do work.
Radiant flux [W]	Φ	Time rate of flow of energy on to, off of, or through a surface.
Radiant flux density Irradiance [W.m^{-2}]	E	Radiant flux incident upon a surface per unit area of that surface.
Radiant exitance [W.m^{-2}]	M	Radiant flux leaving a surface per unit area of that surface.
Radiant intensity [W.sr^{-1}]	I	Radiant flux leaving a small source per unit solid angle in a specified direction.
Radiance or Brightness [W.m^{-2}.sr^{-1}]	B, L	Radiant intensity per unit of projected source area in a specified direction.
Hemispherical reflectance	ρ	Φ reflected/Φ incident for any surface.
Hemispherical transmittance	τ	Φ transmitted/Φ incident for any surface.
Hemispherical absorptance	α	Φ absorbed/Φ incident for any surface.

3.4 Blackbody radiation

All objects, except those at a temperature of absolute zero, emit electromagnetic radiation, commonly called thermal emission. A perfect emitter, a blackbody, is one that emits radiation at the maximum possible rate, a rate that depends only on temperature; this characteristic allows radiometers, instruments that measure radiation, to measure remotely the temperature of an object. Most natural bodies, however, emit at a somewhat reduced rate; and the ratio of the spectral brightness emitted in the direction (θ,ϕ) by a real body relative to that emitted by a blackbody is the spectral directional emissivity, $e(\lambda;\theta,\phi)$

$$e(\lambda;\theta,\phi) = B_\lambda(\lambda;\theta,\phi)/B_\lambda(\lambda)_{\text{blackbody}}$$

In general, e is nearly independent of temperature and depends only on the nature of the substance. But it is this variation in amount of radiation emitted by bodies at the same temperature that allows radiometers to distinguish between types of material—say, between fresh and salt water or between ice and frozen land—provided their temperature can be independently estimated by some other means.

The spectral brightness of a blackbody is described by Planck's equation

$$B_\lambda = \frac{2hc^2}{\lambda^5} \frac{1}{\exp\left[hc/(kT\lambda)\right] - 1}$$

where h is Planck's constant, k is Boltzmann's constant, and T is temperature (table 3.8). Alternatively, the spectral brightness per unit of frequency, B_ν, may be used to describe blackbody radiation. It is calculated from B_λ using

$$\nu\,\lambda = c$$

$$d\nu = -\frac{c}{\lambda^2}\,d\lambda$$

$$B_\nu d\nu = B_\lambda d\lambda$$

Table 3.8 Notation for Describing Blackbody Radiation

B	Brightness [W.sr^{-1}.m^{-2}]
B_λ	Spectral brightness per wavelength band [W.sr^{-1}.m^{-3}]
B_ν	Spectral brightness per frequency band [W.Hz^{-1}.sr^{-1}.m^{-2}]
c	Velocity of light [3×10^8 m.s^{-1}]
e	Emissivity
h	Planck's constant [6.63×10^{-34} J.s]
k	Boltzmann's constant, [1.38×10^{-23} J.K^{-1}]
T	Temperature [K]
λ	Wavelength of emitted radiation [m]
ν	Frequency of radiation [Hz]

where the equation for c results from the expression for the complex wavenumber $k = 2\pi/\lambda$, assuming that the radiation is propagating in free space. With these relations, we obtain

$$B_\nu = \frac{2h\nu^3}{c^2} \frac{1}{\exp\left[h\nu/(kT)\right] - 1}$$

A plot of these functions (fig. 3.4), shows that as temperature increases, the spectral brightness increases rapidly and the peak frequency ν_{max} shifts upward to higher frequencies. But note that B_λ differs from B_ν, and that the shapes of the two functions are slightly different.

The total brightness B, calculated by integrating B_ν over all frequencies, is given by the Stefan-Boltzmann relation

$$B = \frac{2k^4\pi^4}{15c^2h^3}\,T^4$$

$$B = (1.8 \times 10^{-8} \text{ W.sr}^{-1}.\text{m}^{-2}.\text{K}^{-4})\,T^4$$

Integrating this expression over all angles yields the more common expression, the radiant exitance

$$M_B = \pi \, B$$

$$M_B = (5.6 \times 10^{-8} \text{ W.m}^{-2}.\text{K}^{-4}) \, T^4$$

The maximum of the curve for B_ν, is found by solving

$$dB_\nu/d\nu = 0$$

for ν, to obtain

$$\nu_{max} = (0.941)(3k/h) \, T$$

$$\nu_{max} = (5.88 \times 10^{10} \text{ Hz/K}) \, T$$

In a similar way, the wavelength at the maximum in B_λ is found by solving $dB_\lambda/d\lambda = 0$ for wavelength λ. It is important to note, however, that this results in a wavelength λ_{max} having a frequency that differs from ν_{max}.

At lower frequencies, particularly at radio frequencies, the emission of radiation from blackbodies is given by a much simpler relation, the Rayleigh-Jeans Law, which is derived assuming

$$h\nu \, / \, kT \ll 1$$

so that

$$\exp[h\nu/(kT)] = 1 + h\nu/(kT)$$

Introducing this into the expression for B_ν gives

$$B_\nu \approx (2k/\lambda^2) \, T = (2\nu^2 k/c^2) \, T$$

$$B_\nu = (3.07 \times 10^{-40} \text{ W.m}^{-2}.\text{Hz}^{-3}.\text{K}^{-1}) \, \nu^2 T$$

Thus the spectral brightness of a body varies linearly with its temperature; and the signal observed by a radiometer is easily related to the temperature of the body. This is frequently referred to as the *brightness temperature*. The relationship holds as long as

$$\nu \ll kT/h$$

an inequality that is well satisfied for radiation from bodies at room temperature, approximately 300K, observed by radiometers operating at frequencies below 600 GHz. At infrared frequencies the approximation does not apply and the exact equation for brightness must be used.

3.5 Natural sources of radiation

Two natural sources of radiation, the sun and Earth, are of particular importance in remote sensing. The sun, at a mean distance of 149.7 Gm and with a mean radius of 695 Mm, radiates nearly as a blackbody at a temperature of 5900 K. Using these figures, it is easy to calculate that sunlight is confined to a solid angle of 6.77×10^{-5} sr; thus, multiplying the brightness B_λ (in fig. 3.4) by this angle gives the spectral irradiance E_λ of sunlight at the top of the atmosphere (fig. 3.5). This irradiance has a maximum emission per wavelength band at 0.48 μm, and a peak emission per frequency band at 3.5×10^{-14} Hz, a frequency that corresponds to a wavelength of 0.85 μm in the near infrared.

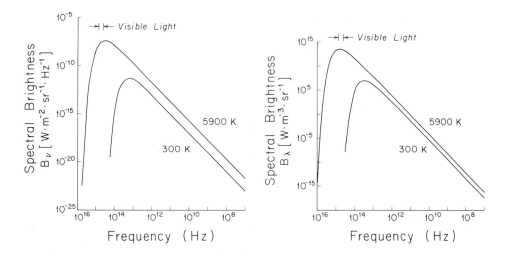

Figure 3.4 The spectral brightness of blackbodies at the indicated temperature as a function of the frequency of the emitted radiation. Left: spectral brightness per frequency band. Right: spectral brightness per wavelength band.

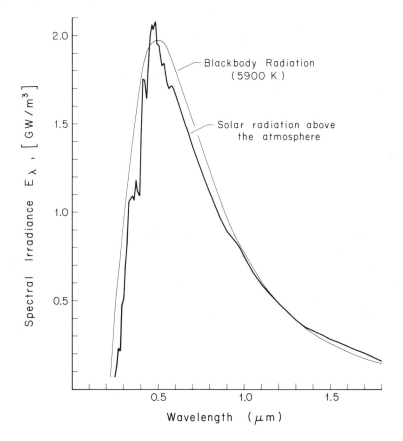

Figure 3.5 Spectral irradiance from the sun according to a proposed standard model (Thekakara, 1972), together with the best-fitting curve of radiation from a blackbody (at 5900 K) the size and distance of the sun (see also fig. 5.8).

The total irradiance, known as the solar constant, is 1376 W.m^{-2} (Hickey et al., 1980), an irradiance that would be produced by a blackbody the size of the sun at the sun's distance and radiating at 5800 K.

Earth radiates very roughly as a blackbody at a temperature near 300 K with a peak emission per hertz at 17 μm. Emission and absorption in the atmosphere and reflected solar radiation greatly complicate the description of Earth's radiation, however, and these topics will be discussed in the next chapter.

At frequencies below 3×10^{10} Hz for the sun, and below 10^7 Hz for the Earth, nonthermal plasma and synchrotron emissions from the solar corona and Earth's ionosphere become increasingly important, and these can radiate many orders of magnitude more energy than thermal emission at the same frequency.

3.6 Kirchoff's Law

The relationship between the rate at which radiation is absorbed by matter and the rate at which radiation is reradiated (emitted) is expressed by Kirchoff's Law. Stated simply, it says that matter in local thermodynamic equilibrium must emit at the same rate that it absorbs; that is, the directional spectral emissivity e is equal to the directional spectral absorptance a:

$$e\,(\lambda\,;\theta\,,\phi) = a\,(\lambda\,;\theta\,,\phi)$$

Thus a perfect emitter, a blackbody, is also a perfect absorber, hence the name.

This simple relationship is central to many applications in remote sensing and reappears time and again in different guises. Thus it is important to realize that e and a are coefficients and not quantities of radiation. The quantity of radiation emitted by an object is the emissivity times the Planck equation for a blackbody at the temperature of the object; and the amount of radiation absorbed by an object is the amount of incident radiation times the absorptance.

To clarify the idea further, consider a simple example. In the open ocean, the sea is a deep cobalt blue and absorbs 98% of any red light incident vertically on the surface. Because no radiation is transmitted out the bottom of the ocean, the transmittance is zero, thus the reflectance must be 0.02. By Kirchoff's Law, the ocean must be an almost perfect emitter of red light. Yet the ocean is almost black because, although a good emitter, it is too cold to radiate according to Planck's equation and thus emits no red light. In fact, more energy is absorbed by the sea during sunny days than is emitted at infrared frequencies, and the imbalance causes the water to warm up. Nevertheless, the process is gradual and the sea remains in local thermodynamic equilibrium. Later, in chapter 6, we will return to this subject and consider in greater detail the balance of radiation near the sea surface.

4

ABSORPTION
IN THE ATMOSPHERE

Sunlight and radar signals, beaming toward Earth, propagate downward through the atmosphere. There they are partly absorbed and scattered before reaching the sea surface. Some energy continues on, penetrates into the sea, and is further attenuated and scattered. Upwelling radiation from within the sea and radiation reflected from the surface are joined by radiation emitted by the sea and the atmosphere, and by radiation scattered within the atmosphere, before passing out of the atmosphere to be captured by satellite instruments (fig. 4.1). The process appears quite complicated, and one wonders how satellites can make accurate surface measurements. The answer lies in the remarkable transparency of the atmosphere to some electromagnetic radiation, notably to light,

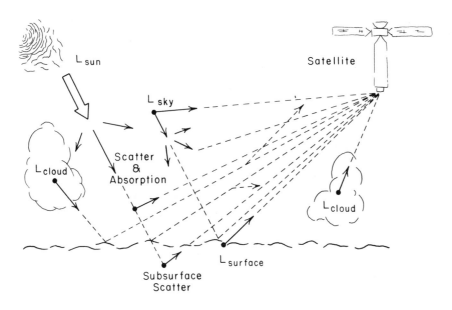

Figure 4.1 Processes influencing radiation observed by satellite.

some infrared, and radio frequencies between 6 and 20 GHz. In these bands absorption, and thus thermal emission, is low, although scatter by clouds, rain and dust can still produce problems. In this chapter we consider the first problem: transmission through a clear atmosphere, one with absorption but no scatter. Later, we will include the additional complications induced by scatter.

4.1 The influence of gases

The primary mechanism for absorption by gases is the exchange of a quantum of energy between a molecule or atom and radiation. The exchange is measured in units of $h\nu$, where h is Planck's constant and ν is the frequency of the radiation in hertz. The highest energies have the highest frequencies and work the greatest changes in the state of the molecule, even to the extent of removing electrons. This latter process, photoionization, mostly requires ultraviolet radiation. Lower frequencies, of importance in remote sensing, work less dramatic changes. In order of decreasing energy, the exchange between radiation and matter has the following influences: (a) visible light raises the energy level of orbital electrons; (b) infrared radiation induces vibration of the molecules; and (c) infrared and radio frequency radiation excites rotational motion of the molecules.

These exchanges of energy between radiation and matter are quantized. Each possible vibration or rotation has a distinct energy associated with it, and only radiation whose energy exactly matches the difference in energy between two states can be absorbed. This quantization results in an absorption spectrum, the absorption as a function of frequency, that is composed of a set of narrow peaks or lines. The frequency ν_0 of each line is determined by the difference in energy levels; the strength S of the absorption line depends, among other factors, on the fraction of molecules in the initial state and on the probability of transition between states, both of which factors depend on temperature. Of course, thousands of levels are possible, and the absorption spectrum of a complex molecule such as water contains tens of thousands of lines ranging from light to radio frequencies. Theoretically, the position and amplitude of all these lines can be calculated from quantum theory with the help of a computer; but in practice a considerable amount of experimental observation is also required.

The computation of the absorption spectrum is made difficult by the necessity of accounting for the widths of the peaks in the spectrum. Widening fills in the region between peaks and can account for most of the absorption of wideband signals. Several processes are involved:

(a) *Natural linewidth* is a function of the probability of transition between energy levels in the molecule, and it varies as the frequency cubed (Derr, 1972, p. 24, sec. 9). The broadening is very small and can almost always be neglected. Typically, the natural linewidth is less than 10^{-4} Hz at UHF.

(b) *Doppler broadening* is due to the thermal motion of the absorbing molecules. Some will be moving toward the source of radiation, others moving away when they encounter radiation, and the frequency of the radiation will be shifted due to the motion. The absorption as a function of wavenumber $\tilde{\nu}$ is (table 4.1; Zuev, 1982:30)

$$k(\nu) = \frac{S}{\pi^{1/2}\gamma_D} \exp\left[-\left(\frac{\tilde{\nu}-\tilde{\nu}_0}{\gamma_D}\right)^2\right]$$

$$\gamma_D = \frac{\tilde{\nu}_0}{c}\left[\frac{2KT}{m}\right]^{1/2}$$

Table 4.1 Notation for Describing Absorption

c	Velocity of light [$3 \times 10^8 \text{m.s}^{-1}$]
L_λ	Spectral radiance [$\text{W.sr}^{-1}.\text{m}^{-3}$]
k	Absorption coefficient [$\text{m}^2.\text{kg}^{-1}$]
K	Boltzmann's constant [$1.38 \times 10^{-23}\text{J.K}^{-1}$]
K	Absorption coefficient [dB.km^{-1}]
m	Mass of absorbing molecule [kg]
p	Pressure [Pa]
p_0	Pressure at standard conditions [Pa]
S	Line strength [m.mol^{-1}] or [m.kg^{-1}]
T	Temperature [K]
T_0	Temperature at standard conditions [K]
x	Exponent
z	Path length [m]
γ_L, γ_D	Line half-width [mm^{-1}] or [m^{-1}]
γ_L^0	Line half-width at standard conditions [mm^{-1}]
κ	Absorption coefficient [m^{-1}]
ν	Frequency [Hz]
$\tilde{\nu}$	Wavenumber [m^{-1}]
$\tilde{\nu}_0$	Line center [m^{-1}]
ρ	Mass density [kg.m^{-3}]

where $\tilde{\nu}_0$ is the wavenumber at the center of the line, $\sqrt{\ln2}\,\gamma_D$ is the half width at the half-power point, c is the velocity of light, K is Boltzmann's constant, T is absolute temperature, m the mass of the absorbing molecule, and S is the strength of the absorption line, normalized such that

$$S = \int_0^\infty k(\tilde{\nu})\,d\tilde{\nu}$$

The Doppler broadening is weak and tends to dominate only in the upper stratosphere where other sources of broadening are even weaker, although it influences line widths at much lower levels. Typically, Doppler linewidths in the visible and infrared vary between 3×10^{-2} and 3×10^{-4} cm^{-1} for atmospheric gases at standard temperature and pressure (Goody, 1964:99). At radio frequencies, under the same conditions, the line widths vary between 10^5 and 10^4 Hz (Waters, 1976:149).

Note that various systems of units are widely used to describe the absorption of radiation. In the infrared and visible bands, linewidth and positions are usually expressed in terms of wavenumber $\tilde{\nu} = \nu/c = \lambda^{-1}$ with units of cm^{-1}, where ν is the frequency in hertz and λ is the wavelength; however, the correct

SI units are mm^{-1} or m^{-1} as stated in table 4.1. In the radio bands, the frequency ν is used instead of $\tilde{\nu}$.

(c) *Collisions* between molecules interrupt the exchange of energy and produce collision broadening. This is a function of the rate at which molecules collide; thus it depends on the size of the molecules (their cross section), their velocity (which depends on temperature), and the number of molecules per unit volume (proportional to gas pressure). The last factor dominates, and the process, often called *pressure broadening*, provides a means for measuring atmospheric temperature as a function of height (pressure) using satellite observations of emitted radiation. The collision linewidth is (Zuev, 1982:30; Houghton, 1977:35; Liou, 1980:16; Farmer, 1974):

$$k(\nu) = \frac{S}{\pi} \left(\frac{\nu}{\nu_0}\right)^2 \left[\frac{\gamma_L}{(\nu-\nu_0)^2+\gamma_L^2} + \frac{\gamma_L}{(\nu+\nu_0)^2+\gamma_L^2}\right]$$

where again γ_L is the half linewidth. For infrared and visible radiation, the last term can be neglected, and the line shape is usually written

$$k(\nu) = \frac{S}{\pi} \frac{\gamma_L}{(\nu-\nu_0)^2+\gamma_L^2}$$

In either case, the linewidth is

$$\gamma_L \approx \gamma_L^0 \frac{p}{p_0} \left[\frac{T_0}{T}\right]^x$$

where p, T are pressure and temperature and γ_L^0 is the linewidth at a standard pressure p_0 and temperature T_0. The exponent x is ½ for most absorption lines in the visible and infrared. For H_2O and CO_2 in the infrared, however, and for absorption lines at radio frequencies, the cross section for collisions tends to be a function of temperature, and $x \approx 0.7 - 1.0$. In general, pressure broadening tends to dominate in the lower atmosphere, and for pressures near 10^5 Pa (1000 mbar), $\gamma_L \approx 0.01 - 0.1$ cm^{-1}.

(d) *Interactions* between lines are possible when the lines are so closely spaced that line broadening causes overlap. The process is not well understood (Zuev, 1982:80), but should be important for some bands of the complex vibrational-rotational spectra of molecules such as water vapor.

The combined influence of Doppler and collision broadening, calculated by convolving the two line shapes, is sufficient for calculating absorption near line centers. It is inadequate, however, for describing absorption in the *far wings* of the lines where $|\nu-\nu_0| \gg \gamma_L$. At these frequencies, various correction factors must be added to the line shapes discussed above (Waters, 1976:150ff.; Zuev, 1982:37). The corrections are especially important because the clarity of the atmospheric windows used by satellite instruments is usually determined by the far wings of strong absorption lines. Fortunately, the corrections are generally known, and the shapes of the important absorption lines can now be calculated accurately.

Various compilations of absorption lines are available based on theory and experiments. These list the five intrinsic variables, those that depend only on the type of molecule, useful for calculating absorption: (a) the line frequency ν_0; (b) the strength S at some standard conditions, γ_L^0; (d) the energy of the lower state E'' of the transition; and (e) the quantum numbers identifying the lines. When used with the extrinsic variables of pressure and temperature, these allow the accurate calculation of absorption by molecules in the atmosphere (fig. 4.2).

A notable compilation of absorption line information is that produced by the Optical Physics Division of the Air Force Geophysics Laboratory (McClatchey et al., 1973, 1978) together with the computer program LOWTRAN to calculate absorption along atmospheric paths (Kneizys et al., 1980). The overall accuracy of the computations of absorption, based on their compilations, is better than 10% over the range 0.25–28.5 μm, with a spectral resolution of 20 cm^{-1}, the largest errors being in the wings of overlapping lines. Similar programs have been assembled by other groups, and present extensions of this work are producing estimates of absorption with greater frequency resolution, extending to lower frequencies and including many atmospheric trace elements. A computer listing (LOWTRAN 5) of parameters for 139,000 absorption lines, covering the absorption by seven principal atmospheric gases and extending from 0.56 μm to 3 cm, is now available in the form of magnetic tape from the NOAA National Climatic Center.

These computer codes yield absorption estimates applicable for bands of radiation typically used for remote sensing, as well as for estimating the influences of gases on the very narrow band radiation from lasers. Usually the estimates are expressed through the absorption coefficient κ, defined to be the fractional loss of spectral radiance L_λ over a short path of length dz:

$$\kappa = \frac{1}{L_\lambda} \frac{dL_\lambda}{dz}$$

This is similar to the attenuation coefficient described in chapter 3, except that here κ refers to the decrease of radiant energy, while before we considered the attenuation of the amplitude of the electric field. Because the radiance is proportional to the square of the electric field amplitude, the coefficient for the attenuation of radiance is twice the coefficient for the attenuation of the electric field.

Typically the absorption coefficient is measured by comparing the radiance L_0 exiting a volume of gas of thickness z with the radiance L_i entering the volume

$$\kappa = \ln(L_0/L_i)/z$$

Thus the absorption coefficient has units of m^{-1}. For radio frequencies, however, this is usually a very small number, and the values for atmospheric absorption are often reported in units of decibels per kilometer. The two are related by

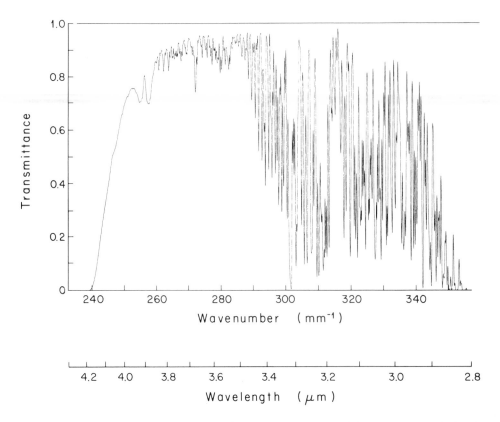

Figure 4.2   The calculated transmittance of the atmosphere in the infrared along a vertical path through the standard atmosphere. The calculation is based on a compilation of absorption lines maintained at the Jet Propulsion Laboratory, and shows the clarity of the atmospheric window centered at $3.7\,\mu$m (from C.B. Farmer, Jet Propulsion Laboratory).

$$K = 10^3 \, d\,(10 \log L/L_0)/dz$$

$$K \,[\text{dB.km}^{-1}] = -4.34 \times 10^3 \,\kappa \,[\text{m}^{-1}]$$

where the units are given in brackets.

For infrared and visible radiation, the mass absorption coefficient k is often used, where

$$k = \frac{1}{\rho L_\lambda} \frac{dL_\lambda}{dz}$$

$$\kappa = k\rho$$

with ρ being the mass of absorbing material per volume defined by the cross sectional area of the beam times the path length dz. Thus the absorption coefficient k is the same as the line shape k discussed above. The mass is expressed either as kilograms or moles of absorber, or in units of centimeter-atmospheres. The latter is the height in centimeters of a column of pure absorbing gas at standard conditions having the same number of molecules of absorbing substance found per unit area along the absorbing path. The conversions are

$$1 \text{ cm}-\text{atm} = 2.69 \times 10^{19} \text{ molecules.cm}^{-2}$$

$$1 \text{ mole} = 6.02 \times 10^{23} \text{ molecules}$$

4.2 The influence of the ionosphere

Radiation with frequencies above 10^9 Hz freely propagates through the ionosphere with only minor influence; at lower frequencies, ionospheric influences become increasingly important, until at frequencies around 10^7 Hz the ionosphere completely blocks the radiation. The primary ionospheric influences important for remote sensing are: (a) the ionosphere slows the group velocity of radiation; (b) it rotates the plane of polarization, a phenomenon known as Faraday rotation; (c) it absorbs radiation; and (d) at frequencies below 10^6 Hz it acts as a mirror, reflecting signals from land out to great distances over the ocean, a condition producing propagation "over-the-horizon."

The equations governing the propagation of radiation in a magnetic, ionized medium with collisions between molecules and electrons are very complicated. But by considering a vertically propagating wave of sufficiently high frequency, it is possible to simplify the discussions considerably and to produce accurate estimates of ionospheric influences.

The key to understanding the propagation of radiation through the ionosphere is the expression for the complex index of refraction n, first derived by Appleton (see Davies, 1966:63) for a plane wave propagating through an ionized medium similar to the ionosphere. For frequencies above roughly 10 MHz, the frequencies important for remote sensing, the index of refraction can be written:

$$n^2 = 1 - \frac{X}{1 \pm Y_L - iZ}$$

using

$$X = f_p^2/f^2$$
$$Y_L = (f_H/f)\cos\theta = f_L/f$$
$$Z = f_c/(2\pi f)$$

with

$$f_p^2 = Ne^2/(4\pi^2 m\epsilon_0) = \alpha N$$
$$2\pi f_H = \mu_0 He/m$$

where f_p is the plasma frequency, f_H is the electron gyrofrequency, f_c the collision frequency, N the electron density, m the electron mass, and H the geomagnetic field intensity (table 4.2).

Table 4.2 Notation for Describing Radio Propagation in the Ionosphere

A	Attenuation [dB]
c	Velocity of light [3.0×10^8 m.s^{-1}]
e	Electron charge [1.6×10^{-19} C]
f	Radiation frequency [Hz]
f_c	Collision frequency [Hz]
f_H	Electron gyrofrequency [Hz]
f_L	Gyrofrequency [Hz]
f_p	Plasma frequency [Hz]
H	Geomagnetic field intensity [A.m^{-1}]
ℓ	Path length [m]
m	Electron mass [9.11×10^{-31} kg]
N	Electron density [m^{-3}]
n	Complex index of refraction
n'	Group refractive index
X, Y, Z	Dimensionless frequencies
α	Constant [80.5 m^3.s^{-2}]
ϵ_0	Permittivity of free space [8.9×10^{-12} F.m^{-1}]
η, χ	Real and imaginary parts of the index of refraction
θ	Angle between wave normal and the geomagnetic field [deg]
κ	Attenuation coefficient [m^{-1}]
μ_0	Permeability of free space [$4\pi \times 10^{-7}$ H.m^{-1}]
Ω	Polarization rotation angle [radians]

The three processes influencing the index of refraction are expressed through the frequencies (f_p, f_H, and f_c) made nondimensional by the radio frequency, thus resulting in the three terms X, Y, and Z. The plasma frequency sets a lower limit on the band of radio frequencies that can propagate through the ionosphere, and the influence is included in X. Radio frequencies higher than f_p propagate into the ionosphere, where the Earth's magnetic field, whose influence is included in Y, splits the radio beam into two components travelling at different velocities, the ordinary and extraordinary rays. Electron motion, excited by the radio field, causes increased collisions between electrons and molecules and extracts energy from the signal, resulting in the Z term. As the radio frequency increases, all three terms become small, n approaches unity, and the ionosphere has a much reduced influence on propagation.

In order to find suitable quantitative approximations to the index of refraction, it is first necessary to estimate the frequency at which X, Y, and Z become small compared with unity. Of the three frequencies, X (and thus f_p) is the largest, and it depends only on the electron density in the ionosphere. The frequency reaches its greatest value at heights near 200 km, where the electron content is roughly $N \sim 10^{12}/m^3$. At this level

$$f_p \sim 9 \times 10^6 \text{ Hz}$$

and rarely exceeds 15 MHz. Because the other frequencies (Y and Z) are smaller than X, the influence of all three is small for radio frequencies much higher than f_p, say, for frequencies exceeding 100 MHz. As a result, the expressions for the index of refraction are greatly simplified for the super high frequencies used by satellite instruments.

Although in general we will not be concerned with frequencies near the plasma frequency, we note that $n = 0$ when $f = f_p$, ignoring the small influences of Y and Z. By applying Snell's Law to a vertically propagating radio signal, it is easy to show that such a wave is totally reflected when $n = 0$. Thus the plasma frequency sets a lower limit on the frequency of radio signals that can penetrate the ionosphere. All radio frequencies below the critical frequency, notably those below 15 MHz, are reflected back.

The total reflection of low frequency radio waves by the ionosphere has great practical importance. It permits "short wave" radio communication because it allows radio waves to propagate to great distances from a transmitter on the Earth's surface. In the early decades of this century, radio provided the only long distance communications to many points on the globe; this practical usefulness of the ionosphere stimulated considerable interest in its properties and its influence on radio propagation. Today, ionospheric reflection allows radars operating at frequencies in the band $10^6 - 10^7$ MHz to observe the sea out to ranges of 4000 km via a single reflection using techniques described in a later chapter. The additional peculiarities of the ionosphere which limit the technique are described there. Essentially, the various layers of free electrons in the ionosphere act as moving distorting mirrors parallel to the Earth, each mirror reflecting a band of frequencies to varying distances.

For very high frequencies and above, X, Y, and Z are all small, leading to substantial simplifications to the expression for the index of refraction. In calculating the influence of the ionosphere, however, it is important to remember that the ionosphere has a large diurnal and a weaker seasonal and long-term variability resulting from solar influences (Davies, 1980). Thus the approximate values of variables describing the ionosphere, such as electron content, can vary by a factor of ten or more, and this variability introduces considerable uncertainty into the final estimates of ionospheric influences (fig. 4.3).

Absorption of radio energy due to collisions between electrons and molecules is most important near the bottom of the ionosphere where the atmosphere is more dense. Recalling that the imaginary part of the index of refraction χ is related to the attenuation, we can write

$$\chi \approx \frac{XZ}{2}$$

assuming

$$X, Y_L \ll 1$$

to obtain the attenuation coefficient for the electric field (see chapter 3):

$$\kappa = \frac{\alpha}{2c} \frac{Nf_c}{f^2}$$

The total attenuation A through the ionosphere is found by integrating κ along a ray path. In units of decibels this integration yields:

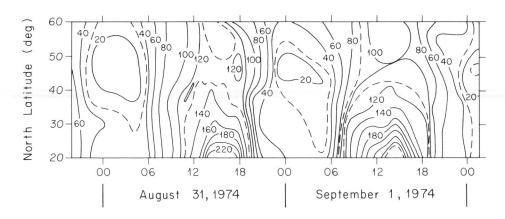

Figure 4.3   Ionospheric electron content on 31 August–1 September 1974, showing typical variability of the ionosphere as a function of latitude and local time. The plot is in units of 10^{15} electrons/m², and was obtained by combining ATS-6 satellite beacon data with Transit satellite observations. The contour levels are in steps of $20 \times 10^{15}/m^2$ (from Davies, Hartmann, and Leitinger, 1977).

$$A = - \frac{20}{\ln 10} \int_0^\infty \kappa \ dz$$

$$A = \frac{1.16 \times 10^{-6}}{f^2} \int_0^\infty Nf_c \ dz$$

Using typical values for Nf_c, Lawrence, Little, and Chivers (1964) estimate that A = 0.05 dB during the day and 0.005 dB at night for a radio frequency of 0.1 GHz; thus, using f in hertz

$$A \approx 5 \times 10^{13}/f^2 \ \text{dB}$$

give or take an order of magnitude, depending on the state of the ionosphere.

The propagation velocity of a radio pulse is important both for satellite altimeters and for radio tracking of satellite positions. The ratio of the pulse velocity to the velocity of light in free space is the group refractive index

$$n' = d(nf)/df = n + f \ dn/df$$

At sufficiently high frequencies

$$n^2 \approx 1 - X = 1 - \frac{N\alpha}{f^2}$$

$$n \approx 1 - \frac{N\alpha}{2f^2}$$

thus

$$n' \approx 1 + \frac{N\alpha}{2f^2}$$

The apparent additional path length $\Delta\ell$ transversed by a pulse, compared with the same path length in a vacuum, is

$$\Delta\ell = \int\limits_0^\infty (n'-1)\,dz = \frac{\alpha}{2f^2} \int\limits_0^\infty N\,dz$$

Thus the group delay is a function only of the total number of free electrons along a ray path. Typically, there are approximately 10^{17} electrons per square meter, so using f in gigahertz

$$\Delta\ell = \frac{4.02\ \text{m}}{f^2}$$

or a length of 4 cm at 10 GHz, give or take an order of magnitude depending on the state of the ionosphere.

The difference in phase velocity between the ordinary and extraordinary ray in the ionosphere, a consequence of the Y_L term in the index of refraction, causes a rotation, the Faraday rotation, in the plane of polarization of the radiation. The difference in the two indices of refraction is

$$\Delta n = n_+ - n_- \approx XY_L = \alpha N f_L / f^3$$

where the subscripts refer to the sign of Y in the expression for n. Again, the additional path length is

$$\Delta\ell = \int\limits_0^\infty \Delta n\,dz = \alpha/f^3 \int\limits_0^\infty N f_L\,dz$$

The degree of rotation Ω of a polarized wave is one half the differential phase path multiplied by 2π divided by the wavelength. Using the definitions of f_L and f_H

$$\Omega = \frac{e^3 \mu_0}{8\pi^2 cm^2 \epsilon_0} \int\limits_0^\infty NH \cos\theta\,dl$$

Evaluating the constants in the expression, using typical values of N and H, for f in hertz

$$\Omega = 7 \times 10^{16}/f^2 \text{ radians}$$

where again the exact value depends on the particular state of the ionosphere. Thus the Faraday rotation is proportional to the total number of free electrons along a ray path times the longitudinal component of the geomagnetic field intensity. Because the geomagnetic field intensity is fairly well known as a function of height, the measurements of Faraday rotation provide an alternate means for estimating ionospheric electron content. The expression is valid for all values of θ except for those within a degree or so of $90°$, at which angle the transverse influences of the magnetic field become important.

In summary, rotation, absorption, and group delay all scale inversely with frequency squared and become increasingly less important as frequency increases.

4.3 The radiative transfer equation

Associated with each ionospheric or molecular absorption mechanism is an absorption coefficient κ that specifies the fractional attenuation over a short segment of path dz in the absorbing medium. Integrating the absorption along the path traversed by the radiation gives the total attenuation of the radiation. At first glance, this fact appears to solve the problem of calculating the influence of the atmosphere on the propagation of radiation. Unfortunately, the problem is more complicated. For infrared or radio energy propagating in the terrestrial environment, the radiation remaining after absorption is usually only a part of the radiation that would be seen by an observer at the end of the path. Every absorber is also an emitter of radiation, and each path segment emits radiation which joins that being absorbed and contributes to the observed signal. We consider here the mathematical formulation for handling this problem following the general approach by Kraus (1966); at the end of the chapter, we will consider the attenuation of signals from Earth's surface to space.

Consider the spectral radiance L seen by an observer viewing a source L_s at the far end of an absorbing path having an absorption coefficient $\kappa(z)$ (fig. 4.4). For simplicity we will consider a straight ray path, with distance z measured along the path. Over an infinitesimal length dz, the signal L will be attenuated by an amount

$$dL = -\kappa(z)\ L(z)\ dz$$

an equation often called Bouguet's Law; and the absorptance along the path is

$$\alpha = \kappa\ dz$$

where κ is the absorption coefficient (table 4.3).

Over the same interval dz, the signal will be increased due to emission of radiation from the material in the volume. By Kirchoff's Law, the emissivity is equal to the absorptance; thus

$$dL = -\ L_B\ \kappa\ dz$$

where L_B is the radiance of a blackbody at the same temperature as the absorbing gas. The net change in radiance is given by the radiative transfer equation

$$dL = -\ L\kappa dz + L_B\kappa dz$$

For most paths, the absorption depends only on the composition, pressure, and temperature along the path (Goody, 1964). That is, we need not consider stimulated emission or the nonlinear interaction of radiation with matter such as occurs in extremely bright laser radiation. Therefore, the absorption coefficient is independent of L and the radiative transfer equation is a linear, first-order differential equation of the form

$$\frac{dL(z)}{dz} + L(z)\kappa(z) = L_B(z)\kappa(z)$$

For a path of length H, the equation has the rather formidable solution:

$$L = L_s\ \exp\ [-\tau(0,H)] + \int_0^H L_B(z)\kappa(z)\ \exp[-\tau(z,H)]dz$$

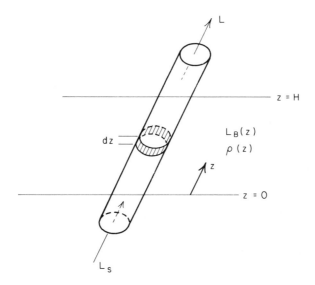

Figure 4.4 Geometry for calculating the radiance exiting an emitting and absorbing atmosphere of thickness H lying above a surface of radiance L_s.

Table 4.3 Notation for Describing Absorption in the Atmosphere

f	Frequency [Hz]
H	Thickness of gas layer [m]
L, L_s	Spectral radiance [W.m^{-3}.sr^{-1}] or [W.m^{-2}.sr^{-1}.Hz^{-1}]
L_B	Spectral radiance of a blackbody [W.m^{-3}.sr^{-1}]
t	Transmittance
T	Brightness temperature [K]
T_B	Temperature of a blackbody [K]
z	Path distance [m]
α	Absorptance
κ	Absorption coefficient [m^{-1}]
τ	Optical depth.

where

$$\tau(z',z) = \int_{z'}^{z} \kappa(z'')dz''$$

is the *optical thickness* (or opacity) of the layer. The first term on the right-hand side of the equation gives the total attenuation of L_s due to absorption; the second term gives the emission $L_B\kappa$ from any point z along the path attenuated by the gas along the remainder of the path.

The general solution for radiative transfer in an atmosphere depends on knowing L_B (or temperature) and κ as a function of height, and the solution is usually calculated numerically by dividing the atmosphere into a number of horizontal layers and summing up the influence of all the layers. A simple

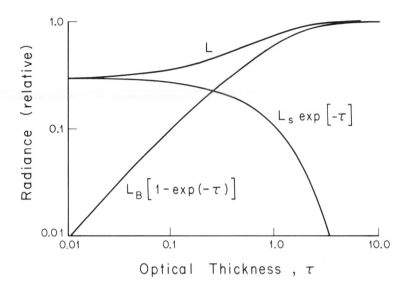

Figure 4.5 Observed radiance L seen by a radiometer observing an emitting absorbing atmosphere of constant radiance L_B and variable optical depth τ lying in front of an object with radiance L_s.

solution that demonstrates the essence of the influence of the atmosphere is found by allowing L_B to be constant. The solution is then (fig. 4.5):

$$L = L_s \exp(-\tau) + L_B [1-\exp(-\tau)]$$

where τ is the optical depth of the medium. For

$\tau = 0$	the atmosphere is transparent,	$L = L_s$
$\tau \gg 1$	the atmosphere is opaque,	$L = L_B$
$\tau = 1$	the atmosphere is partly clear,	$L = 0.37\, L_s + 0.63\, L_B$

In the first case, the source is clearly visible through the atmosphere. In the second case, the atmosphere totally obscures the source and only the atmosphere can be seen. In the last case, some radiation from the source and some from the atmosphere are observed.

The attenuation of spectral radiance and the transmittance t of the atmosphere are both related to the optical thickness of the atmosphere, within the limits of the applicability of Bouguet's Law (see Zuev, 1982:104):

$$t = L/L_s = exp(-\tau)$$

if L_s is the spectral radiance of the surface and τ the optical thickness along a vertical path. For radio frequencies, the transmittance is often expressed in units of decibels, and the transmittance in decibels is related to optical depth through

$$t\ [\text{dB}] = 10 \log t = -4.34\,\tau$$

This expression (fig. 4.6) will be useful later when comparing various estimates of atmospheric absorption.

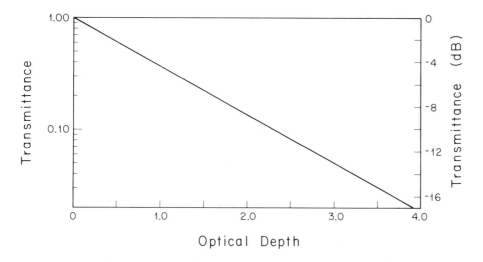

Figure 4.6 Transmittance as a function of optical depth.

At radio frequencies, recall that the blackbody radiance is linearly related to temperature:

$$L_B = 3.07 \times 10^{-4} f^2 T$$

where T is absolute temperature and f the radio frequency. Therefore, radio scientists usually replace L by T :

$$T = T_s \exp(-\tau) + T_B [1 - \exp(-\tau)]$$

and speak of the brightness temperature. That is, for the terrestrial environment, at frequencies above 1 GHz, the radiance received by radiometers sensitive to radio frequencies is directly proportional to the absolute temperature of the medium; and this apparent temperature is referred to as the *brightness temperature*. Note that this differs from the true temperature of the surface below the atmosphere by the factors involving $\exp[-\tau]$, but approaches the true temperature for optically thin gases.

A typical example of the use of the radiative transfer equation in satellite oceanography is provided by Maul et al. (1978). They wished to estimate the accuracy with which it is possible to measure infrared radiation from the Gulf Stream using radiometers on the Goes satellite operating in a geostationary orbit 35,000 km above the equator. The path from the satellite to the sea slants diagonally downward through the atmosphere. Ignoring clouds by looking between them, they estimated the attenuation and emission due to water vapor in the atmosphere (the principal source of attenuation of the radiation they observed) by integrating the radiative transfer equation numerically using measurements from radiosondes of water vapor and temperature as functions of height. The solutions showed that for their particular case, the atmosphere contributed roughly 8°C to the apparent brightness temperature of the sea surface. This atmospheric signal is important, and its role will be discussed further in chapter 8.

4.4 Transmission through the atmosphere

Seven atmospheric constituents are primarily responsible for molecular absorption: (a) carbon dioxide, CO_2; (b) water vapor, H_2O; (c) ozone, O_3; (d) nitrous oxide, N_2O; (e) carbon monoxide, CO; (f) methane, CH_4; and (g) oxygen, O_2. Their influence extends from a strong water vapor absorption line at 22.235 GHz to weak ozone absorption bands in the visible light.

The total absorption of radiation by these gases along a path from the surface to space has been estimated by various workers for different frequency bands. For wavelengths between 0.25 and 28.5 μm, the computer code by Selby and McClatchey (1975) is useful; figure 4.7 shows the transmittance of the atmosphere, for a vertical path in this band, for various typical atmospheric conditions. Clearly visible in the figure are a number of bands with high but somewhat variable transmittance. These are the atmospheric windows through which satellite instruments view the sea surface. Because carbon dioxide, nitrous oxide, carbon monoxide, oxygen, and methane are well mixed, at least in the lower regions of the atmosphere that dominate absorption from the surface to space, the variability in the absorption of visible and infrared radiation arises mainly from the variability of water vapor and to a lesser extent from the variability of ozone in the upper atmosphere.

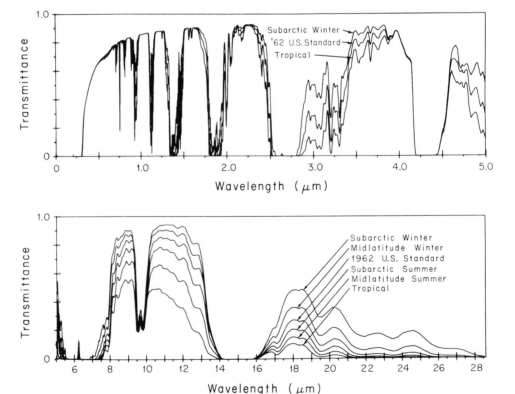

Figure 4.7 Atmospheric transmittance for a vertical path to space from sea level for six model atmospheres with 23 km visibility, including the influence of molecular and aerosol scattering (from Selby and McClatchey, 1975).

Seasat SMMR Water Vapor (gm/cm^2)

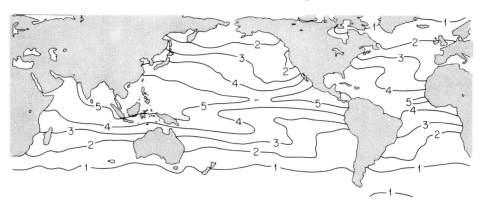

Figure 4.8 The mean atmospheric water vapor during July, August, and September of 1978 as observed by a SHF radiometer on Seasat (from Chelton, Hussey, and Parke, 1981).

Water vapor limits the transparency of the infrared windows, and its concentration depends primarily on the temperature structure of the troposphere. In high latitudes and at high altitudes, the cold atmosphere can hold little water and is therefore dry, so water vapor absorption is low. Conversely, the hot tropical atmosphere can hold much water with correspondingly greater attenuation. Superimposed on these trends are other features of the general circulation of the atmosphere (fig. 4.8) and the day-to-day state of the weather.

Ozone influences the amount of visible and ultraviolet light reaching the surface, and its concentration depends primarily on the mean insolation as a function of latitude and season and secondarily on solar conditions. Thus ozone variability is predominantly seasonal (fig. 4.9), with the variability at 45°N being approximately ± 20% (Waters, 1976).

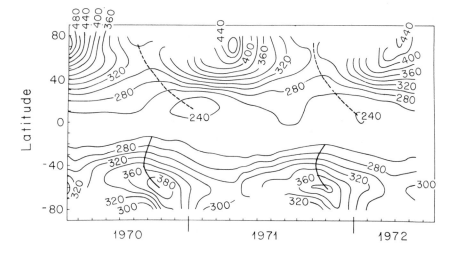

Figure 4.9 The mean zonal integrated ozone content as a function of latitude and season as measured by the backscattered ultraviolet experiment on Nimbus-4. The contours are in Dobson Units, where one Dobson Unit is 21.4 mg.m^{-2} (from Hilsenrath, Heath, and Schlesinger, 1979).

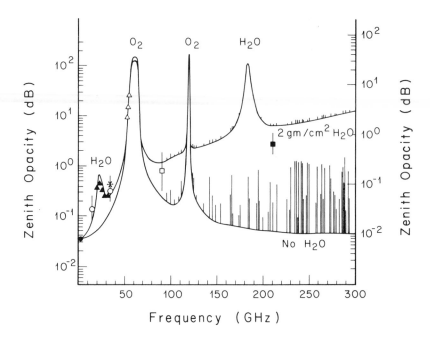

Figure 4.10   Atmospheric zenith opacity for 1962 U.S. Standard Atmosphere model of temperature and pressure for (a) no water and (b) 2 gm.cm^{-2} total water in the atmosphere (10 gm.cm^{-3} at the surface, decreasing exponentially with height with a 2-km scale height) (from Waters, 1976).

Below 300 GHz absorption is primarily due to (a) water vapor lines at 22.235 and 183.31 GHz, (b) an oxygen absorption band around 60 GHz, and (c) an oxygen line at 118.7 GHz. Ozone makes a few small contributions above 67 GHz, but these can be ignored for the most part. Figure 4.10 gives the total zenith optical density (opacity) for frequencies below 300 GHz. Because attenuation is related to emission, the same calculation used to generate opacity can also be used to determine the brightness temperature of the atmosphere, with the additional knowledge of temperature as a function of height. The brightness temperature is given in figure 4.11; but remember that this is the atmospheric contribution only and is the temperature seen by a radiometer looking vertically upward from the sea surface. A downward-looking radiometer will see contributions due to the sea surface, particularly at frequencies with low attenuation, and a slightly different atmospheric contribution because of the asymmetry of the radiative transfer equation.

From these figures, we see that the atmosphere has a number of very clear windows, the major ones being summarized in table 4.4. Recalling that a transmittance of 80% is equivalent to an optical depth of 0.23, or an attenuation of 1 dB, we note that the atmosphere is remarkably transparent in these bands, but not so clear that attenuation can be completely ignored. Or put into other terms, the infrared and radio frequency windows can be even clearer than that for visible light on the clearest days. Of the windows, those in the radio frequency bands, visible light, and the thermal infrared band at 12 μm are the most important. Their properties will be described in the following chapters.

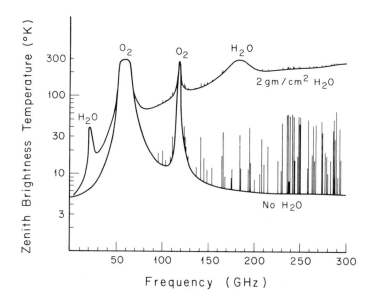

Figure 4.11 Same as figure 4.10, but giving the brightness temperature seen by a zenith looking radiometer, including 3 K cosmic radiation at the top of the atmosphere (from Waters, 1976).

Table 4.4 The Primary Atmospheric Windows

Radiation Band	Variability Due To	Transmittance
$0.4 - 0.7 \, \mu m$	ozone, aerosols, clouds	$0.6 - 0.8$
$2.0 - 2.5 \, \mu m$	water vapor, aerosols, clouds	0.9
$3.5 - 4.0 \, \mu m$	water vapor, clouds	$0.6 - 0.9$
$8.0 - 9.0 \, \mu m$	water vapor, ozone, clouds	$0.5 - 0.9$
$10 - 13 \, \mu m$	water vapor, clouds	$0.2 - 0.9$
$25 - 40$ GHz	water vapor, rain	$0.8 - 0.9$
$0.1 - 20$ GHz	water vapor, rain	$0.9 - 0.99$

5

SCATTER
IN THE ATMOSPHERE

Radiation passing through the atmosphere from the surface to space is absorbed and scattered by particles in the atmosphere. These include the water drops of fog, clouds, and rain and the solid particles of smoke and haze. If the particles are sparse, scatter primarily changes the angle of propagation, thus attenuating the radiation from the surface. If the particles are sufficiently dense, the radiation is scattered again and again, and a portion is scattered back into the original beam, partly obscuring the light from the surface. Examples of dense particles include clouds and haze. Haze is a collection of small particles in the air that scatter sunlight into the field of view of an instrument, obscuring the view through the atmosphere. Clouds tend to be much better at scattering; they strongly scatter visible and infrared light, and the scatter so changes the distribution of radiance that it becomes nearly isotropic. As a result, almost no radiation propagates directly through clouds. Instead, the propagation of radiance is represented more as a diffusion of energy. Thus, on a uniformly overcast day, the bottom of the cloud deck is uniformly bright and the sun cannot be seen directly. Conversely, to an observer looking downward from space, the cloud deck appears white and it completely obscures the surface.

The influence of scatter in the atmosphere is much more difficult to handle than is the influence of absorption. Yet for many radiation bands, the attenuation of radiance by scatter is far more important than the attenuation by molecular absorption described in the last chapter. Molecular absorption merely removes energy from the primary beam and reradiates it uniformly in all directions. Particles also remove energy from the primary beam. But they do so partly by absorption and partly by scattering radiation in different directions, with the latter effect usually dominating (due to the small complex part of the index of refraction for most particulates). The complexity of scatter arises from the strongly directional properties of the scattered radiation; from the polarization of the scattered energy; from the strong dependence of scatter on the size

[64]

of the scattering particle relative to the wavelength of the radiation; and from the need to keep track of the scattered radiation through perhaps two or more encounters with particles (multiscatter).

In calculating the influence of scatter, we can usually assume that the radiation from individual scatterers adds up incoherently at some point far from the particles—the region called the far field. Thus the total scattered field can be represented as a superposition of the power scattered from each individual particle, provided that the following conditions are fulfilled (Derr, 1972:10–22): (a) The scatterers must not be aligned in a regular lattice, otherwise phases from each point will add up in the far field. (b) Individual scattering points must be well separated from each other, in distance measured in wavelengths. This condition requires that $N < 10^3$/unit volume, where N is the number of particles and the unit volume is one wavelength on a side. This is true for gas at atmospheric pressures even for the short wavelengths of red light; however, air molecules produce the Rayleigh scattering of blue light described later in this chapter. (c) Large particles must be ten times their own radius apart, even though there may be few particles per unit volume. That is, each particle must be in the far field of other particles. This is true even for dense fog.

The complexity of the scattering calculations depends on the probability that radiation will be scattered once, twice, or many times. If radiation is scattered once, the scatter can only remove energy from a columnated beam. If the radiation is scattered twice, some of the radiance scattered the second time can reenter the original beam and is not lost (fig. 5.1). If radiation is scattered many times, the radiation must be followed through each scattering step to calculate the radiance ultimately emerging after scattering.

The probability that radiance will be scattered once is nearly equal to the attenuation of a columnated beam traversing the scattering region, provided that molecular absorption can be ignored and that the attenuation is small. The error in this probability estimate is smaller than the square of the attenuation. The probability that the radiance will be scattered a second or a third time is then the square or the cube of the probability of a single scatter, and the probability of multiple scatter is small unless the attenuation is large. Using the relationship between transmittance and optical depth, we can make rough estimates of the importance of scatter as follows:

(a) If $\tau < 0.1$, 10% of the radiance is absorbed, 90% is transmitted, and only single scatter is important.

(b) If $0.1 < \tau < 0.3$, 75% of the radiation is transmitted, and corrections from single-scatter theory must be included to account for small amounts of multiple scatter. (In the primary beam, radiation is lost by scatter from the beam and gained by scatter into the beam.)

(c) If $\tau > 0.3$ multiple scatter dominates and no general solutions are available. This situation can occur for blue light in a clear atmosphere, for the propagation of visible and infrared radiation through clouds, and for the propagation of radio signals through heavy rain.

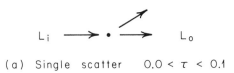

(a) Single scatter $0.0 < \tau < 0.1$

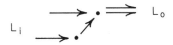

(b) Multiscatter correction to single
     scatter $0.1 < \tau < 0.3$

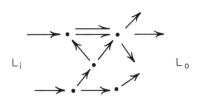

(c) Multiscatter $\tau > 0.3$

Figure 5.1 The influence of scatter on a columnated beam of radiance L. Depending on the probability that the radiance will be scattered one or more times, (a) a portion of the the incoming radiance can be scattered out of the beam; (b) the scattered radiation can be scattered back into the beam to augment the exiting radiance L_0; or (c) the radiance L_i can be scattered many times leading to an isotropic distribution of radiance. The probability of scatter is a function of optical thickness τ and increases exponentially as τ increases.

5.1 Scatter from a single particle: Mie scatter

The theory for the scattering of radiation from spherical particles of finite size with arbitrary dielectric constant is generally associated with the work of Mie (1908), although the problem has a much longer history (Kerker, 1969:54–59). The best recent summaries of the theory of scatter by particles are by Van de Hulst (1957) and Kerker (1969), but they are often difficult to obtain. Other substantial treatments can be found in Born and Wolf (1970), mostly for particles that are perfectly conducting, and in Penndorf (1962).

Consider a spherical particle of radius a and index of refraction n imbedded in space. Radiation incident on the particle is scattered and absorbed by the material of the particle; and the influence is expressed in terms of a cross section. The total *attenuation cross section* σ_E of a particle is defined to be that area, divided by the cross-sectional area (πa^2) of the particle, which yields the ratio of the irradiance Φ_E removed by the particle to the irradiance Φ incident on the particle:

$$\sigma_E / (\pi a^2) = \Phi_E / \Phi$$

Note that various conflicting notations are used in the literature. *Extinction* is a synonym of attenuation, hence the subscript E in our notation; but attenuation is the preferred term. Optical oceanographers recommend that *attenuance* be used instead of either. Occasionally the symbol Q_E is used for cross section, although it generally refers to the efficiency factor $\sigma_E / (\pi a^2)$.

The total attenuation cross section is the sum of two terms, one due to scatter σ_S and the other due to absorption σ_A:

$$\sigma_E = \sigma_S + \sigma_A$$

For spherical particles, the attenuation and scattering cross sections can be expressed as a power series expansion:

$$\sigma_S = \frac{2\pi a^2}{q^2} \sum_{m=1}^{\infty} (2m + 1) \left[|a_m (n, q)|^2 + |b_m (n, q)|^2 \right]$$

$$\sigma_E = \frac{2\pi a^2}{q^2} \sum_{m=1}^{\infty} (2m + 1) \mathrm{Re}[a_m (n, q) + b_m (n, q)]$$

where

$$q = 2\pi a / \lambda$$

is the dimensionless particle size. The other notation is given in table 5.1. The coefficients a_m and b_m depend explicitly on particle size q and index of refraction n; and their functional form is given in the references. In general, they are expressed as ratios of the Ricatti-Bessel functions, and they can be computed using the recursion formulae for these functions.

If $q \ll 1$, the lower-order coefficients in the expansion are:

$$a_1 = -\frac{i}{4\pi} (n^2 - 1)q^5$$

$$b_1 = \frac{2i}{3} \frac{n^2-1}{n^2+2} q^3 \left[1 + \frac{3}{5} \frac{n^2-2}{n^2+2} q^2 - \frac{2i}{3} \frac{n^2-1}{n^2+2} q^3 \right]$$

$$b_2 = \frac{i}{15} \frac{n^2-1}{2n^2+3} q^5$$

If only the lowest-order term b_1 is retained in the expansion; the approximation is called *Rayleigh scatter*:

$$\sigma_S = \frac{128\pi^5 a^6}{3\lambda^4} \left| \frac{n^2-1}{n^2+2} \right|^2$$

$$\sigma_A = \frac{8\pi^2 a^3}{\lambda} \mathrm{Im}\left(-\frac{n^2-1}{n^2+2} \right)$$

Of the two, σ_A is the largest, and

$$\sigma_E \approx \sigma_A = \frac{8\pi^2 a^3}{\lambda} \mathrm{Im}\left(-\frac{n^2-1}{n^2+2} \right)$$

Table 5.1 Notation for Describing scatter

x, y, z	Cartesian coordinates [m,m,m]
r, θ, ϕ	Spherical coordinates [m,−,−]
a	Particle radius [m]
a_m, b_m	Mie scattering coefficients
C	Contrast
D	Particle diameter [m]
H	Scale height of atmosphere [km]
i	$\sqrt{-1}$
L	Spectral radiance [W.m^{-3}.sr^{-1}]
m	Summation index
n	Index of refraction
n_s	Index of refraction of air at standard conditions (STP)
$n(a)$	Particle size distribution [m^{-4}]
N	Number of particles per unit volume [m^{-3}]
N_s	Number of molecules per unit volume at STP [m^{-3}]
$P(\theta)$	Phase function
P_m	Legendre polynomial
P_m^1	Associated Legendre polynomial
q	Relative size of particles
Re, Im	Real and imaginary parts of a variable
S_1, S_2	Scattering functions
V	Volume of particles, per volume of unit volume of air
α, β, γ	Parameters of particle size distribution [m$^{-3}\mu$m$^{-\beta-1}$,−,μm^{-1}]
Φ	Irradiance [W.m^{-2}]
$\kappa_A, \kappa_E, \kappa_S$	Coefficients of absorption, extinction, and scatter [m^{-1}]
λ	Wavelength of radiation [m]
ρ	Constant
π_m, τ_m	Mie scatter coefficients
$\sigma_A, \sigma_E, \sigma_S$	Total absorption, attenuation, and scatter cross sections [m^2]
$\sigma(\Omega)$	Scattering cross section [m^2]
Ω	Solid angle [sr]
τ	Optical depth

Thus, when the particles are much smaller than the wavelength of the radiation, the scatter is small and proportional to λ^{-4}, and the extinction cross section is proportional to the volume of the scattering sphere. The latter has an important practical consequence: the total attenuation is proportional to the volume of scattering material in the beam, and at super high radio frequencies this will be shown to be the liquid water content of the atmosphere.

The range of validity of the Rayleigh equations can be estimated by comparing σ_E (Rayleigh) with σ_E (Mie). Gunn and East (1954) found that when $q < 0.05$, for liquid water at radio frequencies, the Rayleigh approximation is accurate to ~20% (fig. 5.2); and that the approximation becomes increasingly inaccurate as q increases. Thus the range of validity of the Rayleigh equations is quite small and Mie theory must be used for most practical calculations. For example, 3 GHz radiation has a wavelength of 10 cm, and $q < 0.05$ requires

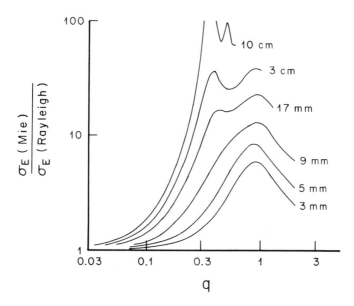

Figure 5.2 The ratio of Mie attenuation cross section σ_E (*Mie*) to the small particle approximation of Rayleigh σ_E (*Rayleigh*). The curves are for the scatter of various wavelengths by pure water drops at 18°C. (The wavelength of the radiation is indicated on the curves.) The ratio indicates that the Rayleigh approximation is valid only for very small particles such that $q \ll 0.05$, where $q = 2\pi a/\lambda$, a is the radius of the drops, and λ is the wavelength (from Gunn and East, 1954).

that $a < 1.6$ mm, a value satisfied only by small raindrops. At 30 GHz, however, $a < 0.16$ mm, and the theory is limited to non-raining clouds. For infrared, $\lambda = 10\,\mu$m, and $a < 0.16\,\mu$m is the limit, a limit exceeded by most particulates in the atmosphere. For light frequencies, the Mie theory must be used.

Scattered radiance, in contrast with absorbed radiance, is not lost, but exits the scattering volume at various angles. Part is scattered into the half-plane in the direction of propagation of the original radiation; this is forward scatter. Part is scattered back toward the source of radiation; if the scatter angle is 180°, that is, if scatter is directly back toward the source, this portion is called backscatter. Thus for many studies we require the directional distribution of radiance scattered from particles.

The directional distribution of the scattered radiation is a function not only of q and n but also of the polarization. To simplify the discussion, we will consider natural (unpolarized) radiation scattered from spherical particles. The more complicated examples of the scatter of polarized radiation from ice crystals and other non-spherical particles is described in Liou (1980).

The directional distribution of scattered spectral radiance L_s is given by the *scattering cross section* $\sigma(\Omega)$ defined by

$$\sigma(\Omega)/(\pi a^2) \equiv 4\pi L_s/\Phi$$

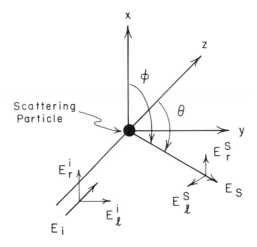

Figure 5.3 Geometry for describing scattered radiation. The incident E_i and scattered E_S radiation define a reference plane y,z, and the scattering angle θ lies in this plane. The two components of the electric field of the radiation are defined with respect to the reference plane. E_r is perpendicular to the plane, and E_l is parallel to the plane.

This is related to the total scattering cross section through

$$\sigma_S = \frac{1}{4\pi} \int_{4\pi} \sigma(\Omega)\,d\Omega$$

In optics; the cross section is usually written as the product of the total cross section times a *phase function* $P(\Omega)$:

$$\sigma(\Omega) = \sigma_S P(\Omega)$$

normalized such that

$$\int_{4\pi} P(\Omega)\,d\Omega = \int_0^{2\pi}\int_0^\pi P(\theta,\phi)\,\sin\theta\,d\theta\,d\phi = 4\pi$$

Thus the phase function specifies the fraction of the radiation scattered in the directions θ, ϕ, where the scattering angle θ (fig. 5.3) is measured in a reference plane defined by the incident and scattered radiation and ϕ is measured in a plane perpendicular to the reference plane. For polarized radiation incident on a particle, the scattered radiation will depend on both θ and ϕ. For unpolarized radiation, the electric field vector of the incident radiation is randomly distributed in ϕ and the scatter depends only on θ. But note that in either case the scattered radiation is partly polarized.

For spherical particles, the phase function can be written as

$$P(\theta) = \frac{\lambda^2}{2\pi\sigma_S}\,[|S_1(\theta)|^2 + |S_2(\theta)|^2]$$

where the two components S_1 and S_2 of P describe the directional distribution of the incident and scattered radiation with electric field parallel and perpendicular to the reference plane. From Mie theory, they are:

$$S_1(\theta) = \sum_{m=1}^{\infty} \frac{2m+1}{m(m+1)} \ [a_m \pi_m (\cos\theta) + b_m \tau_m (\cos\theta)]$$

$$S_2(\theta) = \sum_{m=1}^{\infty} \frac{2m+1}{m(m+1)} \ [b_m \pi_m (\cos\theta) + a_m \tau_m (\cos\theta)]$$

where

$$\pi_m (\cos\theta) = \frac{1}{\sin\theta} \ P_m^1(\cos\theta) \qquad \tau_m = \frac{d}{d\theta} \ P_m^1(\cos\theta)$$

where $P_m^1(\cos\theta)$ are the associated Legendre polynomials derived from the Legendre polynomials $P_m(\cos\theta)$ through

$$-\frac{\partial}{\partial\theta} \ P_m (\cos\theta) = P_m^1(\cos\theta) \qquad P_0^1 (\cos\theta) = 0$$

If $q \ll 1$, the polynomials reduce to $\pi_1 = 1$ and $\tau_1 = \cos\theta$, and the components of the phase function for Rayleigh scatter are:

$$S_1(\theta) = 3/2 \ b_1 \cos\theta \qquad S_2(\theta) = 3/2 \ b_1$$

thus

$$\sigma_S = 6\pi a^2 |b_1|^2/q^2$$

and together these give

$$P(\theta) = 3/4 \ (1 + \cos^2\theta)$$

Thus Rayleigh scatter is polarized and nearly isotropic. For example, for $\theta = 90°$, $S_1 = 0$ and only radiation with an electric field perpendicular to the reference plane is scattered into this angle. For larger q, the scatter is mostly in the forward direction; for yet larger q, the same is true, but the forward lobe is broken into many narrow beams. In general, the width of the forward lobe is proportional to q^{-1}.

In the far field, Born and Wolf (1970:659) and others show that for $q \gg 1$, $\sigma = 2D$ where D is the geometrical cross-sectional area of the particle. That is, the scattering cross section is twice the geometrical cross section. This surprising result is due to diffraction around the edge of a particle, and arises from the assumption that scatter into very small forward angles is lost from the beam. Van de Hulst (1957) points out that a flower pot resting on a window blocks light equal to that falling on its cross sectional area, but the same pot in space seen through a telescope blocks light equal to that falling on twice its cross-sectional area.

5.2 Collections of particles

The *particle-size distribution* $n(a)$ is defined to be the number of drops per unit volume having radii in the range a to $a + da$, such that

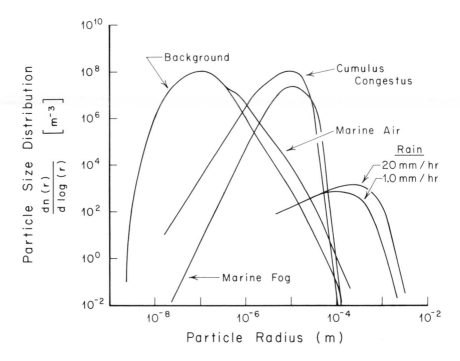

Figure 5.4 Typical distributions of particle sizes in the atmosphere. While these curves are representative of the various types of particles, individual observations of the particle-size distribution often differ significantly from these idealized representations. The curves are based on data in SMIC (1971); Falcone, Abreu, and Shettle (1979); and Marshall and Palmer (1948).

$$N = \int_0^\infty dN = \int_0^\infty n(a)\,da$$

where N is the total number of drops per cubic meter of the medium. Because particles usually have a very wide range of sizes, it is convenient to plot the number of drops per interval on a logarithmic scale (fig. 5.4). For such a plot:

$$dN = 2.3\,an(a)\,d\log a$$

Many different empirical drop size distributions have been proposed with various degrees of success. A few of these are illustrated in figure 5.4. The clearest atmosphere is found in the middle and high troposphere, but even this air has a few particles; their distribution is labeled "background" in the figure. At lower levels over the ocean, this background is augmented by salt particles produced by breaking waves, and their distribution is labeled "marine air." Although these distributions are shown extending to sizes as small as 0.01 μm, the number of particles smaller than 0.1 μm is poorly known and is based more on inference than on direct observation (SMIC, 1971).

For larger particles of fog, clouds, and rain, a family of distributions is frequently used (Falcone, Abreu, and Shettle, 1979):

$$n(a) = \alpha a^\beta exp(-\gamma a)$$

Some typical values of the parameters of these curves, together with the units appropriate for *a* in micrometers, are listed in table 5.2. In general, these equations are idealized representations of the actual distributions, and individual observations of drop sizes can differ significantly from the representations. For example, the distribution for rain drops suggested originally by Marshall and Palmer (1948) has been observed on many occasions. These observations show that the empirical constants depend somewhat on the type of rain, whether from stratus or cumulus clouds; and that the values given in the table are only rough approximations to the actual values observed in any particular rainstorm.

Table 5.2 Drop-size Distribution Parameters

Particle type	$\alpha \; [m^{-3}\mu m^{-\beta-1}]$	β	$\gamma \; [\mu m^{-1}]$
Marine fog	2.7×10^4	3	0.3
Stratus clouds	2.7×10^7	2	0.6
Cumulus clouds	1.4×10^6	2	0.328
Rain	8.0	0	$0.0082\,R^{-0.21}$ *

* R is rain rate in mm.hr^{-1}.

Most particles in the atmosphere occur in the lower levels of the atmosphere, with two major exceptions. Thin bands of ice particles (cirrus clouds) frequently occur at heights of 3 to 6 km; and volcanoes occasionally inject many small particles into the stratosphere at heights of 15 to 20 km (fig. 5.5).

The rate at which spectral radiance L is lost from a beam is given by the attenuation coefficient κ_E:

$$\kappa_E = -\frac{1}{L}\frac{dL}{dz}$$

Using the definitions of the particle size distribution together with the attenuation cross section for each size particle, the coefficient can be rewritten

$$\kappa_E = \int_0^\infty n(a)\sigma_E(a)\,da$$

This is the local rate of change of radiance due to a collection of particles of different sizes. In the next section we will use this expression in the equation of radiative transfer to obtain the rate of change of radiation in a beam propagating through a scattering atmosphere. Here, however, we note one simplification of the above integral: for super high frequency radio waves propagating through rain and clouds, the Rayleigh approximation is accurate, and

$$\kappa_E = \kappa_A = \frac{8\pi^2}{\lambda}\,\mathrm{Im}\left[-\frac{n^2-1}{n^2+2}\right]\int n(a)\,a^3\,da$$

$$\kappa_A = \text{constant } V$$

where V is the volume of water per volume of space. Thus the absorption of SHF radio signals by rain is proportional to the volume of water in the beam.

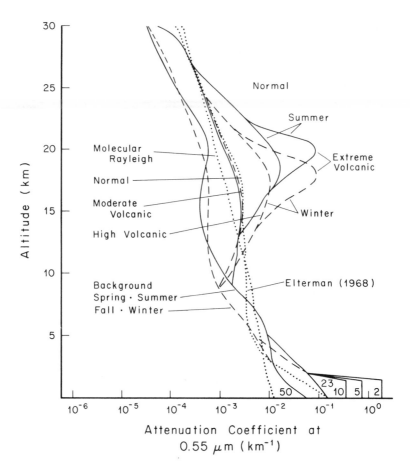

Figure 5.5 The vertical distribution of particles that scatter light in the atmosphere. The numerical values at lower right give the visibility in miles (from Selby and McClatchey, 1975).

(In using this expression, note the distinction between the particle-size distribution $n(a)$ and index of refraction n.)

5.3 Radiative transfer with particulates

Given the sizes and types of particles and how they scatter radiation it is possible, in principle, to calculate the influence of particles on radiation propagating through the atmosphere. In practice, the calculations are often exceedingly difficult, and many techniques have been proposed for finding approximate solutions of the radiative transfer equation. Fortunately, we are interested in the propagation of radiation through nearly clear atmospheric windows. Thus multiple scatter tends to be unimportant, and the equations can often be greatly simplified. Here we will consider the radiative transfer equation with particulates. In later chapters we will investigate particular solutions to the equation.

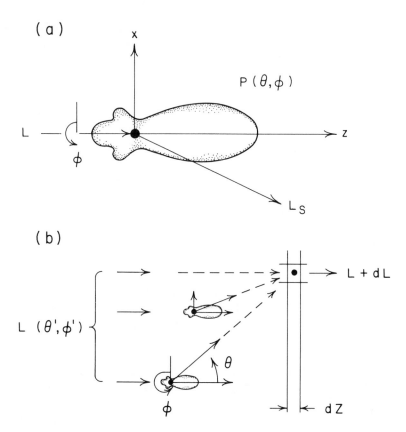

Figure 5.6 Geometry for discussion of radiative transfer through an atmosphere with particles. (a) Particles at arbitrary positions scatter radiation as a function of angle θ,ϕ with the scattered intensity determined by the phase function $P(\theta,\phi)$. (b) Radiation from many particles can be scattered into the beam, augmenting the radiation in the beam (L).

Using the phase function described in section 5.1 and the geometry in figure 5.6, the equation of radiative transfer is

$$dL(\theta,\phi) = -\kappa_E L(\theta,\phi)dz + \kappa_A L_B(\theta,\phi)dz + \kappa_S L_S(\theta,\phi)dz$$

where

$$L_S = \frac{1}{4\pi}\int_0^{2\pi}\int_0^{\pi} P(\theta,\phi,\theta',\phi')L(\theta',\phi')\sin\theta'\ d\theta'd\phi'$$

and

$$\kappa_A = \int_0^{\infty} n(a)\sigma_A(a)da$$

$$\kappa_S = \int_0^{\infty} n(a)\sigma_S(a)da$$

The first term on the right-hand side gives the attenuation of radiance due to absorption and scatter; the second term gives the increase in radiance due to emission (from Kirchoff's Law, using Planck's equation for the blackbody radiance L_B); and the third term is the integrated influence of the scatter of radiation into the beam. Note that L, L_B, κ_A, and κ_S all depend on position z along the path traversed by the radiation.

The equation is often rewritten in an alternate form:

$$dL(\theta,\phi) = (-L(\theta,\phi) + (1-\omega_0)L_B(\theta,\phi) + \omega_0 L_S)d\tau$$

with

$$d\tau = \kappa_E dz$$

$$\omega_0 \equiv \kappa_S/\kappa_E$$

where τ is the optical depth defined earlier and ω_0 is the *single scattering albedo*.

If only single scatter is important, the last term (L_S) in the radiative transfer equation can be neglected. If the double scatter correction to single scatter is required, then the integral for L_S need be calculated only once at each level in the atmosphere. If multiple scatter is important, radiation must be tracked through many scattering events at each level and the integral for L_S evaluated for each event.

Note that the expression for L_S is only a simplified notation for a much more difficult problem. After one scatter, natural radiation is polarized and must be described by four functions: the Stokes parameters, which give the amplitudes, phases, and direction of rotation of the two components of the polarization. Each component is scattered differently upon encountering a second particle, and the quantities $L(\theta,\phi)$ and $P(\theta,\phi)$ are really vectors that describe the scatter of polarized radiation (see Chandrasekhar, 1960:40; Liou, 1980).

5.4 The integrated influence of particulates

Various estimates of the propagation of radiation through the atmosphere from the surface to space have been made using the theory of radiative transfer together with estimates of the distribution of natural particles. Because the particle size distributions are often poorly known, and because the particles are not always the uniform spheres assumed by the Mie theory, these calculations tend to be inexact. Nevertheless, they provide an indication of the importance of particles in the atmosphere.

For the shortest wavelengths of visible light, Rayleigh scatter from molecules dominates the transmission of radiation, and it is important to estimate the Rayleigh scatter with very good accuracy in order to interpret observations made using visible light. Penndorf (1957) points out that many widely used equations were derived by making approximations that greatly reduce their accuracy, but that other equations are more accurate. These are given here.

The scattering coefficient κ_S and phase function P for atmospheric molecules is (Penndorf, 1957; Peck and Reeder, 1972):

$$\kappa_S = \sigma_S(\lambda)N = \frac{8\pi^3(n_S^2-1)^2 N}{3\lambda^4 N_S^2}\left[\frac{6+3\rho}{6-7\rho}\right]$$

$$P(\cos\theta) = 0.7629\,(1+0.932\cos^2\theta)$$

with

$$\rho \approx 0.035$$

where the index of refraction n_S and the number of molecules per unit volume N_S are evaluated at standard atmospheric temperature and pressure (15°C and 1.013×10^5 Pa), and N is the number of molecules per unit volume at any temperature and pressure. The term involving ρ accounts for the influence of anisotropic molecules (Young, 1981); but it is not known precisely. The value given here is from Penndorf (1957), but Young (1981) recommends $\rho = 0.028$, a slightly smaller value.

The index of refraction of air is

$$(n_S-1)\times 10^8 = \frac{5791817}{238.0185-(1/\lambda)^2} + \frac{167909}{57.3562-(1/\lambda)^2}$$

where λ must be in units of micrometers, a function tabulated in Penndorf (1957) as a function of wavelength and temperature.

The optical thickness, or the optical depth of the atmosphere due to molecular scatter, is:

$$\tau_S = \int_0^\infty \kappa_S\,dz$$

This integral is easily evaluated for an isothermal atmosphere by noting that $N = N_S \exp(-z/H)$, $\kappa_S = \sigma_S(\lambda)N$ to obtain $\tau_S = \sigma_S(\lambda)N_S H$, where the scattering cross section is a function only of the wavelength of the radiation, $N_S = 2.687 \times 10^{25}$ molecules.m^{-3}, and $H = 7.995$ km is the scale height of a standard atmosphere. This function is plotted in figure 5.7. If more accurate values are needed, the tables in Penndorf (1957) should be used.

The influence of particles is important for visible and infrared radiation propagating through even the clearest air, and the particles produce a total optical thickness that varies from 0.01 to 1.0 depending on wavelength (fig. 5.7). Recalling that multiple scatter is important when $\tau > 0.3$, it becomes obvious that even clear air is not really clear when quantitative measurements of the propagation of radiance are required. Of course, all but the thinnest clouds completely block the transmission of these bands of radiation.

The influence of particles depends on their size, number, and kind. Over the ocean, they are produced mostly by breaking waves and are hygroscopic. Thus the optical thickness of marine atmosphere is a function of wind speed and relative humidity. Near land, dust is also important, and over the tropical Atlantic Saharan dust is sometimes carried from Africa to the Caribbean. Occasionally volcanoes spew particles into the stratosphere, and the eruption of El Chichon in April 1982 produced a stratospheric haze that hampered infrared and visible observations of the ocean.

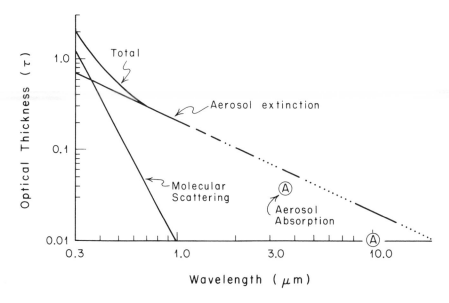

Figure 5.7 Optical thickness for a vertical path through the atmosphere, showing contributions by molecules and aerosols. The latter contribution varies with wind, humidity, and aerosol concentration; the curve is a rough estimate for clear air and 80% relative humidity. Note that molecular absorption dominates the aerosol extinction in the infrared except in the atmospheric windows. (From data in Nilsson, 1979; Penndorf, 1957; and Villier, Tanre, and Deschamps, 1980.)

The cumulative effect of particulates, molecular scattering, and weak ozone absorption on solar radiation propagating through a clear atmosphere is shown in figure 5.8. Clouds further reduce the total solar radiation. Averaged over the globe, 30% of the solar radiation is reflected and scattered to space by clouds (24%) and particles (6%); 17% is absorbed by clouds (3%) and atmospheric constituents (14%); 22% reaches the earth as diffuse skylight; and only 31% reaches the surface as direct-beam solar radiation (Sellers, 1965).

At microwave frequencies (3–300 GHz), the primary source of scatter is rainfall and dense thick clouds; and their influence is greatest at the highest frequencies. No single estimate of total absorption can be given, because it depends on rain rate, on the distribution of drop sizes, and on the height of freezing level. Ice, whose dielectric constant is small, only weakly attenuates and scatters radiation and can usually be neglected; however, a rough estimate of attenuation by rain is given in figure 5.9. Further discussion of attenuation by rain is included in chapter 10.

At all frequencies, particles not only reduce the radiance in a beam propagating through the atmosphere but also scatter radiance into the beam, obscuring the signal from the surface. The latter leads to a loss of contrast between objects at the sea surface viewed from space. The *contrast* is defined as

$$C \equiv \frac{L_t - L_s}{L_s}$$

where L_t is the radiance of the object and L_s is the background radiance. At the top of the atmosphere this becomes

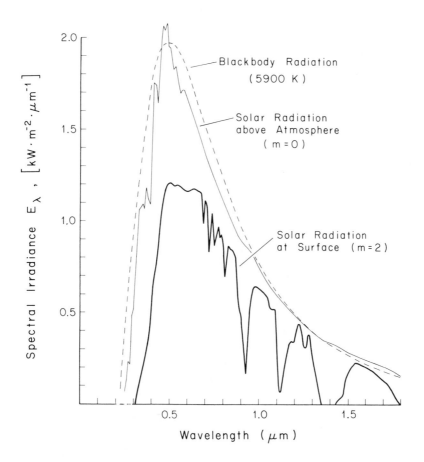

Figure 5.8 Spectral irradiance (E_λ) of direct sunlight before and after it passes through a clear atmosphere, together with the best-fitting curve of radiation from a blackbody (at 5900 K) the size and distance of the sun (see also fig. 3.5). The number of standard atmospheric masses is designated by m. Thus $m = 2$ is applicable for sunlight when the sun is 30° above the horizon (from Moon, 1940).

$$C = \frac{L_{to} - L_{so}}{L_{so} + L_A \, e^\tau}$$

where τ is the optical depth of the atmosphere and L_A is the light scattered into the beam viewed by the receiver. The latter is frequently called the airlight. As the optical depth of the atmosphere increases, so does the airlight, and the contrast decreases. Fig. 5.10 gives the result of the calculation in Fraser (1964) of the contrast viewed at the top of a Rayleigh atmosphere for two optical thicknesses, where the contrast C has been divided by its value at the surface. This is the *contrast transmittance* given by:

$$\frac{C}{C_0} = \frac{1}{1 + \dfrac{L_A}{L_s \, e^{-\tau}}}$$

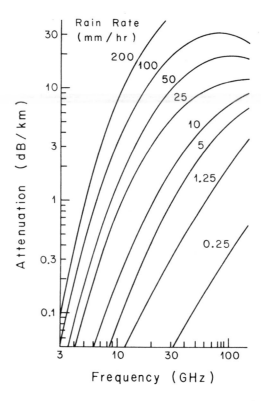

Figure 5.9 Horizontal path attenuation at various rain rates, compiled from data in the literature (from Schanda, 1976:196).

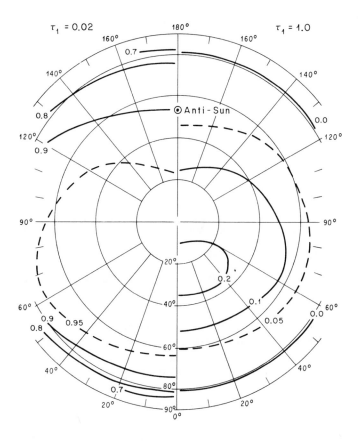

Figure 5.10 Contrast transmittance computed for a model of a Rayleigh atmosphere and a Lambert surface ($\rho = 0.25$). The optical thickness is 0.02 on the left and 1.0 on the right, and the sun is 53.1° from the vertical (from Fraser, 1964).

6

INTERACTION OF
RADIATION WITH SURFACES

Remote sensing of the sea depends in a fundamental way on an understanding of the interaction of radiation with surfaces, for it is this interaction that enables satellite instruments to infer properties of the surface, such as its temperature or roughness. In this chapter we discuss the reflection and emission of radiation from smooth and rough surfaces, as well as the influence of the surface on solutions of the equations of radiative transfer. In later chapters, we will consider particular aspects of the problem pertinent to specific frequency and wavelength bands.

Electromagnetic radiation impinging upon a surface is partly reflected and the remainder is transmitted through the surface. Both the reflected and transmitted radiation are important. The reflected radiation is sensed by radars. The transmitted portion is ultimately absorbed within the ocean, and can be related to the emissivity of the surface through Kirchoff's Law. The reflectance, transmittance, and emissivity of the surface are all functions of the viewing angle, the polarization of the radiation, and the dielectric constant of the surface. Later, these factors will be related to wind speed, wave height, and surface temperature. The precise relationships among the electromagnetic properties of the surface are not simple, but they can be explicitly stated in the case of a plane dielectric surface.

6.1 The Fresnel reflection coefficients

The *reflectance* ρ of a surface is defined to be the ratio of the irradiance (exitance) M reflected from the surface to the irradiance E incident on the surface.

$$\rho \equiv \frac{M}{E}$$

In general, it is not useful to discuss the reflectance of radiance because most surfaces change the solid angle of the beam of radiant energy. (Think of the reflection from a wavy mirror or from a polished ball.) Thus reflectance would be a function of the shape of the surface as well as its composition. If the surface is a smooth plane, however, the reflection is specular. In this case, the reflection of spectral radiance can be defined:

$$\rho(\lambda;\theta_r,\phi_r-\pi) = \frac{L_r(\lambda;\theta_r,\phi_r)}{L_i(\lambda;\theta_r,\phi_r-\pi)}$$

where λ is the wavelength of the radiation and θ, ϕ are angles in spherical coordinates (table 6.1).

Table 6.1 Notation for Describing Reflectance

a,b	Empirical functions
e	Emissivity
E	Exitance [W.m^{-2}]
f	Frequency [Hz]
h	Height of surface roughness [m]
i,r	Subscripts denoting incident and reflected radiation
L	Radiance [W.m^{-2}.sr^{-1}.μm^{-1}]
M	Irradiance [W.m^{-2}]
n	Index of refraction
p,q	Reflectance functions
S	Salinity [$^{\circ}$/oo]
t	Transmittance
T	Temperature [$^{\circ}$C]
U_{10}	Mean wind speed 10 m above the sea [m.s^{-1}]
V,H	Subscripts denoting vertical and horizontal polarization
α	Empirical constant
ϵ',ϵ''	Components of the dielectric constant
ϵ_∞	Dielectric constant at infinite frequency
ϵ_r	Dielectric constant
ϵ_s	Static dielectric constant
ϵ_0	Permittivity of free space [8.85$\times$10^{-12} F.m^{-1}]
λ	Wavelength [m]
η,χ	Components of the complex index of refraction
θ,ϕ	Spherical coordinates
θ	Incidence angle
θ_B	Brewster's angle
ρ	Reflectance
σ	Ionic conductivity [S.m^{-1}]
τ	Relaxation time [s]
τ	Optical depth
ω	Frequency [rad.s^{-1}]

A variety of notation and terms is used to discuss reflection. Reflectance, denoted by ρ, is the accepted term and symbol; but reflectivity and R are sometimes used to describe the reflectance from a plane dielectric. *Albedo* is an astronomical quantity denoting the ratio of solar energy reflected from a body to solar energy incident on a body; but the term is sometimes used (incorrectly) as a synonym for reflectance.

The spectral reflectance of a plane dielectric surface is given by the *Fresnel reflection coefficients*. These are obtained by considering a plane electromagnetic wave incident on the surface, applying the appropriate boundary conditions, and solving for the reflected and transmitted wave. The details are given in many textbooks (see Born and Wolf, 1970, 38ff. 615ff.). We use the expressions in Schanda (1976:208) for the reflectance:

$$\rho_H(\theta) = \frac{(p - \cos\theta)^2 + q^2}{(p + \cos\theta)^2 + q^2}$$

$$\rho_V(\theta) = \frac{(\epsilon'\cos\theta - p)^2 + (\epsilon''\cos\theta + q)^2}{(\epsilon'\cos\theta + p)^2 + (\epsilon''\cos\theta + q)^2}$$

where ρ_H, ρ_V are the reflection coefficients for horizontally and vertically polarized components of the radiation and $\epsilon = \epsilon' - i\epsilon''$ is the complex dielectric constant of the surface. The relations p and q are:

$$p = \frac{1}{\sqrt{2}} \left\{ [(\epsilon' - \sin^2\theta)^2 + \epsilon''^2]^{1/2} + [\epsilon' - \sin^2\theta] \right\}^{1/2}$$

$$q = \frac{1}{\sqrt{2}} \left\{ [(\epsilon' - \sin^2\theta)^2 + \epsilon''^2]^{1/2} - [\epsilon' - \sin^2\theta] \right\}^{1/2}$$

The angle θ is the *incidence angle*, the acute angle between the line of propagation of the radiation and the normal to the surface. When $\theta = 0$, the radiation is perpendicular to the surface and the equations reduce to:

$$\rho_H = \rho_V = \rho = \frac{(p - 1)^2 + q^2}{(p + 1)^2 + q^2} = \frac{(\eta - 1)^2 + \chi^2}{(\eta + 1)^2 + \chi^2}$$

with

$$p = \eta = \frac{1}{\sqrt{2}} \left[(\epsilon'^2 + \epsilon''^2)^{1/2} + \epsilon' \right]^{1/2}$$

$$q = \chi = \frac{1}{\sqrt{2}} \left[(\epsilon'^2 + \epsilon''^2)^{1/2} - \epsilon' \right]^{1/2}$$

where the complex index of refraction is

$$n = \eta + i\chi$$

Recalling that

$$n^2 = \epsilon' - i\epsilon''$$

we obtain the usual form of the reflection coefficient at normal incidence:

$$\rho = \left| \frac{n - 1}{n + 1} \right|^2$$

Thus the reflectance depends essentially on the contrast between the index of refraction of the surface and that of the air, which is the contrast between the velocities of light in the two media.

The reflectance is zero when the numerator of the Fresnel equations is zero.

For nonabsorbant materials ($\epsilon'' = 0$) two cases are possible: (a) the trivial instance requires $\epsilon' = 1$, and there is no contrast in dielectric constant, hence no discontinuity in the velocity of light at the interface; and (b) $\rho_V = 0$ at *Brewster's angle θ_B*. This requires

$$\tan^2 \theta_B = \epsilon'$$

$$\tan \theta_B = \eta$$

Because only $\rho_V = 0$, the reflected radiance at Brewster's angle is horizontally polarized. If $\epsilon'' \neq 0$, then $\rho_v \neq 0$ at the Brewster angle, but it does reach a minimum.

The spectral emissivity of a surface is related to its spectral reflectance through Kirchoff's Law. Consider an optically thick ocean, one that is so absorbant that no transmitted radiance is reflected back out from subsurface particles or from the bottom. Then all radiation transmitted through the surface is ultimately absorbed, and the transmittance t is equal to the absorptance A. By Kirchoff's Law, the spectral emissivity e must be equal to the transmittance. But

$$\rho + t = 1$$

Therefore,

$$e_V(\theta) = 1 - \rho_V(\theta)$$

$$e_H(\theta) = 1 - \rho_H(\theta)$$

where the subscripts refer to the polarization of the radiation.

In applying this result to a real surface, some care must be taken. The emitted radiation comes not from the surface, but from a layer several skin depths thick just below the surface. Some transmitted radiation could be reflected from subsurface particles or from the bottom, thus violating the assumptions used to relate emissivity to reflectance. Sea water strongly absorbs the radio frequency and infrared radiation used in remote sensing, and subsurface reflection is rarely a problem; however, snow and ice only weakly absorb radiation, and subsurface reflections and scatter are important. For these materials, the equations of radiative transfer, applied to that portion of the radiation transmitted through the surface, can be solved to calculate the scattered, reflected, and emitted radiation propagating back up toward the surface. A second application of the reflection coefficients then gives the amount of this radiation that is transmitted through the surface.

The reflectance and emissivity of sea water as a function of incidence angle (fig. 6.1) illustrates the properties of a typical surface. For visible and infrared radiation

$$n \approx 1.2 - 1.3$$

so

$$\rho \approx 0.02 - 0.01$$

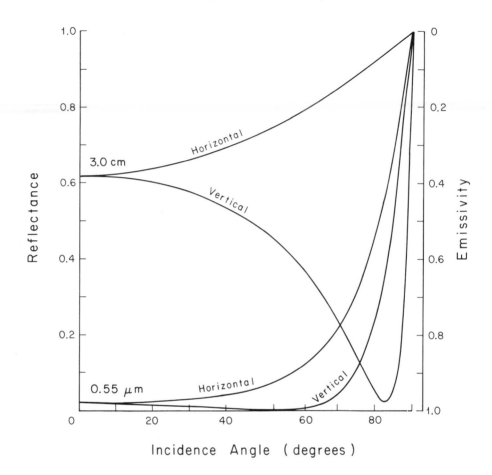

Figure 6.1 Reflectance and emissivity of a plane sea-water surface as a function of incidence angle (angle from the normal to the surface), for two different wavelengths of radiation. Horizontal and vertical refer to the polarization of the radiation.

and

$$e(0) \approx 0.98 - 0.99$$

Thus water is almost a blackbody at visible and infrared frequencies. For vertically polarized radiation incident at Brewster's angle ($\approx 53°$), all radiation is absorbed, and water is a perfect emitter. For 3 cm (10 GHz) radio waves and 20°C sea water, the dielectric constant, reflectance, and vertical emissivity are:

$$\epsilon_r \approx 52 - 37\,i$$

$$\rho \approx 0.61$$

$$e(0) \approx 0.39$$

Thus sea water is a poor emitter of 3 cm radio waves.

6.2 Dielectric constants of surfaces

In order to specify further the spectral reflectance and emissivity of typical surfaces using the Fresnel reflection coefficients, we need to know either the complex dielectric constant or the index of refraction of the surface as a function of frequency. For some materials, such as pure or salt water, the function is accurately known. For other substances, such as snow and ice (in polar regions), the dielectric constants are known mostly empirically, and only for limited bands of frequency or wavelength.

The materials of interest in satellite oceanography are pure water (rain), ice, salt water, and sea ice. Both sea ice and ice are found in many forms, such as ice in clouds, snow, hail, first-year sea ice, and multi-year sea ice, each with different emissivities and dielectric constants depending on structure and amount of salt inclusions. These differences are important, and they enable satellite instruments to contribute to our understanding of polar regions.

The complex index of refraction of pure water and ice has been investigated over a very wide band of wavelengths (figs. 6.2 and 6.3) by many workers; the results are summarized by Ray (1972), but see also Downing and Williams (1975). For pure water, note that the real part of n is nearly 1 in the visible and infrared, and emissivities are large; that the complex part is small at visible wavelengths; and that it increases rapidly to a value 100 times larger at infrared wavelengths near $10 \, \mu$m. Thus absorption is weak at light frequencies, but large at infrared. Pure ice differs from water primarily at radio frequencies. Ice is nearly transparent, while water is a good absorber.

For conducting materials (and all but the driest desert sand has enough water to be conducting), and for frequencies away from absorption bands (radio frequencies), the dielectric constant is usually represented by the Debye equation

$$\epsilon_r = \epsilon_\infty + \frac{\epsilon_s \, \epsilon_\infty}{1 + (i\omega\tau)^{1-\alpha}} - i \, \frac{\sigma}{\omega\epsilon_0}$$

where $\omega = 2\pi f$ is the frequency of the radiation, ϵ_∞ is the dielectric constant at infinite frequency, ϵ_s is the static dielectric constant, τ is the relaxation time, σ is the ionic conductivity, and α is an empirical constant. While simple in form, the equation is actually quite complex because ϵ_s, τ, and σ are all functions of the material.

The conductivity of sea water increases the complex part of the dielectric constant according to Debye's equation. This is important only at radio frequencies. Light and infrared have very high frequencies (ω); consequently, the conductivity term is small. The only influence of the ions in sea water is through their changes in the absorption bands of the liquid. The dielectric constant of sea water was first investigated, with great care, by Lane and Saxton (1952), and their work has been the standard against which later measurements have been compared. Using their equations for the behavior of sea water, Wilheit (1972) gives an overview of both the dielectric constant (fig. 6.4) and the emissivity of sea water (fig. 6.5) at SHF frequencies (1–50 GHz). From these figures, note that:

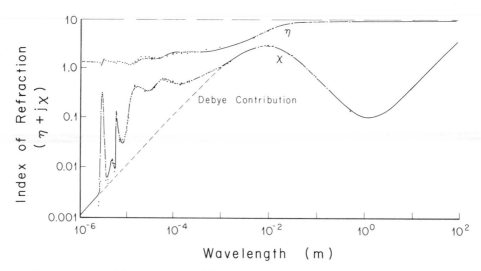

Figure 6.2 Real (η) and imaginary (χ) components of the index of refraction of pure water as a function of wavelength (from Ray, 1972).

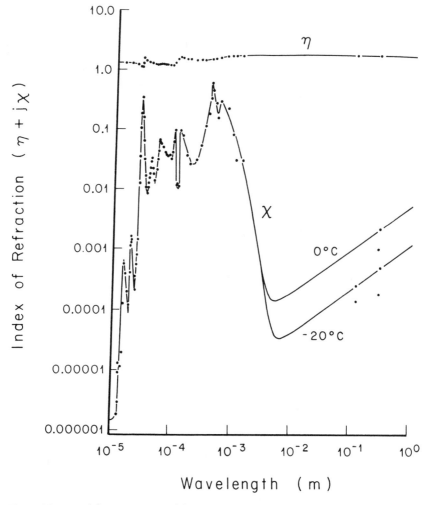

Figure 6.3 Real (η) and imaginary (χ) components of the index of refraction of pure ice as a function of wavelength (from Ray, 1972).

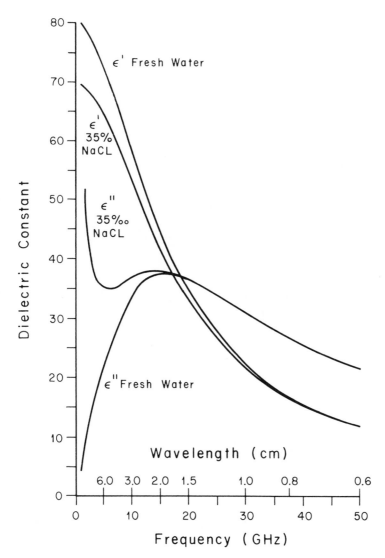

Figure 6.4   Dielectric constant of pure and salt water at 20°C as a function
of radio frequency (from Wilheit, 1972).

(a) The influence of salt on emissivity is small, except at frequencies below 5 GHz, where it lowers the emissivity.

(b) The emissivity of both salt and fresh water is very low, approaching 0.36 at frequencies below 5 GHz, and that it remains, for the most part, below 0.5.

(c) The complex part of the dielectric constant is large, and exceeds the real part for frequencies above 20 GHz.

The influence of temperature is more complicated. The emissivity of sea water for frequencies near 1.5 and 20 GHz is inversely proportional to temperature, so the brightness temperature of sea water is independent of surface tem-

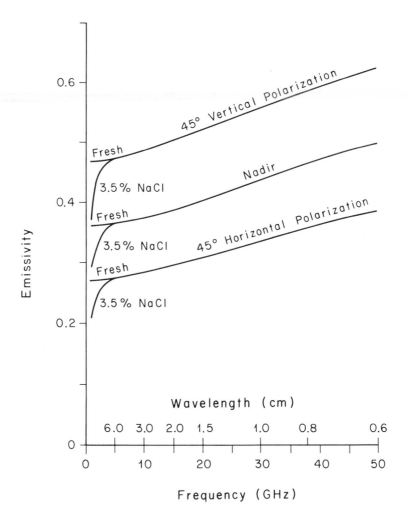

Figure 6.5   The emissivity of pure and salt water at 20°C as a function of radio frequency
(from Wilheit, 1972).

perature (fig. 6.6) at these two frequencies. This fact has a very useful consequence. From figures 6.5 and 6.6, it is clear that a dual-frequency radiometer,
operating close to the sea surface to eliminate the influence of the atmosphere,
can independently measure salinity and temperature if it uses one frequency
near 1.0 GHz, and the other near 5.0 GHz.

This overview of radio-frequency emissions from sea water is useful for
estimating the influences of the surface, but precise estimates require careful
measurements of the dielectric constant as a function of temperature, salinity,
and frequency if the error in knowing the emissivity of sea water is to be less
than the 0.1 K error of modern radiometers. Klein and Swift (1977), reviewing
the literature, give analytic expressions for the parameters in the Debye equation. They propose:

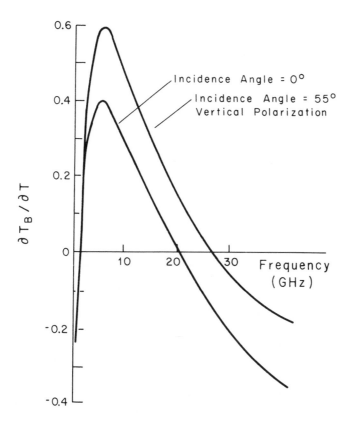

Figure 6.6   Change in sea surface brightness temperature per change in surface temperature over the range 0-30°C, as a function of radio frequency (from
Wilheit, 1978).

$$\alpha = 0$$

$$\epsilon_s(T,S) = \epsilon_s(T)a(S,T)$$

with coefficients

$$\epsilon_s(T) = 87.134 - 1.949 \times 10^{-1}T - 1.276 \times 10^{-2}T^2 + 2.491 \times 10^{-4}T^3$$

$$a(T,S) = 1.000 + 1.613 \times 10^{-5}ST - 3.656 \times 10^{-3}S + 3.210 \times 10^{-5}S^2$$
$$- 4.232 \times 10^{-7}S^3$$

and

$$\tau(T,S) = \tau(T,O)b(S,T)$$

with coefficients

$$\tau(T,O) = 1.768 \times 10^{-11} - 6.086 \times 10^{-13}T + 1.104 \times 10^{-14}T^2 - 8.111 \times 10^{-17}T^3$$

$$b(S,T) = 1.000 + 2.282 \times 10^{-5}ST - 7.638 \times 10^{-4}S - 7.760 \times 10^{-6}S^2$$
$$+ 1.105 \times 10^{-8}S^3$$

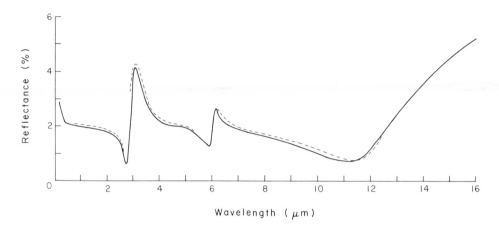

Figure 6.7   Reflectance of pure water (solid line) and sea water (dashed line) for radiation incident normal to a plane water surface (from Friedman, 1969).

plus

$$\epsilon_\infty = 4.9 \pm 20\%$$

where T is temperature in degrees centigrade and S is salinity in parts per thousand.

Dissolved salts in sea water influence the infrared reflectance by shifting the H_2O absorption bands, by adding a weak absorption band at 9 μm due to the polyatomic sulfate ion, and by increasing the refractive index (Friedman, 1969). Kropotkin and Sheveleva (1975) observed the strong influence of the sulfate ion, especially $MgSO_4$, in strong ionic solutions. Hobson and Williams (1971) note that if the salt concentrations are typical of sea water (3%), most influences are too small to be measured, except in the region of 8–14 μm, where the reflectance of sea water can differ from that of pure water by up to 10%. Friedman (1969) gives accurate values for the reflectance of sea water over the full infrared band (fig. 6.7), and his measurements, together with the above studies, provide a firm foundation for use of infrared to study sea surface temperatures. For exact estimates of the reflectance, his published tables should be used.

The reflectance and emissivity of snow and ice, in contrast to those of sea water, are much more complicated and less well understood. For SHF radio waves interacting with sea ice, subsurface scatter tends to be small and the variability of reflectance depends primarily on the variability of absorptance. Both salt in the form of brine inclusions and melt water on the surface and within the ice have much greater absorptance than ice. The concentrations of water and brine depend on the history of the surface: brine depends on the long-term history of melting and freezing of the ice, and tends to be less in old ice and greater in newly formed ice; water concentration depends on the recent thermal history of the ice, and water exists whenever temperatures exceed freezing.

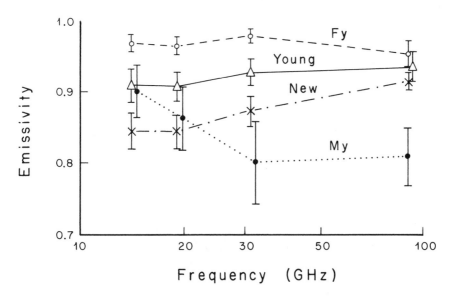

Figure 6.8 Microwave emissivity at vertical incidence for four types of sea ice as a function of frequency. The vertical bar shows one standard deviation of the estimate; FY designates first-year ice, MY multiyear ice. All data are at the same frequency, but some points have been slightly displaced to avoid overlap (from Troy et al., 1981).

Because of the complexity of these processes, the only reliable estimates of reflectance and emissivity come from studies of ice in nature. In general, the emissivity of ice is relatively high compared with that of water, and first-year Arctic ice differs from multi-year ice. Both differ from snow. First-year ice, in the Arctic, has an emissivity that is near 0.95 and independent of frequency (fig. 6.8). Multi-year ice has a decrease in emissivity at higher frequencies, although the exact value for either type of ice is not well known (Cavalieri, Gloersen, and Campbell, 1981; Svendsen et al., 1983). Snow cover strongly influences the emissivity of sea ice (Campbell et al., 1978), and snow-free ice has higher emissivities than does Arctic ice in its natural state. In summer, thin pools of melted water form on the surface of the ice, increasing the emissivity to the point that first-year and multi-year ice cannot be differentiated (Gloersen et al., 1978).

Pure snow weakly absorbs visible, infrared, and SHF radiation; radiation penetrates deeply, and subsurface scatter is important. Calculations of subsurface propagation of radiation (Choudhury and Chang, 1979; Chang et al., 1976) indicate that emissivity is a function of grain size and that emissivity can be estimated from Mie theory. The grain size is a function of time: grains sink slowly and grow in size, and the average grain size near the surface is a function of the rate of snow accumulation. Thus emissivity can be used to measure the accumulation rate of snow and ice on glaciers.

The reflectance of snow and ice greatly decreases as the wavelength increases from the visible to the infrared (fig. 6.9). The absorptance of light is low, and

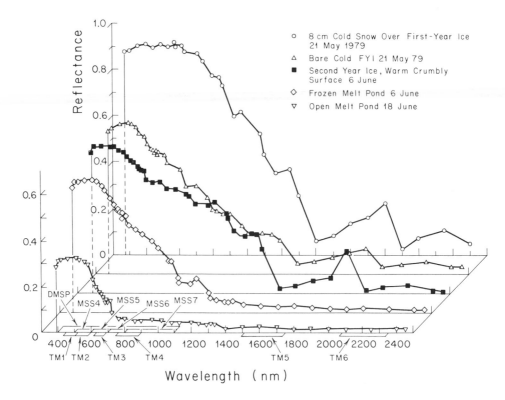

Figure 6.9   The reflectance of various types of snow and ice as a function of wavelength of the radiation. Also indicated on the ordinate are the wavelength bands observed by the Multi Spectral Scanner (MSS) on Landsat, by the Thematic Mapper (TM) on Landsat 4; and by instruments on the Defense Meteorological Satellites (DMSP). Figure supplied by T.C. Grenfell.

subsurface scatter reflects radiation just as does foam on the sea surface; thus snow and ice are white. Longer wavelengths penetrate less deeply and scatter is less important. Ultimately, for wavelengths near $10\,\mu$m, frozen water is a good absorber. For these wavelengths reflectivity is small, emissivity is large, and snow and ice are nearly blackbodies. The wavelength at which scatter ceases to be important is a function of snow or ice type, and near infrared radiation can be used to differentiate among the various types (Dozier, Schneider, and McGinnis, 1981).

These various estimates of the emissivity of natural surfaces are summarized in table 6.2. The values of emissivity can be placed into perspective by noting that a change in emissivity of 0.01 produces a change in emitted radiance that is equivalent to a change in surface temperature of 0.7°C at $11\,\mu$m and 2.7°C at 10 GHz for surfaces near room temperature. Thus emissivities must be known with an accuracy of 0.1% in order to make useful measurements of sea surface temperature. At the same time, radiometers have typical resolutions of a few tenths of a degree Kelvin.

Because the emissivity of a surface is a function of the material within a few

Table 6.2 Vertical emissivities of typical surfaces

Frequency	Substance	Emissivity
Light	Fresh & Salt water	0.98
	Ice	0.35
Infrared	Fresh & Salt water	0.99
$(10-12\mu)$	Ice	0.95
Radio (1 GHz)	Fresh water	0.36
	Salt water	0.30
	First-year sea ice	0.98
	Multi-year sea ice	0.90
Radio (11 GHz)	Fresh or salt water	0.38
	First-year sea ice	0.95
	Multi-year sea ice	0.90

skin depths of the surface, estimates of these depths are useful. Some typical values are: (a) 75 m for 0.4 μm blue light propagating through pure sea water; (b) 3 m for 0.7 μm red light propagating through pure sea water; (c) 3 μm for 10 μm infrared indication incident on sea water; and (d) 0.5 cm for 5 GHz radio waves incident on sea water. Less exact figures are available for frozen water, but 19 GHz radio waves tend to propagate a few millimeters into first-year ice, a few centimeters into multi-year ice, and a few meters into dry snow and pure ice (Zwally and Gloersen, 1977).

6.3 Radiative transfer in the presence of a boundary

Surfaces, being partial reflectors, influence the solution of the radiative transfer equation. Radiation emitted by or propagating through the atmosphere is partially reflected by the surface, and the reflected radiation joins the upwelling radiation emitted or scattered by the atmosphere (fig. 4.1). To clarify the concept, consider a simple example: the transfer of radiation in an atmosphere of constant temperature T_A and optical depth τ, above a sea of temperature T_s, with emissivity e. What is the temperature seen by a downward looking SHF radiometer flying above the atmosphere (fig. 6.10)?

To calculate T, break the problem into a number of small pieces. The brightness temperature of radiation incident on the surface due to atmospheric emissions is

$$T_A (1 - t)$$

where $t = \exp [-\tau]$. The amount reflected is

$$(1 - e) T_A (1 - t)$$

and the amount emitted is eT_s, so the total brightness temperature of upwelling radiation at the sea surface is

$$eT_s + (1 - e) (1 - t)T_A$$

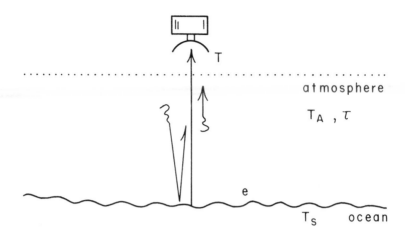

Figure 6.10   Transfer of radiation in the presence of a surface.

This radiation is attenuated and joined by additional atmospheric emission, so that the total brightness temperature seen by the radiometer is

$$T = T_A (1 - t) + t [eT_s + (1 - e)(1 - t) T_A]$$

$$T = et [T_s - T_A (1 - t)] + T_A (1 - t^2)$$

Note that if $\tau = 0$, $t = 1$, then $T = eT_s$ as it should be. If $\tau \gg 0$, $t \approx 0$, then $T = T_A$; again, as expected.

Now do a simple calculation. Assume the radiometer is observing 5 GHz radiation emitted from the sea surface and propagated through a typical clear atmosphere with attenuation and brightness given by figures 4.9 and 4.10. Of course, this is not quite a correct use of the data in figure 4.10 because an upward looking radiometer does not measure exactly the same brightness temperature for the atmosphere as does a downward looking radiometer, but the point is not critical for our argument. From the figures, the attenuation is 0.035 dB, or $t = 0.99197$, $T_A = 2$ K (after removing the 3 K cosmic background from the figure). From figure 6.4, $e = 0.36$; and if we assume $T_s = 300$ K, then:

$$T = (0.36)\,(0.99197)\,[300\text{ K} - 2.0\text{ K}\,(0.008)] + 2.0\text{ K}\,(0.016)$$

$$T = 107.16\text{ K}$$

so the observed surface temperature is:

$$T/0.36 = 297.66\text{ K}$$

Therefore, the radiometer measures a sea surface temperature that is 2.34 K colder than it really is due to the influence of the atmosphere and its reflection in the sea. This value can be broken down as follows: The atmosphere reduces the radiation emitted by the sea surface by 2.408 K, but direct atmospheric emissions add 0.044 K and reflected atmospheric emissions add another 0.028 K (-2.33 K $= -2.41$ K $+ 0.04$ K $+ 0.03$ K).

This simple example illustrates several ideas: (a) the atmosphere was remarkably clear, yet it could not be ignored; (b) on the other hand, the corrections were small enough to be estimated accurately; (c) the radiation from the atmosphere reflected in the surface was almost as large as the direct contribution; and (d) the atmospheric absorption was much greater than the emission. The latter is not surprising, because radiation from a warm body (the sea) must warm up the colder atmosphere, so more energy is absorbed than radiated. Remember, Kirchoff's Law only gives the rate at which energy is exchanged relative to a blackbody of the *same* temperature.

6.4 Rough surfaces

Up to this point we have considered the specular reflection from plane dielectric surfaces. But the ocean surface is seldom smooth, and is usually covered with waves. To understand the reflectance of the sea, we need both a criterion for smoothness and techniques for calculating the reflectance of a rough surface.

A *smooth* (specular) surface is one with

$$h \cos \theta_i < \lambda/8$$

where h is the height of the roughness and λ is the wavelength of the radiation. Conversely, a *rough* surface is one that does not meet this criterion. Between these two extremes is yet a third category often used for describing the sea: a *composite* surface is one that is locally smooth, but with significant variation in slope due to long waves. Thus the surface can be decomposed into smooth facets, each with a local incidence angle that differs from the mean incidence angle averaged over a large area. The reflectance of each facet is calculated from the Fresnel coefficients, and the average reflectance of the surface is obtained by integrating over the influence of all facets. Despite the simplification resulting from assuming a composite surface, calculations of the scatter or emission from the sea surface often require additional assumptions, the most common requiring that either the slopes or the local radius of curvature of the surface be small. Complete and detailed discussions of scatter from rough surfaces in general, and the sea in particular, can be found in Beckmann and Spizzichino (1963), Bass and Fuks (1979), Ishimaru (1978), and the references in chapters 9 and 10.

The reflectance of light and infrared radiance can be calculated assuming a composite surface with a distribution of surface slopes given by Cox and Munk (1954). In fact, these slope statistics were obtained using observations of sunlight reflected from the sea. Because the facets required by the composite-surface assumption can easily be smaller than a millimeter on a side, there is no doubt that they are locally smooth surfaces and produce specular reflection. The calculation indicates that sunlight reflected from rough seas can be seen in a cone of angles 20–30° about the expected point of specular reflection; and that the size of the cone increases with increasing wind speed. The reflected sunlight, called sun glitter, can be useful in itself or can be a source of noise, depending on the particular measurement.

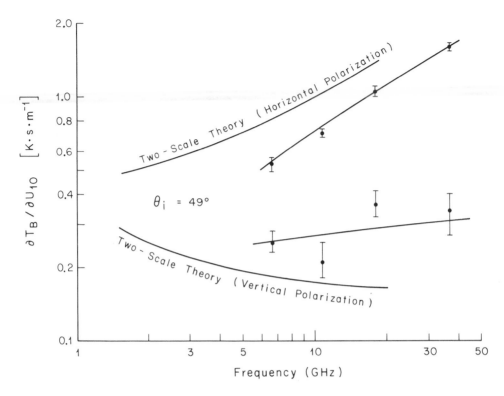

Figure 6.11   Sensitivity of brightness temperature of the sea surface to wind speed at a height of 10 m, as a function of frequency, at an incidence angle of 49°. The observed sensitivity (points) is explained almost entirely to variations in surface slopes due to ocean waves (from data in Wentz, Christensen and Richardson, 1981).

The vertical reflectance of SHF radio waves (fig. 6.11) can be calculated in a similar way (Wilheit, 1979a). The reflectance of wavelengths shorter than 1 cm (30 GHz) is accurately calculated using the Cox and Munk statistics, but longer wavelengths respond as though the sea were somewhat smoother than this (Hollinger, 1971). The apparent smoothness results from the capillary and short gravity waves on the surface not being high enough to violate the criterion of smoothness, and thus not contributing to the slope of the local facets. More detailed calculations, including atmospheric emission scattered by the small-scale roughness within each facet, have been reported by Wentz (1975), but they do not substantially change the conclusions.

The reflectance of SHF radio waves at large incidence angles involves coherent scatter from waves with wavelengths matching the projected radio wavelength. This is *Bragg scatter*, discussed more fully in chapter 10.

Foam on the sea surface, produced by breaking waves, alters the reflectance of light and SHF radiation. The influence depends on the amount of foam and foam streaks. These increase rapidly as the wind speed exceeds 7 m.s^{-1}, and eventually completely cover the surface when winds exceed roughly 40 m.s^{-1}. The foam and foam streaks are white when viewed with visible light, in con-

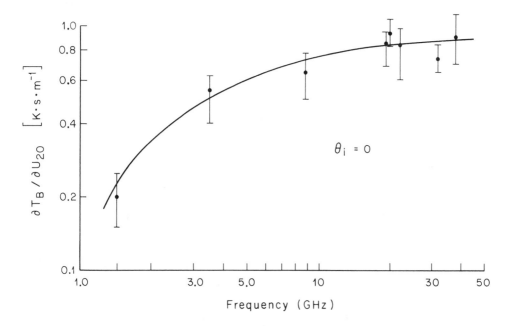

Figure 6.12 Sensitivity of brightness temperature of the sea surface to wind speed at a height of 20 m, as a function of frequency, at vertical incidence angle. This sensitivity is due almost entirely to foam produced from breaking waves generated by the wind (from data in Basharinov and Shutko, 1980).

trast with the low reflectance of calm water; thus they increase the reflectance. Because infrared radiation is strongly attenuated by water, foam has little influence, and both foam and calm water have nearly the same reflectance.

The influence of foam on SHF radiation is known primarily through experiments. These indicate that (a) the change in emissivity due to foam results mainly from the distortion of the sea surface at the interstices between bubbles (Williams, 1971), rather than by the bulk properties of the foam; (b) the emissivity increases linearly with wind speed, the increase being greatest at the highest frequencies; (c) the influence of foam tends to be more important than the emissivity at vertical incidence (fig. 6.12); and (d) conversely, at incident angles near 50°, the influence of surface slopes tends to dominate the influence of foam until the wind speed exceeds roughly 20 m.s^{-1} (Wentz, Christensen, and Richardson, 1981).

7

OBSERVATIONS
USING VISIBLE LIGHT

Observations of natural radiation in the visible, infrared, and radio frequency bands are used to study the sea from space. Each band has its inherent advantages and disadvantages; each is exploited to measure particular phenomena; and each has been observed by succeeding generations of instruments of increasing sensitivity and usefulness. In general, instruments that observe light or infrared have greater spatial resolution but are limited to cloudless, clear atmospheres. Instruments observing radio frequencies have lower spatial resolution but can see through clouds. The complexity of the instrument increases with wavelength, and longer wavelengths have been exploited more recently. Light was observed by the first instruments beginning in 1961; infrared was developed over the following decade and the radio bands still later. The end is not yet in sight. Newer, still more complex spacecraft and instruments are being prepared. Lasers are expected to measure currents, winds, and perhaps sea surface temperature. Very precise radars and improved cameras are being developed for use later in this decade. With this chapter we start a series that surveys the various frequency bands, their advantages and disadvantages, and their particular instruments and techniques. We begin with visible light, progress to infrared- and radio-frequency radiometers, and end with a discussion of radar scatter from the sea and radars.

7.1 The instruments

The first images of Earth from space were made by Tiros-1 and showed sunlit clouds. The pictures were crude, with great geometric distortion and low range of sensitivity. But they were soon followed by photographs taken by orbiting astronauts, which showed Earth in stunning detail. These spurred the development of remote automatic systems able to view Earth with high resolution and sensitivity. Out of this work emerged the Earth Resources Technology Satellite (ERTS-1), soon renamed Landsat 1.

[100]

Landsat 1 observed the Earth with a resolution of 80 m in seven wavelength bands ranging from light to near infrared, using two relatively sensitive instruments, the multispectral scanner and the return-beam vidicon. Landsat 1 was soon followed by two more satellites, Landsat 2 and 3. These Landsats are nearly the same, the primary differences being listed in tables 7.1 and 7.2 (see also Rabinove et al., 1981). Of the two instruments carried on the satellites, the multispectral scanner has proved to be the most useful and durable. Its three versions have produced tens of thousands of pictures of the world. In contrast, only a few images were made by the return-beam vidicon. The success of the Landsats has led to new systems and instruments. Landsat 4, launched in 1982, carried a completely different instrument, the thematic mapper (Salomonson et al., 1980); in the future, France plans to fly a high resolution visible instrument on the Spot satellite (Chevrel, Courtois and Weil, 1981).

The land-observing satellites were designed to view scenes with a wide range of brightness. Thus their multispectral scanners were poorly suited to observe the subtle variations in the hue of the oceans that indicate the distribution of

Table 7.1 Multispectral Light Sensors

Satellite	Instrument	Acronym	Swath (km)	Resolution (m)	Incidence angle	Operational
Landsat 1,2	Return Beam Vidicon	RBV	185	79	±5.8°	1972
	Multispectral Scanner	MSS	185	79	±5.8°	1972—
Landsat 3	Return Beam Vidicon	RBV	98+98	24	±5.8°	1978—
	Multispectral Scanner	MSS	185	79 / 237 (band 8)	±5.8°	1978—
Nimbus 7	Coastal Zone Color Scanner	CZCS	1600	800	±40°	1978—
Dynamics Explorer 1	Scanning Auroral Imager	SAI	348— 14,000	2.9—117 (km)	±15°	1981—
Landsat 4	Thematic Mapper	TM	185	30 / 120 (band 7)	±5.8°	1982—
Spot 1	High Resolution Visible	HRV	60+60	10—20	±33°	1986—
MOS 1	Multispectral Electronic Self Scanning Radiometer	MESSR	100+100	50	±3.2°	1986—

Table 7.2 Spectral Response of Light Sensors

Satellite Instrument	Band No.	Wavelength (μm)	Full Scale (W.m^{-2}.sr^{-1}.μm^{-1})	Sensitivity (W.m^{-2}sr^{-1}.μm^{-1})
Landsat 1,2,3	1	0.475—0.575	200	10
Return Beam	2	0.580—0.680	225	(analog
Vidicon	3	0.698—0.830	188	signal)
(RBV)	1	0.550—0.750†	82	
Landsat 1,2,3	4	0.5—0.6	248	3.9—1.3
Multispectral	5	0.6—0.7	200	3.2—1.0
Scanner	6	0.7—0.8	176	2.8
(MSS)	7	0.8—1.1	153	2.4
	8	10.4—12.6†	340(K)	1.5(K)
Nimbus 7	1	0.433—0.453	115—54	0.21
Coastal Zone	2	0.510—0.530	76—35	0.14
Color Scanner	3	0.540—0.560	62—29	0.11
(CZCS)	4	0.660—0.680	29—13	0.05
	5	0.700—0.800	29	0.11
	6	10.5—12.5	320(K)	0.22(K)
Dynamics Explorer 1 Scanning Auroral Imager (SAI)	7	0.471—0.493	3×10^{-6}	2×10^{-9}
Landsat 4	1	0.45—0.52	143-40	0.16
Thematic	2	0.51—0.60	314-27	0.10
Mapper	3	0.63—0.69	225—22	0.08
(TM)	4	0.76—0.90	214—14	0.05
	5	1.55—1.75	30—4	0.02
	7	2.08—2.35	16—1.7	0.006
	6	10.4—12.6	320(K)	0.5(K)
Spot 1	1	0.50—0.59	350	0.86
High Resolution	2	0.61—0.68	355	0.80
Visible	3	0.79—0.89	262	0.48
(HRV)	4	0.51—0.73	344	0.92
MOS-1	1	0.51—0.59	288	3.1
Multispectral	2	0.61—0.69	213	2.4
Electronic	3	0.72—0.80	200	2.3
Self Scanning Radiometer (MESSR)	4	0.80—1.10	180	2.0

†Landsat 3 only.

riverborne sediments, growing plankton, and shallow bottoms. The Coastal Zone Color Scanner (CZCS) carried on Nimbus-7 was specifically designed for these tasks. It has good sensitivity for observing slight changes in radiance from the sea, a number of channels to observe different colors of light, and a broad field of view capable of observing wide areas of the ocean. This too has proven to be a popular and useful instrument, and similar ones are planned for future satellites. The first will be the multispectral electronic self scanning radiometer on the Japanese Marine Observing Satellite (MOS-1), and another may be flown on a U.S. meteorological satellite.

In parallel with the development of multispectral scanners for viewing the land and sea was the development of the visible and infrared radiometers for viewing clouds in the atmosphere. In general, the meteorological sensors produce poorer quality images than those from the multispectral scanners. The former have a resolution of only a kilometer or so and observe only a single wide band of light. Still they have been useful, particularly for studying polar regions and cloud-level winds. The characteristics of the meteorological sensors will be described in the next chapter, along with those of the infrared radiometers, with which they are closely combined. In this chapter we will be content to describe just their contribution to oceanography.

Several instruments have been especially designed to observe the sea at night. The first, carried on the Applications Technology Satellite ATS-4, failed to orbit. A second and different instrument, the Scanning Auroral Imager, was launched aboard the Dynamics Explorer in 1981 (Frank et al., 1981). It has one channel that responds to the blue light of bioluminescence, and sufficient sensitivity for observing the signal from space. An airborne system for observing bioluminescence, operated by Zapata Fisheries Development Corporation (Bulban, 1979), uses a low-light-level television camera in a light aircraft flying at 2000 m. It is used routinely to search for schools of fish in coastal regions. To view bioluminescence, these instruments must be very sensitive to light. Luminous patches of organisms produce light with an upwelling brightness of only around 10^{-5} W.m^{-2}.sr^{-1}.μm^{-1}. In contrast, sunlit waters are 100 million times brighter, and starlit waters on a moonless night are only one one-hundredth as bright.

The satellite-borne instruments generally look vertically downward, that is, they are *nadir viewing*. As the satellite moves along, a photocell (or equivalent sensing element) scans from side to side along a line nearly perpendicular to the subsatellite track (fig. 7.1). The width of the observed area is the *swath* width, and the diameter of the smallest discernible spot on the sea surface is the *resolution* or instrument field of view (IFOV). The image produced by the instrument consists of lines of points or *pixels*, each giving the brightness of an area on the surface the size of the instrument field of view.

Such scanning imagers are not always ideal. The scanning element can jam; and the photocell must view from nadir out to the edge of the swath and back in the time it takes the satellite to move the distance of one resolution element. Landsats move roughly 8 km.s^{-1}, and have a resolution of 0.08 km, so they must complete one scan in 0.01 s. If the swath is wide and the IFOV small, the photocell may not gather enough photons of light from each point on the surface for the instrument to respond properly — this is equivalent to setting too short an exposure time on a camera.

Satellite Imagers

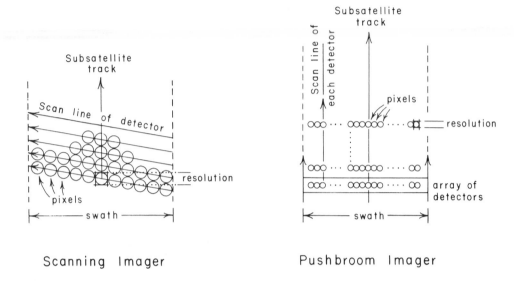

Scanning Imager   Pushbroom Imager

Figure 7.1 Two techniques used by visible-light imagers to observe wide areas. The most common scans a single detector from side to side as the satellite moves along, building up an image much like a television picture. A second technique, planned for newer instruments, has an array of stationary detectors that view the surface as it passes beneath the satellite.

The solution is to combine hundreds or even thousands of photocells in a linear array, each cell viewing along a line parallel to the subsatellite track. This is the *pushbroom* approach, and it allows each detector to view the surface for sufficiently long periods of time to provide a sensitive response to variations in the brightness of the surface.

These descriptions of instruments are brief and not meant to be a practical guide. Complete details are given in the various publications produced by the agencies that developed the satellites and by the companies that built them. For example, General Electric publishes user guides to the Nimbus-7 and Landsat systems, and these are available through the data-distribution centers listed in chapter 16. Other descriptions of optical instruments are given by Slater (1980).

Two types of publications must usually be consulted: (a) technical descriptions of the instrument, the voltages or signals it produces in response to a specified input, steps taken to process the signals, and the format of the data; and (b) descriptions of the scientific basis for the measurement and its interpretation. The former information is given in the various user guides, the latter in the bulk of the papers referenced here.

The technical descriptions of the instruments yield a bewildering mass of detail important to those who must deal with particular instruments, but it is information that is quickly outdated. The MSS on Landsat 1 was succeeded by a nearly identical yet different MSS on Landsat 2, with a concomitant change in the format of the data. Furthermore, most instruments operate in several

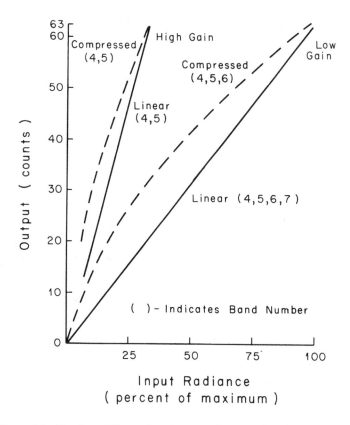

Figure 7.2 The four different functional relationships (gains) relating input to output of the Landsat Multispectral Scanner. Similar flexibility in the operation of other instruments requires users of satellite data to be well informed of the characteristics of each particular instrument.

modes (fig. 7.2) with different sensitivities to radiance or different functional relationships between input and output. The scientific basis for the MSS measurements is more general and applies to all MSS data regardless of the particular source of data.

Because the exact values of the technical information listed in tables 7.1 and 7.2 depends on the details of the satellite orbit, on the status of the instrument, and on other unspecified information, the values are only a guide to the performance of the instrument. The orbit slowly changes, the performance of the instrument degrades with time, the resolution depends on the contrast of the target, and the lifetimes are inexact. As the instruments age, sometimes one or another frequency band quits but the instrument continues to operate, and some instruments are more noisy and less useful than others in their class.

7.2 Observations of oceanic features

The wide range of oceanic phenomena observable from space were first catalogued in the photographs made by the Gemini and Skylab astronauts, and were reproduced in a number of NASA Special Publications (NASA SP). A few of the important works are:

(a) NASA, 1967. *Earth photographs from GEMINI III, IV, and V.* NASA SP-129. Washington: U.S. Government Printing Office.

(b) NASA, 1968. *Earth photographs from GEMINI VI through XII,* NASA SP-171. Washington: U.S. Government Printing Office.

(c) Nicks, O.W., ed. 1970. *This island earth.* NASA SP-250. Washington: U.S. Government Printing Office.

(d) NASA Lyndon B. Johnson Space Center, 1977. *SKYLAB explores the Earth.* NASA SP-380. Washington: U.S. Government Printing Office.

(e) Short, Nicholas M., Lowman, Paul D., Freden, Stanley C., and Finch, William A. 1976. *Mission to Earth: LANDSAT views the world.* NASA SP-360. Washington: U.S. Government Printing Office.

These beautiful picture books show swirls and eddies in the ocean, sediment plumes, shallow underwater banks and reefs, cloud streets, and many other phenomena of interest (plates 2, 3, 4, and 7). But all are unique photos, with no continuous, synoptic coverage showing how these features are generated and how they evolve. Only Landsat and Nimbus-7, together with the meteorological satellites, provide the continuous coverage needed for such studies.

Before turning to the more difficult quantitative measurements of the oceanic color, we should first consider the simpler class of observations that distinguish oceanic features by their shape and the studies based on these observations. Included in the class are observations of the extent of polar ice, ice-free leads, ice motion, surface manifestations of internal waves, boundaries of surface currents, and cloud motion (wind). Each type of observation depends on a more or less binary distribution of brightness (an ice edge exists or it does not; a wave exists or it does not), rather than on a quantitative measurement of a continuous distribution of brightness characteristic of measurements of chlorophyll-a and its relation to plankton concentrations. Oceanic features distinguished by shape can be observed even by simple instruments. The Landsat multispectral scanner is particularly useful for its high resolution and repeated coverage of the polar regions, and the visible light scanners (VISSR) on the geostationary meteorological satellites are particularly useful for their repeated coverage of tropical and mid-latitude regions.

A number of studies have used images from the Landsat multispectral scanner to study the distribution and movement of ice in the Arctic and Antarctic (plate 6; fig. 7.3). These, together with studies of ice using the coarser imaging radiometer on the meteorological satellites, have provided much new and important information on ice in remote areas as well as data with which to test theories of ice dynamics. McClain (1974) catalogued the satellites capable of providing information, together with typical images from them. Muench and Ahlnas (1976) reported the distribution of ice in the Bering Sea during the spring of 1976 using images from the meteorological satellites Noaa 2 and 3 (fig. 7.4). Sobczak (1977) used Landsat images to study the movement of individual icebergs and leads (fig. 7.5), then compared the

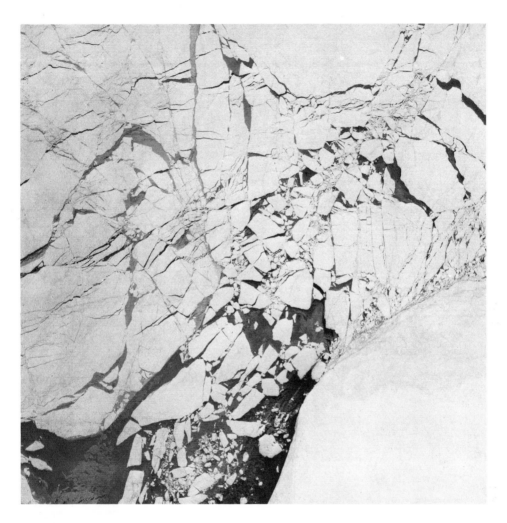

Figure 7.3 Landsat Multispectral Scanner image of sea ice near Point Barrow Alaska on 22 March 1977 using band 7 (0.8–1.1 μm). Images such as this are used to monitor sea ice extent and to study ice dynamics.

observed motion with that predicted by theories of ice dynamics using measured surface winds. Both DeRycke (1973) and Swithinbank, McClain and Little (1977) tracked the motion of bergs around Antarctica to study the circulation in the Southern Ocean (fig. 7.6).

In general, observations of the Arctic and Antarctic using sunlight are hampered by winter darkness and persistent summer clouds. Only spring and autumn tend to be clear. Clouds are difficult to distinguish from ice, and snow-covered thin ice resembles thick ice. To avoid some of these problems,

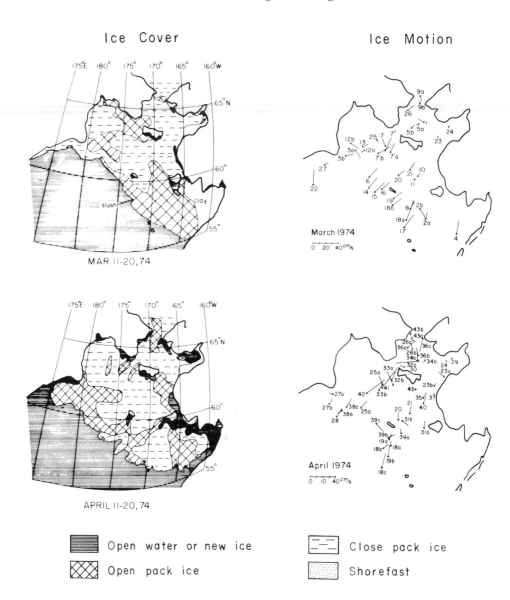

Figure 7.4 Ice distribution (left) and motion (right) in the Bering Sea during March and April 1974, from observations made by instruments on Noaa 2 and 3 satellites (from Muench and Ahlnas, 1976).

infrared images are often combined with the visible images. Infrared does not need daylight, and snow, ice, and clouds all have different brightness depending on wavelength (fig. 6.9).

The Landsat images do not usually show features in the open ocean, but special processing to increase the contrast of faint features allows more to be seen

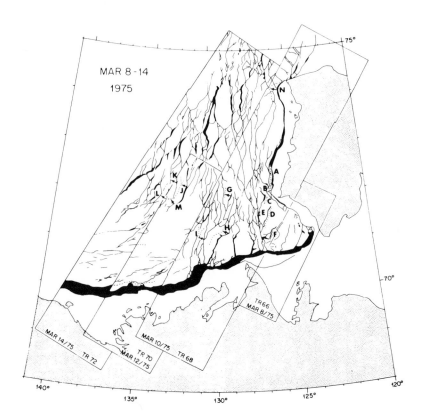

Figure 7.5 Pattern of sea-ice leads observed during the period of 8–14 March 1975 by the multispectral scanner on Landsat-1. Lettered areas denote places where ice movement was measured. Solid and dashed lines note the location of the same prominent feature between adjacent satellite passes (from Sobczak, 1977).

(fig. 7.7). Apel et al. (1975) have used Landsat images to study wavelike patterns along the edge of the continental shelf (fig. 7.8). They conclude these are surface slicks associated with internal waves generated by tidal currents. The images allow them to infer the motion of the waves and to study their wavelength. In fact, so many waves have been seen, in so many places, that the observations have been summarized in an atlas (Sawyer and Apel, 1976).

Sequential observations of tropical and mid-latitude clouds made every 30 minutes by the visible and infrared radiometers on the geostationary meteorological satellites GOES, METEOSAT, and GMS have been particularly useful for observing cloud motion and hence winds at cloud heights. Such information is useful for meteorology, but surface winds are needed for oceanography. Uncertainty in the extrapolation of winds from the lowest cloud levels to the sea surface is the primary limitation of the technique, but the absence of clouds in large mid-latitudes is also a concern.

The extrapolation from cloud-level to surface winds is based on statistical

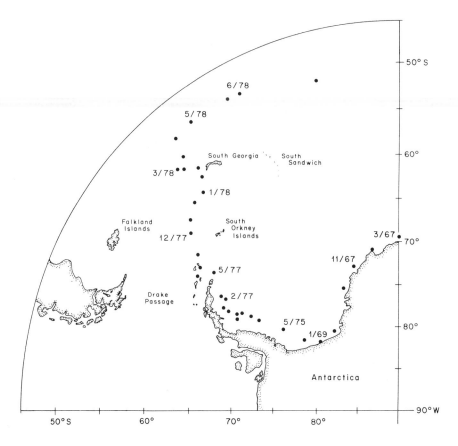

Figure 7.6   Positions of the Trolltunga, a very large tabular iceberg, observed over a period of nearly eleven years by Landsats and meteorological satellites (data from McClain, 1978).

comparisons between winds measured by ships and winds calculated from cloud motion. The wind tends to decrease from cloud levels toward the surface, and to change direction as predicted by Ekman (1905), but the exact form of the variation depends on atmospheric conditions. Different relationships are often used in different oceanic regions and for different wind directions. Thus cloud level winds from the north in the Arabian Sea have a different relationship with surface winds than do cloud level winds from the south in the same region (fig. 7.9).

Once useful empirical relations have been derived for particular regions and seasons, 3—5 days of data can be averaged together to produce basin-wide maps of surface wind velocity (fig. 7.9). Both ship and satellite data are used. The ship data tend to be more accurate and help fill in cloud-free areas, while clouds give denser but somewhat less accurate measurements of wind in

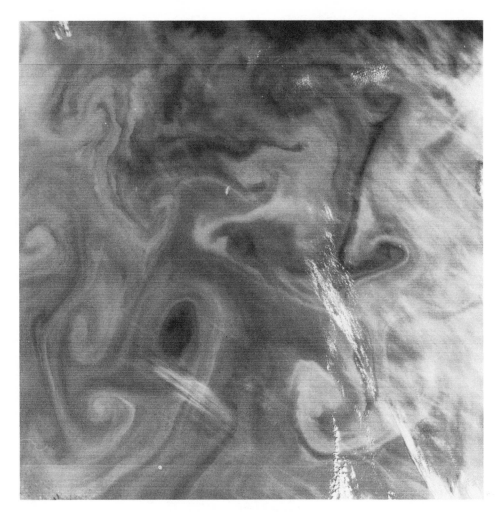

Figure 7.7 Landsat Multispectral Scanner (MSS) image of a plankton bloom south of Iceland on 19 June 1976 using band 4 of the instrument (0.5–0.6 μm). Although not as sensitive as the Coastal Zone Color Scanner, the MSS has been useful for studying dense plankton patches in coastal regions and at high latitudes.

regions remote from shipping lanes. Such maps of the average wind velocity are accurate to about ± 2–3 m.s^{-1} in speed and $\pm 30°$ in direction in tropical regions. Unaveraged data are much less accurate, and Halpern and Knox (1983) found little coherence between winds calculated from cloud motion and winds measured by a buoy in the equatorial Pacific for frequencies greater than 0.2–0.3 cycles/day, but very good coherence at lower frequencies.

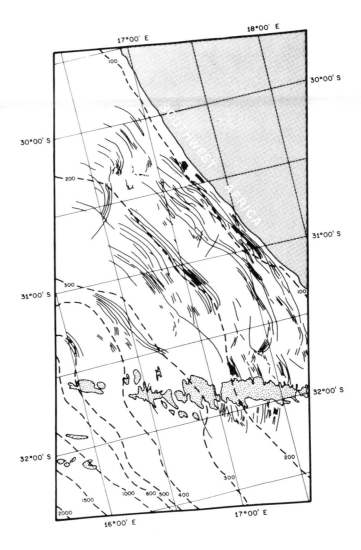

Figure 7.8 Line drawing of slicks associated with internal waves off the coast of South Africa observed by the Multispectral Scanner on Landsat, together with bathymetry (from Apel et al., 1975).

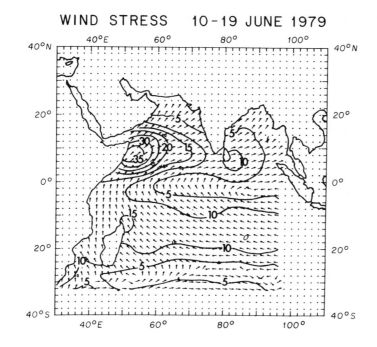

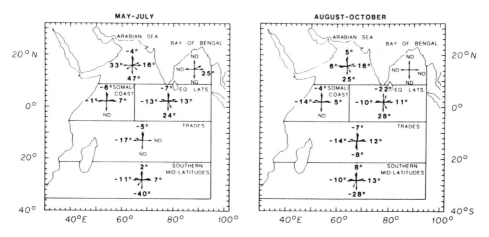

Figure 7.9   Upper: average wind stress in centi-Newtons per square meter on the Indian Ocean during 10–19 June 1979, calculated from low-level cloud velocities measured by the VISSR on Goes-1 and by ships (from Wylie and Hinton, 1982b). Lower: average veering angle (cloud direction minus surface wind direction) in the Indian Ocean. The differences are due to regional and seasonal variations in the structure of the atmospheric boundary layer (from Wylie and Hinton, 1982a).

7.3 Ocean color and chlorophyll concentrations

Primary productivity in the ocean results from one-celled plants, the phyto-plankton, converting nutrients into plant material by using sunlight with the help of chlorophyll. Many different species contribute, although only a few species, occurring in great numbers, are found at any one time or place. The mix of species slowly changes with the seasons and location. The individual plants live at various depths, from the surface to nearly 100 m, but preferably in the sunlit regions having light sufficient for photosynthesis, the *euphotic zone*. Usually this is defined to be the zone whose depth is the level at which the irradiance is 1% of that at the surface.

The chlorophyll pigments in the plants absorb light, and the plants them-selves scatter light. Together, these processes change the color of the ocean as seen by an observer looking downward into the sea. Very productive waters appear blue-green or sometimes red (the red tide) in contrast with the deep-blue, almost black color of very pure water. On clear days, the color can be seen from space and provides a means of mapping the distribution of phyto-plankton over large areas. In the following paragraphs we consider, in reverse order, the chain of ideas linking satellite observations of radiance to ocean color; color to chlorophyll concentration; and concentration to productivity, using the notation in table 7.3. This discussion lays the foundation for inter-preting images from the Coastal Zone Color Scanner that views oceanic radi-ance at 443 nm, 520 nm, 550 nm, 670 nm, and 750 nm, as well as for similar instruments specially designed to measure ocean color.

The relationships between total chlorophyll in the euphotic (lighted) zone and productivity were explored by Smith and Baker (1978a). Using data representative of a wide variety of oceanic conditions and seasons, they found that total productivity is related to the average concentration of chlorophyll-like pigments in the surface layers of the water (fig. 7.10), where the surface layer is defined to be that layer from the surface down to one attenuation depth. An attenuation depth is the level where downwelling irradiance is $1/e$ its value just below the surface. In terms of the attenuation coefficient for irradiance, K_T, the depth at the bottom of the surface layer is K_T^{-1}.

The reason for the correlation was also apparent in the data. The chlorophyll in the euphotic zone is correlated with the chlorophyll in the surface layer (fig. 7.10). This correlation results from the plants tending to move closer to the surface as their density and productivity increase (Smith, 1981). Note that chlorophyll-a, a particular type of chlorophyll, is highly correlated with all other chlorophyll and chlorophyll-like pigments (phaeopigments) in the water column; thus the total pigment concentration can be used as a substitute for chlorophyll-a concentration. Because it is usually difficult to separate chlorophyll-a from other pigments, the ability to relate total pigment concentra-tion to chlorophyll-a concentration is very useful.

Table 7.3 Notation for Describing Visible Images of the Sea

a	Radius of aerosols [m]
a,b	Constants [mg Chl-a.m^{-3}, $-$]
a,b	Absorption and backscatter coefficients [m^{-1}, m^{-1}]
a_w, b_w	Absorption and backscatter by water molecules [m^{-1}, m^{-1}]
a_p, b_p	Absorption and backscatter by plants [m^{-1}, m^{-1}]
$c_1 \cdots c_6$	Constants
C	Chlorophyll concentration [mg Chl.m^{-3}]
C_K	Average chlorophyll concentration in surface layer [mg Chl.m^{-3}]
C_T	Average chlorophyll concentration in euphotic zone [mg Chl.m^{-3}]
$E(\lambda)$	Spectral irradiance [W.m$^{-2}.\mu$m^{-1}]
$E_0(\lambda)$	Spectral solar irradiance outside the atmosphere [W.m$^{-2}.\mu$m^{-1}]
$g(\lambda)$	Ratio of aerosol optical thickness
K_T	Attenuation coefficient for diffuse irradiance [m^{-1}]
L	Spectral radiance [W.m$^{-2}.$sr$^{-1}.\mu$m^{-1}]
L_g	Sun glitter ["]
L_r	Skylight reflected from the sea ["]
L_T	Spectral radiance observed by spaceborne radiometer ["]
L_u	Upwelling spectral radiance just below the sea surface ["]
L_w	L_u after propagation through the surface ["]
L_0	Upwelling spectral radiance just above the sea surface ["]
L^*	Path radiance ["]
m	Angstrom's exponent
n	Number of aerosol particles per unit volume [m^{-3}]
N	Number of samples used to determine a correlation
r	Correlation coefficient
$R(\lambda)$	Subsurface spectral reflectance
T_A	Transmittance of the atmosphere
T	Transmittance, water to air
α, β	Parameters of particle size distribution [m$^{-3}\mu$m$^{\beta-1}$, $-$]
λ	Wavelength of light [m]

Chlorophyll concentration in the surface layer can be related to ocean color either empirically, using oceanic observations, or theoretically, using the theory of radiative transfer in sunlit waters. In either case, color is determined by the spectral reflectance $R(\lambda)$ of the water observed near vertical, at angles that exclude reflected sunlight (sun glitter).

The observations of Clarke, Ewing, and Lorenzen (1970) are typical of the experimental approach. They used a spectrometer on a low-flying airplane to measure reflected light as a function of chlorophyll concentration. They found that the curves of reflectance as a function of wavelength λ have a characteristic shape that changes with chlorophyll concentration (fig. 7.11). Usually the reflectance near 0.500 μm is relatively insensitive to chlorophyll concentration and this is a "hinge point" of the curves. The reflectance of blue light decreases

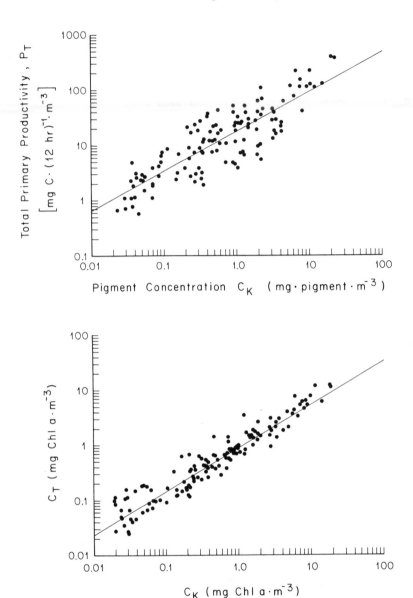

Figure 7.10   Upper: the total average productivity P_T within the euphotic zone as a function of the chlorophyll-like pigment between the surface and a depth of one attenuation length C_K. The best-fitting line through the data is log $P_T = 1.254 + 0.728$ log C_K ($r = 0.855$, $N = 126$). Lower: average total chlorophyll-a concentration C_T in the euphotic zone, as a function of the average chlorophyll-a concentration C_K between the surface and one attenuation depth. The best-fitting line through the data is log $C_T = -0.020 + 0.788$ log C_K ($r = 0.955$, $N = 140$; from Smith and Baker, 1978a).

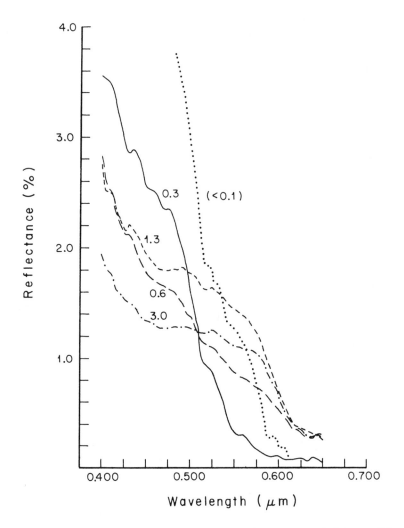

Figure 7.11 Spectral reflectance of sea water observed from an aircraft flying at 305 m over waters of different colors in the NW Atlantic. The numerical values are the average chlorophyll concentration in the euphotic zone in units of $mg.m^{-3}$. The reflectance is for vertically polarized light observed at Brewster's angle of 53°. This angle minimizes reflected skylight and emphasizes the light from below the sea surface (from Clarke, Ewing, and Lorenzen, 1970).

with concentration and eventually saturates at high concentrations. The reflectance of yellow light increases with concentration and is a more sensitive indicator of high concentrations. Together, these observations suggest that measurements of the ratio of reflectance in one band to that in another can be used to measure chlorophyll concentration (fig. 7.12). This is the approach used for most remote observations of chlorophyll.

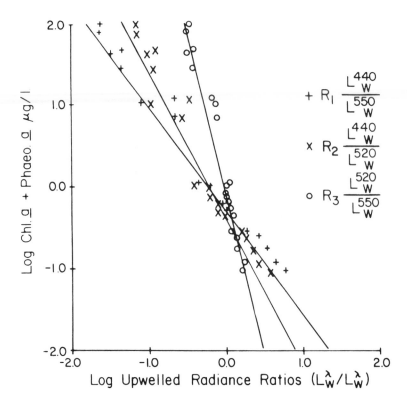

Figure 7.12 Ratios R of upwelling radiance just above the sea surface between pairs of light bands, as a function of the chlorophyll and phaeopigment concentration at the surface. The superscript on L refers to the wavelength in nanometers (from Gordon and Clark, 1980).

The observations of the spectral reflectance of productive waters can be predicted theoretically, thus providing a firmer basis for interpreting the oceanic observations. The reflectance of light occurs within a volume just below the surface, and 90% comes from water shallower than one attenuation depth. For simplicity, we can consider the reflectance to occur at a fictitious surface just below the real water surface (to avoid additional problems introduced by the real surface). The subsurface reflectance is defined to be

$$R(\lambda) = \frac{E_{up}(\lambda)}{E_{down}(\lambda)}$$

where $E(\lambda)$ is the spectral irradiance propagating either upward or downward. A spectrometer on an aircraft or satellite, however, measures radiance above the water surface. Thus $E_{up}(\lambda)$ must be converted to upwelling radiance L_u knowing the angular distribution of radiant flux, then propagated through the surface to obtain the vertically propagating radiance L_w (fig. 7.13).

From oceanic observations and from theory

$$E_{up}(\lambda) = 5.0\, L_u(\lambda)$$

$$L_w(\lambda) = \frac{T}{n_w^2}\, L_u(\lambda)$$

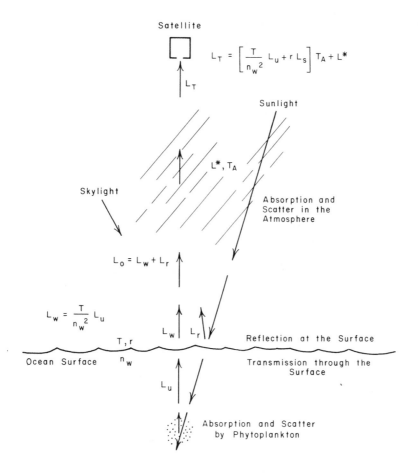

Figure 7.13   Sketch of the various contributions to the light observed by a satellite
instrument observing sunlit seas (from Austin, 1979).

Therefore,

$$L_w(\lambda) = \frac{T}{5.0n_w^2} R(\lambda) E_{\text{down}}(\lambda)$$

The factor of 5.0 accounts for the directional distribution of undersea radiation being slightly peaked away from the vertical when the sun is high (Austin, 1979). If the subsurface radiation were independent of angle, the factor of 5.0 would be replaced by π. The factor n_w^{-2} accounts for the change in solid angle when radiance crosses from water with an index of refraction $n_w \approx 1.33$ into air; and $T \approx 0.98$ is the transmittance of the interface. The problem now reduces to estimating $R(\lambda)$ knowing the properties of the water.

Morel and Prieur (1977) show that Duntley's approximation (also proposed by Gumburtsev)

$$R(\lambda) \approx 0.33 \frac{b(\lambda)}{a(\lambda)}$$

is supported by theory and observations. Here $b(\lambda)$ is the total backscatter coefficient for radiance and $a(\lambda)$ the absorption coefficient for radiance. For most open ocean areas, termed Case 1 waters, the coefficients can be expanded into

$$a(\lambda) = a_w(\lambda) + a_p(\lambda)C$$
$$b(\lambda) = b_w(\lambda) + b_p(\lambda)C$$

where the subscripts w refer to absorption and scatter by water molecules, the subscript p refers to plant material, and C is the concentration of chlorophyll. In coastal waters, termed Case 2 waters, suspended sediments further influence the propagation of light in the water, but the analysis of observations from these areas will not be discussed here. Both $a(\lambda)$ and $b(\lambda)$ have been measured as a function of wavelength and chlorophyll concentration for a number of different phytoplankton species in the ocean. From these, $R(\lambda)$ can be calculated with good accuracy (Smith and Baker, 1978b; Morel and Prieur, 1977). Because both $a(\lambda)$ and $b(\lambda)$ can vary independently, at least two different ratios of reflectance should be used to estimate chlorophyll concentration.

Several equations have been proposed relating chlorophyll concentration C to ratios of the spectral radiance. These relationships are known as *bio-optical algorithms*. The most common are of the form (Smith and Baker, 1982; Gordon and Clark, 1980; Gordon et al., 1983)

$$\log C = \log a + b \log [L_w(\lambda_1)/L_w(\lambda_2)]$$

or

$$C = a\,[L_w(\lambda_1)/L_w(\lambda_2)]^b$$

Smith and Baker (1982) propose $a = 1.26$ mg.m^{-3}, $b = -2.589$ as the best fit to 999 observations ($r = 0.95$) in the Southern California Bight. But other constants give better fits to other sets of data. Gordon and Clark (1980) found $a = 0.505$ mg.m^{-3}, $b = 1.269$ to fit 21 observations from a variety of geographic regions with a correlation coefficient of $r = 0.99$ for the fit; other values are given in Gordon et al. (1983). Figure 7.12 is an example of the observations in Gordon and Clark (1980). They suggest, however, that one pair of wavelengths should be used for low concentrations of chlorophyll ($C < 1$ mg.m^{-3}) and a different pair for high concentrations. Wilson and Austin (1978) use a more elaborate scheme and propose

$$C = \cfrac{c_1 + c_2\,\dfrac{L_w(\lambda_3)}{L_w(\lambda_2)} + c_3\,\dfrac{L_w(\lambda_1)}{L_w(\lambda_2)}}{c_4 + c_5\,\dfrac{L_w(\lambda_3)}{L_w(\lambda_2)} + c_6\,\dfrac{L_w(\lambda_1)}{L_w(\lambda_2)}}$$

where the constants $c_1, c_2 \cdots c_6$ are determined by theory or experiment.

7.4 Influence of the atmosphere on observations of ocean color

To relate the radiance L_T seen by a satellite-borne instrument to the radiance L_w at the sea surface, it is necessary to consider the influence of the atmosphere. This is not easy. The light from the intervening atmosphere L^*, the *path radiance*, is far brighter than L_w (fig. 7.14), and it must be estimated with

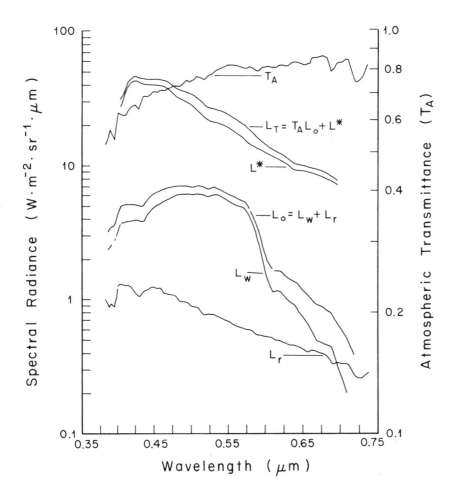

Figure 7.14 Computed spectral radiances that contribute to the light seen by a satellite instrument viewing typical sunlit seas. Note that the upwelling radiance from the sea, L_w, is much smaller than the light reflected and scattered by the atmosphere, L^*. Also shown is the vertical transmittance T_A of the atmosphere (from Wilson and Austin, 1978).

an accuracy of around 1% in order to estimate L_w with an accuracy of 10%. This is the *atmospheric correction* problem.

The radiance at satellite height is the sum of the attenuated surface radiance L_0 and the path radiance L^*:

$$L_T = L_0 T_A + L^*$$

where T_A is the transmittance of the atmosphere. Furthermore, the surface radiance is composed of sun glitter L_g, reflected skylight L_r, and the signal we wish to measure, the radiance L_w just above the surface which has come from below the surface:

$$L_0 = L_g + L_r + L_w$$

The sun glitter is large and generally avoided. The reflected skylight is lumped together with the path radiance so that:

$$L_T = L_w t + L^*$$

Thus the atmospheric correction reduces to calculating L^* and the transmittance T_A.

The atmospheric correction is partly calculated and partly observed. The technique has been guided by comparisons with accurate solutions to the equations of radiative transfer through a real, scattering atmosphere. Such solutions have encompassed a variety of atmospheric models, with varying aerosol types, and have included multiscattering and the influence of the polarization of the scattered light. From this work has emerged a set of approximations that simplify the computations without unduly compromising accuracy.

The technique described here is a somewhat simplified solution to the problem, and is along the lines proposed by Gordon and Clark (1980) and first used by the Nimbus-7 Coastal Zone Color Scanner Team. But other methods are being explored, and a good recent review of the subject is given by Gordon et al., (1983). The simplified technique is essentially a linear approximation to the problem that works well for clear atmospheres. It assumes: (a) the scatter by molecules and aerosols can be calculated separately; (b) the aerosol phase function and single-scatter albedo are nearly independent of wavelength; (c) the scattered light from aerosols is proportional to the optical thickness of the aerosols; and (d) L_w (670 nm) = 0 (this latter assumption need not be true, but merely serves as a good initial assumption for beginning the computation).

For simplicity in discussing the idea of an atmospheric correction, the angular dependence of the upwelling and downwelling radiance is ignored here. A more complete statement of the problem is given by Gordon et al. (1983) and by Viollier, Tanre, and Deschamps (1980).

The first three assumptions result in

$$L_T = L_w \exp\left[-(\tau_{\text{molecules}} + \tau_{\text{aerosol}} + \tau_{\text{ozone}})\right] + L_{\text{molecules}} + L_{\text{aerosol}}$$

where τ is the optical thickness due to scatter from air molecules and aerosols and absorption by ozone; $L_{\text{molecules}}$ is the spectral radiance of light scattered from molecules; and L_{aerosols} is the spectral radiance scattered from aerosols.

The molecular and ozone contributions to L_T are readily determined. The scatter of sunlight from molecules and their optical depth are calculated using the equations of radiative transfer outlined in section 5.3, together with the estimates of the scattering coefficient and phase function given in section 5.4. The ozone absorption is calculated using the computer code of Selby and McClatchey (1975) augmented by the estimates of latitudinal and seasonal variability of ozone (see fig. 4.9). The solar irradiance is taken from standard models such as that proposed by Thekakara (1974), plotted in figure 3.5, or the more accurate values by Neckel and Labs (1981). Altogether, the relative importance of these various terms is sketched in figure 7.15. With the molecular and ozone contributions calculated, only L_{aerosol} and $T_{\text{aerosol}} = \exp(-\tau_{\text{aerosol}})$ remain to be determined. The former dominates, and the latter can be approximated with little error by $T_{\text{aerosol}} \approx 1$.

The influence of the aerosol radiance L_{aerosol} is the most difficult process to estimate. The fourth assumption above, that L_w (670 nm) = 0, enables the influence of aerosols to be measured at a wavelength of 670 nm.

$$L_{\text{aerosol}}(670) = L_T(670) - L_{\text{molecules}}(670)$$

where L_T is observed and $L_{\text{molecules}}$ is calculated as noted above. The problem

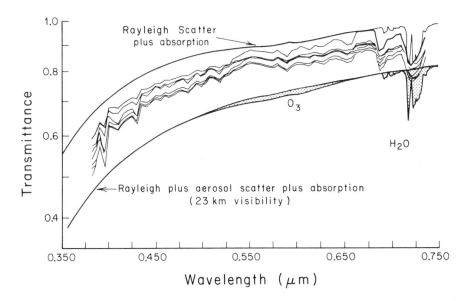

Figure 7.15 Transmittance of irradiance through typical very clear atmospheres, showing the influence of Rayleigh scatter and absorption by ozone and water vapor. The smooth curves are calculated using an early version of the Selby and McClatchey (1975) computer code for an atmospheric visibility of 23 km; the irregular curves are transmittance measured by the Visibility Laboratory of Scripps Institution of Oceanography.

now reduces to estimating L_{aerosols} at shorter wavelengths.

The aerosol radiance can be determined three ways:

(a) It can be measured by specially equipped ships in the field of view of the satellite. This method is accurate, but not very practical.

(b) Alternatively, the influence can be measured directly by the satellite if there is an area in the image where the water has a very low chlorophyll concentration, assuming only that the aerosol concentration but not the type varies across the image. Gordon and Clark (1981) show that the optical properties of clear water with $C < 0.25$ mg.m^{-3} are predictable and can be used to estimate $L_w(\lambda)$ for $\lambda = 520$ nm, 550 nm, and 670 nm, but not blue light at $\lambda = 443$ nm. The latter must be calculated from longer-wavelength observations. Such low-chlorophyll areas are found in most images, or nearby images, and the technique is useful.

(c) If there are no low-chlorophyll areas, or as a first approximation to finding such areas, the aerosol radiance at short wavelengths can be calculated using an extrapolation from L_{aerosol} (670 nm). Assuming the aerosol single-scatter albedo and phase function are nearly independent of wavelength, then

$$L_{\text{aerosol}}(\lambda) = g(\lambda) L_{\text{aerosol}}(670) E_0(\lambda)/E_0(670)$$

$$g(\lambda) = \tau_{\text{aerosol}}(\lambda)/\tau_{\text{aerosol}}(670)$$

where $E_0(\lambda)$ is the solar irradiance outside the atmosphere at a wavelength λ. (Gordon and Clark (1980) use the symbol ϵ instead of g, but ϵ is already used in too many other contexts. They also use the term Rayleigh scatter for scatter from molecules. The latter term is used here to avoid confusion with the Rayleigh approximation for scatter from aerosols.) Usually it is assumed that

$$g(\lambda) = (670 \text{ nm}/\lambda)^m$$

where m is Angstrom's exponent, and where the value of m can be estimated from measurements of aerosol types and sizes. Bullrich (1964:121) showed that spherical aerosols with a particle size distribution

$$dN/da = \alpha a^{-\beta}$$

give

$$\tau_{\text{aerosol}} \sim \lambda^{-(\beta-3)}$$

where the notation is that of section 5.2. From figure 5.4 $\beta \sim 4$, thus

$$\tau_{\text{aerosol}} \sim \lambda^{-1}$$

and Angstrom's exponent is approximately unity. Viollier, Tanre, and Deschamps (1980) found that the average value of the exponent, measured over a period of a year at Lajes in the Azores, was $m = 0.93 \pm 0.3$. However, the wide variability of atmosphere aerosols should yield different values of m at different areas and times.

If a low-chlorophyll area is found in the image, then m can be calculated at $\lambda = 520$ and 550 nm, and then extrapolated to $\lambda = 443$ nm.

The errors involved in estimating the atmospheric correction have been discussed in considerable detail by others. The results indicate that the correction is accurate enough to allow chlorophyll to be estimated within a factor of 2 over a wide range of values from 0.01 to 10 mg.m^{-3} (fig. 7.16). In particular, Sorensen (1979) notes the following accuracies are required in order to estimate L_T with an accuracy of 1%:

(a) The geometrical configuration — 1°;
(b) $E_0(\lambda)$ — 1%;
(c) Ozone — 0.5 g.m^{-2};
(d) Water vapor (750 nm) — 1gm.cm^{-2};
(e) Whitecap coverage — 0.02 to 0.5%;
(f) τ_{aerosol} — 14% at 440 nm, 1% at 750 nm;
(g) $\tau_{\text{molecules}}$ — air pressure with an accuracy of 1% (10 mbar).

The errors introduced by inaccurate estimates of aerosol properties have been investigated by Curran (1972), Viollier, Tanre and Deschamps (1980), Gordon (1978), and Gordon and Clark (1980) among others. The latter authors note that (a) it is better to underestimate the value of m rather than to overestimate it, as long as $C < 0.2$ mg/m^3; and (b) as pigment concentration increases, the value of m must be more accurate. The assumption that L_{aerosol} is proportional to τ_{aerosol} has been tested both experimentally and theoretically by Griggs (1975) and found to be a good approximation (fig. 7.17).

A relaxation of the assumption that $L_w (670) = 0$ was suggested by Smith and Wilson (1981). They note that $L_w (443)/L_w (670)$ is highly correlated with $L_w (443)/L_w (550)$ ($r = 0.971$, 101 points). Thus once the latter ratio is computed using the techniques just described, the value of $L_w (670)$ can be estimated from the correlation. If it is not small or zero, the value is used to correct the computations and the ratio is recomputed in an iterative fashion. The iteration is finished when the upwelling radiance at 670 nm is consistent with the radiance at 443 nm.

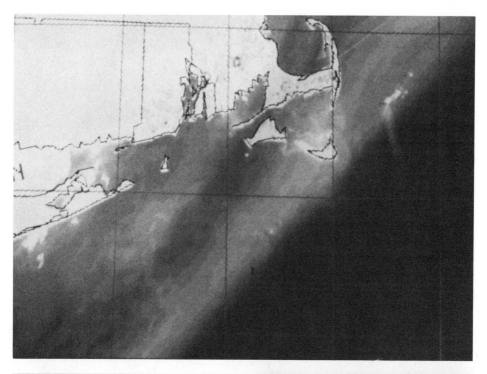

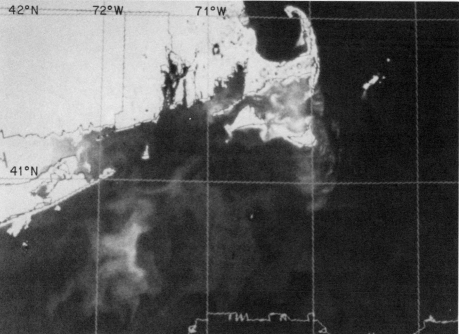

Figure 7.16 Upper: Coastal Zone Color Scanner image of an area just south of Rhode Island on 10 June 1979 using band 3 of the instrument (0.433–0.453 μm) with no atmospheric correction. Lower: the same image with atmospheric light removed to show upwelling radiance at the sea surface. Pairs of such corrected images are used to calculate chlorophyll concentration in surface waters of the ocean (from Gordon et al., 1983).

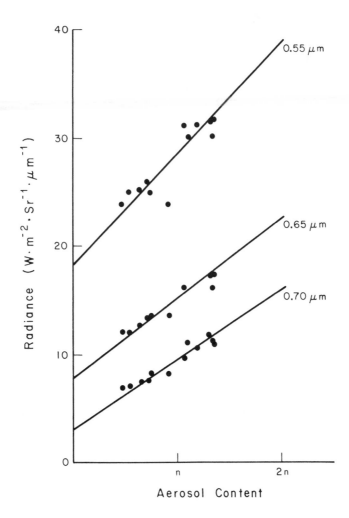

Figure 7.17 Radiance outside the atmosphere in different wavelength bands due to sunlight scattered from aerosols in the atmosphere as a function of aerosol content. The radiances were measured by the Multispectral Scanner on Landsat 2, and are for nadir viewing normalized to a sun zenith angle of 63° (from Griggs, 1977).

Finally, Gordon et al. (1983) show that any instrument used to measure $L_T(\lambda)$ in space must be calibrated with a high degree of consistency with the instrument used to measure $E_0(\lambda)$. This should be done by observing $E_0(\lambda)$ and $L_T(\lambda)$ using the same instrument on the spacecraft.

7.5 Satellite observations of ocean color

Of the several instruments capable of measuring L_T from space, only the Coastal Zone Color Scanner (CZCS) and the Landsat Thematic Mapper have been specifically designed to observe ocean color. The CZCS, carried aboard the Nimbus-7, can make quantitative measurements of oceanic radiance in five bands in the visible spectrum as well as one in the infrared. Viewing geometry, sun angle, orbit, and accuracy were all optimized for this task. Compared with the Landsat Multispectral Scanner (MSS), the CZCS has more bands (table 7.1)

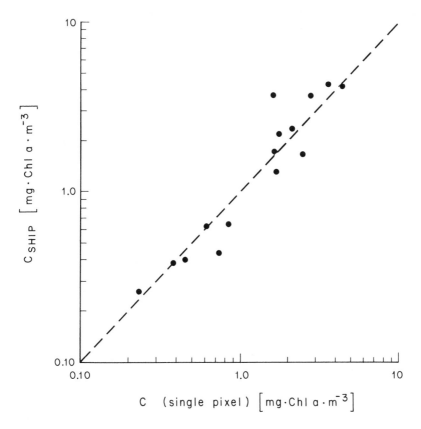

Figure 7.18 Comparisons between chlorophyll concentrations measured by a ship and those estimated from Coastal Zone Color Scanner images of an area offshore of Southern California on 6 March 1979, using L_w (443 nm)/L_w (550 nm).

and greater sensitivity (Hovis et al., 1980). The instrument scans 0.7 rad (40°) on either side of the sub-satellite point (nadir) crosswise to the direction of travel of the satellite; and the scan track can be tilted ±0.35 rad (20°) fore or aft to avoid sun glint. All six spectral bands are precisely coregistered and calibrated. The spatial resolution is 800 m, and the instrument views a swath of ocean 1600 km wide. Data from the instrument can be obtained directly as they are observed, or they can be recorded on the satellite for later transmission to the ground.

The observations from the coastal zone color scanner are now being used to measure chlorophyll and to study oceanic productivity over large regions (plate 5). The accuracy of the observations of chlorophyll has been assessed by comparing surface observations of chlorophyll with simultaneous (or nearly so) estimates calculated from satellite measurements of radiance (fig. 7.18). The comparisons show that the instrument is accurate within 1/3 unit on a logarithmic scale (3 dB), when satellite observations are combined with a single surface observation of skylight (Gordon, et al., 1980; Smith and Baker, 1982). If only satellite data are available, the accuracy is slightly less. Thus satellite images make possible the mapping and study of oceanic productivity over large regions (Smith, Eppley, and Baker, 1982).

8

OBSERVATIONS
USING INFRARED RADIATION

Measurements of both the thermal balance of Earth and the temperature of clouds have long been a primary concern of meteorologists, who have flown a wide variety of infrared radiometers in space. The first were designed to study the Earth as a whole, but soon the area viewed by the radiometer decreased to around 10 km^2, while the sensitivity increased to around 1K. Present radiometers view an area around 1 km in diameter and measure incoming radiation with an accuracy of around 0.3K. Although the instruments were primarily designed to study the atmosphere, the more recent ones are able to measure the patterns of sea surface temperature associated with fronts and eddies in the ocean. With further processing to remove errors introduced by the atmosphere, the signal from the instruments may also be used to measure the true temperature of the sea surface, not just its spatial variability.

8.1 The instruments

Radiometers useful for oceanography have been flown on a variety of satellites, but especially on the Nimbus, Noaa, Geos, and Dmsp series of satellites. The instruments — known by their acronyms, HRIR, VHRR, SR, AVHRR, and VISSR (table 8.1) — all look at the sea through the 3–4 μm or 10–12 μm atmospheric windows, which have low absorption. Most have fields of view sufficiently narrow to peer through the clouds and to provide detailed maps of thermal radiation from the sea. Most also have an additional channel for viewing visible light, used to note the position of clouds or to map snow or ice as discussed in the last chapter.

By far the most important radiometers have been those on the various meteorological satellites. The instruments have good sensitivity and resolution, have flown on a succession of satellites, and cover the globe. The format of the data has remained unchanged for long periods of time, and the data are readily available either directly from the satellites or from the meteorological agencies that record and process global sets of data.

[128]

Table 8.1 Infrared Sensors for Viewing the Sea

Satellite	Instrument	Acronym	Swath (km)	Operational
Nimbus-1,2,3	High Resolution Infrared Radiometer	HRIR	2700	1966—1972
Nimbus-5	Surface Composition Mapping Radiometer	SCMR	800	1972
Noaa-2,3,4,5	Very High Resolution Radiometer	VHRR	2900	1972—1979
Noaa-1,2,3,4,5 Itos-1	Scanning Radiometer	SR	2900	1970—1978
DMSP	Optical Line Scanner	OLS	3000	1976—1980
SMS-1,2 Goes-1,2,3,4 GMS-1,2	Visible Infrared Spin-Scan Radiometer	VISSR	13,000 (full disc)	1974—1981
HCCM	Heat Capacity Mapping Radiometer	HCMR	720	1978—1980
Tiros-14	Advanced Very High Resolution Radiometer	AVHRR/1a	2700	1978—1979
Noaa-6	Advanced Very High Resolution Radiometer	AVHRR/1b	2700	1979—1981
Noaa-7,8	Advanced Very High Resolution Radiometer	AVHRR/2	2700	1981—
Tiros-14 Noaa-6,7,8	Tiros Operational Vertical Sounder	TOVS	2250	1978—
Goes-5	VISSR/Atmospheric Sounder	VAS	13,000 (full disc)	1981—
MOS-1	Visible and Thermal Infrared Radiometer	VTIR	1,500	1986— (planned)

The most widely used radiometers are those on the satellites operated by the U.S. National Oceanic and Atmospheric Administration (NOAA). Observations from the Scanning Radiometer (SR) on the early Noaa series of polar-orbiting satellites were stored in analog form on board the satellite for later transmission to a central receiving station. These observations were global, and for many years were used to construct global images of cloud cover. The companion Very High Resolution Radiometer (VHRR) on the same satellites provided higher density, more accurate regional observations of areas within a few thousand kilometers of ground stations able to receive data directly from the satellites (table 8.2). Both instruments were replaced by the Advanced Very High Resolution Radiometer (AVHRR) on the latest satellites in the Noaa series. This instrument observes more infrared bands and operates in two modes: (a) high density images with 1 km resolution are transmitted directly to local stations on the ground; and (b) these same observations are averaged

Table 8.2   Characteristics of the Infrared Sensors

Name	Acronym	Resolution (km)	Wavelength (μm)	Sensitivity NEΔT (K)
High Resolution Infrared Radiometer	HRIR	8	3.4–4.2	1–4
Very High Resolution Radiometer	VHRR	0.9–1.5	0.6–0.7 10.5–12.5	1.0–4.0
Surface Composition Mapping Radiometer	SCMR	0.6	0.8–1.1 8.3–9.3 10.2–11.2	0.17
Scanning Radiometer	SR	4–13 4–13 8–19	0.40–1.10* 0.52–0.73 10.5–12.5	1.0
Optical Line Scanner	OLS	0.6–2.8	0.4–1.1 8–13	1.56
Visible Infrared Spin-Scan Radiometer	VISSR	1 8	0.55–0.70 10.5–12.6	0.3
Heat Capacity Mapping Radiometer	HCMR	0.5 0.5	0.5–1.1 10.5–12.5	1.0*** 0.4
Advanced Very High Resolution Radiometer	AVHRR/1a	1.1(HRPT)† 4.0(APT) all bands	0.55–0.68** 0.55–0.90 0.725–1.10 3.55–3.99 10.5–11.5	0.12 0.12 0.13
Advanced Very High Resolution Radiometer	AVHRR/2	1.1(HRPT) 4.0(APT) all bands	0.58–0.68 0.725–1.10 3.55–3.93 10.3–11.3 11.5–12.5	0.12 0.12 0.13
VISSR Atmospheric Sounder	VAS	1 8 8 8	0.55–0.70 10.5–12.6 9–13 3.2–4.2	0.3
Visible and Thermal Infrared Radiometer	VTIR	0.9 2.7 2.7 2.7	0.5–0.7 6–7 10.5–11.5 11.5–12.5	0.5 0.5 0.5

*Noaa 3,4,5 only   †HRPT = high resolution picture transmission
**AVHRR/1b only   APT = automatic picture transmission
***W.m^{-2}. sr^{-1}.μm^{-1}

together to obtain a value representative of an area 4 km in diameter, and are then digitized and recorded on the satellite to provide global coverage.

The Visible and Infrared Spin Scan Radiometer (VISSR) on the Goes series of geostationary satellites observed the Earth every 30 minutes with a resolution of 8 km in the infrared. Viewing from an altitude of roughly 35,000 km above the equator, the radiometers see the full disc of the Earth. This is about 120° in diameter, so the observations are most useful in the tropical and

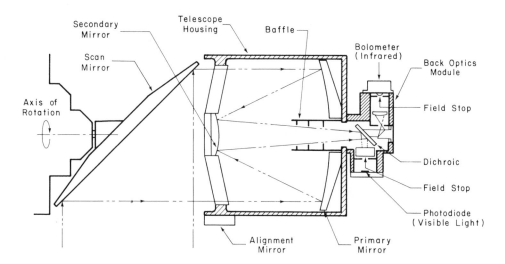

Figure 8.1 A cross section of the Scanning Radiometer used for many years on the Noaa weather satellites. Most other visible and infrared instruments, including the Multispectral Scanner and Coastal Zone Color Scanner discussed in the last chapter, operate in a similar way. The scanning mirror reflects radiation from the Earth into a telescope, which then focuses the radiance onto a detector. Two detectors are used in this instrument, but more are used on the more complex instruments.

sub-tropical areas. The instrument has now been replaced by the VISSR/Atmospheric Sounder (VAS), which continues the capability of the VISSR but combines the data with observations from within the atmosphere. Instruments very similar to the VISSR are also carried on geostationary satellites operated by Japan (the GMS satellites) and by the European Space Agency (the Meteosats).

The sensitivity of radiometers depends on the ratio of their internally generated signal to that produced by incoming radiation. The power available to the instrument is

$$\Phi_{\text{in}} = A \; d\Omega \; d\lambda \; L(\lambda)$$

where A is the aperture of the instrument, $d\Omega$ is the angular extent of the area viewed by the instrument, $d\lambda$ is the bandwidth, and $L(\lambda)$ is the spectral radiance of the area viewed. The power is usually focused on a detector that converts radiant power into electrical power (fig. 8.1). This is then amplified and either recorded or transmitted to the ground. The output of the detector Φ_{out} is the sum of the incoming signal Φ_{in} plus the noise $\Delta\Phi$ generated within the transducer:

$$\Phi_{\text{out}} \equiv \Phi_{\text{in}} + \Delta\Phi$$

$$\Phi_{\text{out}} = A \; d\Omega \; d\lambda \; L(\lambda) + \Delta\Phi$$

The signal-to-noise ratio, S_n, of the instrument is then

$$S_n = \frac{\Phi_{\text{in}}}{\Delta\Phi} \approx \frac{A \; d\Omega \; d\lambda \; L(\lambda)}{\Delta\Phi}$$

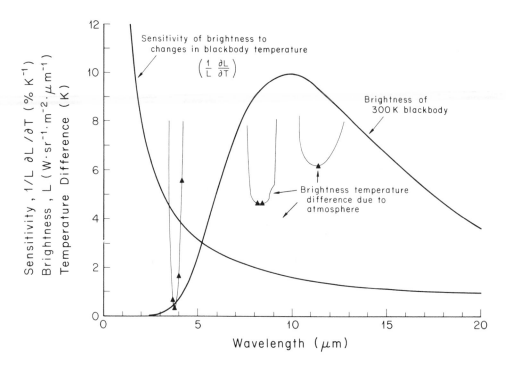

Figure 8.2 Factors influencing the choice of wavelength for observing the temperature of the sea from space. Shown are: (a) the brightness from a 300 K blackbody (the sea); (b) the percent change in radiance for a 1° change in surface temperature; and (c) the difference between the temperature observed by a spaceborne radiometer and the true surface temperature, due to the influence of the clear atmosphere with 5.3 gm.cm^{-2} of water vapor (data from Chahine, 1981).

and the designer wishes to maximize this function. One way is to observe the same area many times, N, with the same instrument, assuming L does not change meanwhile. The average of these observations converges toward Φ_{in} as N increases. If the noise in different observations is uncorrelated, then on average

$$S_n = \frac{\sqrt{N}\ A\ d\Omega\ d\lambda\ L(\lambda)}{\Delta\Phi}$$

Another way to maximize the function is to increase each term in the numerator to its greatest practical value. The aperture A can be increased indefinitely, but this causes the instrument to become larger and more costly. The field of view $d\Omega$ can be increased, but the spatial resolution must be balanced against the increase in $d\Omega$. For this, lower satellite orbits help. The bandwidth $d\lambda$ can also be increased, but only to the width of the natural atmospheric windows. Lastly, $\Delta\Phi$ can be reduced by improving the detector; for example, by cooling it to very low temperatures. The ultimate limit is the quantization of the radiation Φ_{in}, but almost all instruments have internal noises considerably larger than the inherent quantum limit.

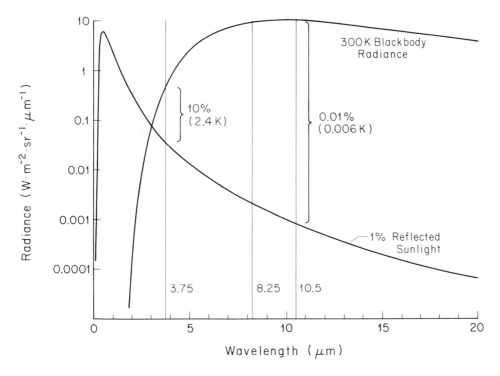

Figure 8.3 The influence of reflected solar radiance on infrared observations. Shown is the radiance from a 300 K blackbody contrasted with solar radiation reflected from a Lambert surface that reflects 1% of the incoming irradiance, with the ratio calculated at wavelengths commonly observed from space.

Often, the instrument noise figure is inserted into the Planck equation and converted to an apparent change in temperature, ΔT. This is referred to as the Noise Equivalent ΔT, NEΔT, and is given in table 8.2 for typical instruments.

The selection of an operating wavelength is an important part of the design of a radiometer. Several constraints must be considered (fig. 8.2): (a) the width of the atmospheric windows, which determines $d\lambda$; (b) the clarity of the windows, which influences the accuracy of the observations; (c) the sensitivity of the emitted radiance L to changes in temperature T of the surface, $1/L \, \partial L/\partial T$, which enables the instrument to measure subtle changes in sea surface temperature; and (d) the intensity of the radiance L, which influences the choice of detectors.

The 3.7 μm window is clearly best for observing the sea. The radiance at this wavelength is very sensitive to changes in surface temperature; it is easily measured; and the window is clearer than at 8.5 and 11 μm. This wavelength has been avoided by meteorologists because solar radiation reflected from clouds renders the observations nearly useless during the day (fig. 8.3). For oceanography, the problem can be avoided. Incidence angles for observing the sea can be chosen to avoid reflected sunlight, or only nighttime observations can be used.

8.2 Sources of error within the atmosphere

The accuracy of satellite observations of sea surface temperature depends on the ability of radiometers to view the sea with little error introduced by the atmosphere. Yet the error can be substantial, even in relatively clear air. In this section we will consider first the errors influencing observations in the 10.5 μm window, the wavelength used by most radiometers, followed by a consideration of those in the 3.7 μm band. In the next section, we will take up the subject of reducing or avoiding the influence of the atmosphere on the observations.

Clouds and water vapor are the primary sources of error. Water vapor partly absorbs the signal and reradiates energy at the colder temperatures typical of the atmosphere a few kilometers above the surface. Dense, widespread clouds completely block radiation from the sea, and although they restrict the number of surface observations they introduce no errors. Their existence is obvious. Thin clouds and clouds smaller than the instrument's field of view are the major concern: both lower the apparent sea temperature, but not so much that it is obviously wrong. The problem is to detect their presence and then to correct or reject the erroneous measurements.

The thin clouds include high cirrus and some very low stratus clouds. Either can be so thin that they are nearly invisible in the images of the Earth made by either the infrared or visible sensors. The cirrus clouds are particularly important. They are much colder than the surface and even a few small clouds can contribute large errors. Being invisible, the clouds remain undetected; and being thin, they do not reduce the apparent temperature of the sea to unrealistically low values.

The small unresolved clouds are usually cumulus. Typically, trade wind cumulus are less than 1 km in diameter and fall into this category, as do thin, scattered clouds.

The influence of water vapor on observations in the 10–12 μm band was investigated by Maul et al. (1978). They used radiosonde observations of water vapor in the clear atmosphere, together with the equation for radiative transfer, to calculate the brightness seen by an infrared radiometer in space looking obliquely at the sea surface through the atmosphere. They then estimated the ability of the VISSR to measure sea surface temperature accurately and to observe the gradient of sea temperature along the northern boundary of the Gulf Stream. They found: (a) Radiosondes must be selected to avoid those that pass through clouds. Failure to detect the increase in humidity due to clouds leads to 2–5K errors in the calculations. (b) The clear atmosphere reduces the apparent temperature of the surface by 4–8K, with the greatest difference in regions and seasons with high temperature and humidity (warm air can hold more water vapor than can cold air). The reduction in the apparent surface temperatures at any particular time shows a spatial variability on scales of 1000–2000 km typical of regional changes in water vapor. (c) The temperature contrast across a current boundary is reduced by approximately

$$\nabla T_0 = t \, \nabla T_s$$

where ∇T_s is the temperature contrast, ∇T_0 is the contrast seen by the satellite, and t is the transmittance of the atmosphere. They found that $0.4 < t < 0.8$ for Goes observations, and that the contrast in surface temperature is reduced sufficiently that the VISSR may not be able to see the edge of the Loop Current in the Gulf of Mexico in summer. Similar degradations in satellite infrared observations have been reported by others.

Atmospheric particles and gases other than water vapor reduce the observed sea surface temperature only slightly at 10.5 μm. The latter contribute reductions on order 0.1K. Jacobowitz and Coulson (1973), using crude estimates of the influence of aerosols, report that these are expected to reduce temperatures by 0.1−2.0K. They note that the calculated error depends strongly on the number of large aerosols and that these produce the 2.0K error.

Atmospheric radiation reflected by the surface introduces another small error. Typically this is less than a degree Kelvin. Maul (1981) estimated the influence at 10.5 μm for atmospheric conditions typical of the Gulf of Mexico, and found that the reflected radiation contributed an error of 0.2−0.7K.

Atmospheric errors, except those produced by clouds and aerosols, tend to be smaller in the 3.7 μm window than in the longer-wavelength windows. Accurate calculations of the radiance from the well-mixed atmospheric gases and from the far wings of the water-vapor lines have been reported by Chahine (1981) and by others. They have found that the mixed gases contribute an error of 0.1K in narrow windows near 3.7 μm, and that 5.3 gm.cm^{-2} of water vapor contributes an additional error of only 0.3−0.6K. Aerosols should be considerably more important. Recalling that the extinction of radiation by small aerosols varies as λ^{-4} in the Rayleigh approximation, we might expect their influence will be 60 times greater at 3.7 μm than at 10.5 μm. Of course, the approximation is not quite valid because some aerosols are nearly as large as the wavelength of the radiation, but its use is not too far wrong when the air is clear and aerosols small. During the day, solar radiation is backscattered from the aerosols, and this too must be considered.

The relative importances of the various errors contributed by the atmosphere are summarized in table 8.3.

Table 8.3 Relative importance of the atmosphere on radiometric observations of sea-surface temperature

3.7μm band		10.5μm band	
undetected clouds	0−10K	undetected clouds	0−10K
aerosols	0.3−5K	water vapor	1−8K
water vapor	0.3−1K	aerosols	0.1−2K
other gases	0.1K	reflected sky	0.2−0.7K
		other gases	0.1K

8.3 Removing atmospheric influences

The atmospheric errors are too large to be ignored, yet small enough that there is some hope that they can be accurately accounted for. The problem is not simple; to be comparable with conventional sea surface observations, infrared observations of sea surface temperatures must have an accuracy of around 0.5K. Yet many processes introduce errors of this magnitude. If the goal is an accuracy around 1K, then the problem is much simplified and may be solvable. The most recent work tends to indicate that the error estimates may be pessimistic and that errors of less than 1K can be expected in practice, provided great care is taken in all aspects of the collection, recording, and processing of infrared observations.

The accurate determination of sea surface temperature begins with the calibration of the radiometers. Not all are calibrated, and not all can make observations with an accuracy of 0.5K or better. Maul (1981) reports that the VISSR has a constant offset of 2−4K. The AVHRR is better calibrated (it views an internal reference standard once each scan) and should have an accuracy comparable to its noise.

The accuracy of satellite observations is also influenced by the recording of data. Observations from the VISSR, AVHRR, and OLS are digitized on the satellite before they are recorded and transmitted to the ground. Thus the observations are not further degraded by these steps. In contrast, observations from earlier instruments were recorded and transmitted in analog form, and this degraded the accuracy of the measurements by 2−4K.

Clouds dominate the errors introduced by the atmosphere, and various techniques seek to detect and eliminate pixels containing clouds (Henderson-Sellers, 1982). Once cloud-free pixels are located, other techniques then seek to correct for the residual influences of water vapor. Methods for detecting clouds include the following:

(a) *Maximum temperature.* The simplest and earliest techniques for correcting for the influence of clouds assumed that ocean-surface features persist much longer than clouds; that clouds can only lower the observed temperature; and that occasionally a pixel will be completely cloud-free (Smith et al., 1970). All observations of a small surface area, say 0.1° on a side (11 km), made over a period of several days, are examined, and the highest temperature in the set is retained as the best estimate of the temperature in the area. This value is then compared with nearby values to determine if it is reasonable, that is, if it does not deviate more than that expected from knowledge of typical gradients of surface temperature. If it appears reasonable, it is accepted as true. This technique cannot find persistent thin or unresolved clouds, however.

(b) *Two-wavelength infrared.* The second technique relies on the nonlinear relation between temperature and brightness at two different infrared wavelengths, usually 3.7 and 10.5 μm, as stated by the Planck equation for blackbody radiation. If a scene is composed of scattered but unresolved clouds above a warm surface, the image will give two different apparent temperatures in the two bands. If the scene is composed of a uniform cloud deck or sea sur-

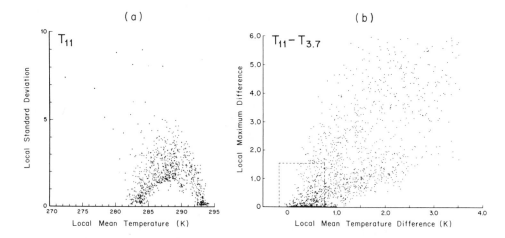

Figure 8.4 The influence of clouds on infrared observations. (a) The standard deviation of the radiance from small, partly cloudy areas, each containing 64 pixels. The feet of the arch-like distribution of points are the sea surface and cloud-top temperatures (from Coakley and Bretherton, 1982). (b) The maximum difference between local values of $(T_{11}-T_{3.7})$ and the local mean values of the same quantity. Values inside the dashed box indicate cloud-free pixels. T_{11} and $T_{3.7}$ are the apparent temperatures at 11.0 and 3.7 μm (data from K. Kelly, Scripps Institution of Oceanography).

face, the apparent temperatures will be the same; however, a uniform cloud deck is usually apparent in the visible image of the scene. As an example, consider a radiometer viewing an area composed of 1/3 clouds at 200K, 1/3 clouds at 250K, and 1/3 sea at 300K. For these conditions the radiometer will measure an apparent temperature of 277.5K at 3.7 μm and 259.9K at 11 μm, a difference of 17.6K (Smith et al., 1974). Unfortunately, the temperature differences produced by low, warm clouds is small, and under some circumstances there is no difference at all. If 20% of a 300K ocean surface is covered by 284K clouds, then the difference is undetectable, yet the error in estimates of sea-surface temperature introduced by the clouds is slightly greater than 2K.

(c) *Infrared variability.* A third technique notes that the apparent temperature of clouds tends to be much more variable in space than the temperature of the sea surface. Thus an oceanic scene is divided into areas each containing 25–64 pixels. The standard deviation of the radiance from each area is plotted as a function of the mean radiating temperature of the area. Coakley and Bretherton (1982) show that the points fall along an arch (fig. 8.4a) and that the feet of the arch determine the sea surface and cloud-top temperatures. A more stringent test uses the maximum difference between pixel radiance and the mean radiance (fig. 8.4b). In either case, all areas having a small deviation from the mean, and with a brightness temperature close to that expected of the sea in the region, are accepted as good (fig. 8.5).

(d) *Two-wavelength visible-infrared.* The fourth technique uses reflected sunlight to detect clouds, on the assumption that the sea is much darker than

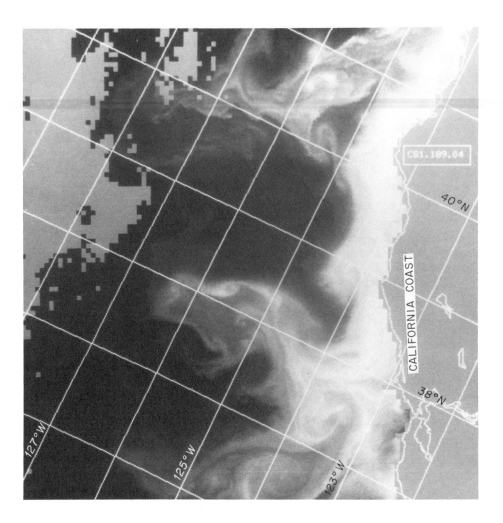

Figure 8.5   Application of a cloud detection algorithm. Observations of 3.7μm radiation made by the AVHRR on NOAA-6 on 7 July 1981 have been processed by computer to remove cloudy areas using a technique based on infrared variability within 5×5 pixel areas. The data were remapped onto an equal-area projection, converted to apparent sea-surface temperature, then displayed as grey tones proportional to sea-surface temperature in the range of 6—13°C, with warm areas being darkest (from K. Kelly, 1983).

clouds. Maul (1981) calculates a discrimination function based on the joint probability of radiance in the visible (0.55—0.7 μm) and infrared (10.5 μm) observed by the VISSR. This is used with a computer routine that recognizes the particular form of the discrimination function characteristic of cloud-free regions, to find pixels with low radiance in the visible and high radiance in the infrared. In a similar but somewhat simpler way, Bernstein (private communication) uses visible light (0.7—1.1 μm) observed by the AVHRR. In either case, the technique has two relatively minor flaws. The sensitivity of the visible-light radiometers is not quite good enough to distinguish clear from cloudy areas; and clouds in the shadow of other clouds have low radiance similar to that of the surface. The lack of sensitivity of the VISSR is such that one count in the visible channel is associated with a 0.5K error in the infrared when a few puffy cumulus are in the field of view (Maul, 1981).

Once cloud-free pixels have been identified, the infrared radiance of these pixels must be corrected for the influence of water vapor in the atmosphere in order to obtain accurate values of sea surface temperature. Again, a number of techniques have been proposed:

(a) *Single-wavelength corrections.* Many instruments observe only a single infrared wavelength, usually 10.5μm. Thus no independent information useful for estimating the influence of water vapor is available from the spacecraft. In this circumstance, either climatological estimates of water vapor or regional radiosonde observations can be used. Maul (1981) used radiosonde observations from stations around the Gulf of Mexico to calculate the regional influence of water vapor. By further comparing values of surface temperature calculated from the infrared data with a few ship or buoy reports of temperature in the Gulf, the residual error due to instrument calibration, reflected atmospheric radiation, and surface influences was reduced to a standard error of ± 0.5K. This is perhaps the best accuracy that might be achieved over a region.

(b) *Two-wavelength corrections.* Because radiation at 10.5μm is much more sensitive to water vapor than is radiation at 3.7μm, the former can be used to correct the latter (Deschamps and Phulpin, 1980). The difference in apparent temperatures at the two wavelengths $(T_{3.7} - T_{10.5})$ is used to estimate the correction for the influence of water vapor. Bernstein (1982) proposes

$$T_s = 1.0726 \, T_{3.7} + 0.31 \, (T_{3.7} - T_{10.5}) - 18.11 \quad \text{(daytime)}$$

based on 48 observations of radiance over the Pacific made by the AVHRR which were compared with surface temperature T_s observed by expendable bathythermographs (XBT) or drifting buoys (fig. 8.6); the subscripts in the equation refer to the wavelength in microns and the temperatures are in degrees Celsius. Similar corrections have also been proposed by McClain (1981) and by Prabhakara, Dalu, and Kunde (1974).

On 17 November 1981, Noaa replaced the older GOSSTCOMP with a two- and three-wavelength technique, the Multi-Channel Sea Surface Temperature MCSST analysis, using

$$T_S = 1.0574 \, T_{3.7} + 0.447 \, (T_{3.7} - T_{10.5}) - 14.6 \text{ (nighttime)}$$

$$T_S = 1.035 \, T_{10.5} + 3.046 \, (T_{10.5} - T_{12}) - 10.78 \text{ (daytime)}$$

but slightly different equations have been used since September 1982 (Strong and McClain, 1984).

Although the technique is simple and accurate, it appears to succeed for the wrong reasons. The influence of water vapor in the 3.7μm window should be roughly 1K, according to theoretical calculations of water vapor absorption coefficients, yet the correction accounts for nearly a 3K range in temperature.

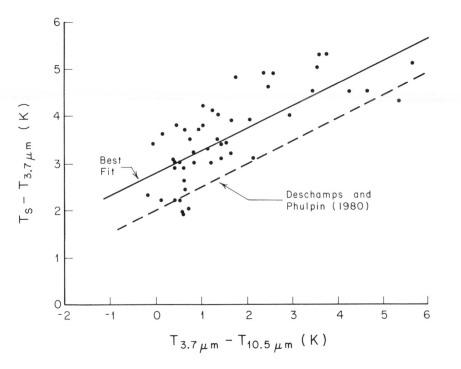

Figure 8.6 The two-wavelength correction for the influence of atmospheric water vapor. The apparent sea surface temperature measured at a wavelength of $10.5\,\mu$m $(T_{10.5})$ is used to correct the observation of surface temperature observed at a wavelength of $3.7\,\mu$m $(T_{3.7})$ to obtain the true sea-surface temperature T_s. The infrared observations were made over the Pacific in daytime by the AVHRR, and the surface temperatures were measured by surface instruments at times and positions close to the satellite observations (from Bernstein, 1982).

Atmospheric aerosols may account for part of the effect. Marine aerosols tend to be hygroscopic, and a change in relative humidity from 75% to 99% causes the number of large aerosols to increase by a hundredfold (Hanel, 1976:136). Thus both the scatter of sunlight and the nighttime emissions from aerosols should covary with water vapor, or more precisely with relative humidity; and $10.5\,\mu$m observations of water vapor lead to useful corrections at $3.7\,\mu$m. But this may not be the full explanation. Kelly (1983) has used buoy and AVHRR data from offshore of California and the data in Bernstein (1982) to investigate the ability of the multichannel AVHRR data to estimate sea surface temperature. Using the statistical analysis of Davis (1977), Kelly showed that combining $3.7\,\mu$m and $10.5\,\mu$m data from the AVHRR is no more useful in calculating surface temperature than is the $10.5\,\mu$m data alone, for cloud-free regions. The single-channel observations and the two-channel technique, after careful screening for the influence of clouds, both give surface temperature with comparable accuracy of less than 1K (0.63K for Bernstein's data and 0.8K for California coastal data less rigorously screened for clouds). Thus either appears to be able to measure sea surface temperature accurately, but the nature of the errors in the measurement are not yet clearly understood.

(c) *Multi-wavelength corrections.* Because clouds are so common, techniques to correct for their influence may be more useful than techniques to detect and eliminate cloudy pixels. Chahine has proposed and tested a method that uses many wavelengths to measure the temperature and humidity profiles in the atmosphere as well as the percent cloud cover and the cloud temperatures (Chahine, 1974; Chahine, Aumann, and Taylor, 1977). The information is then used to estimate the atmospheric radiance at 3.7 μm, the correction necessary to convert observations of radiance to surface temperature. Instruments such as the Tiros Operational Vertical Sounder (TOVS), carried along with the AVHRR on the new Noaa series of satellites, are useful for the task (table 8.4;

Table 8.4 The Tiros Operational Vertical Sounder (TOVS)

High Resolution Infrared Radiation Sounder (HIRS)				
Channel	Wavelength (μm)	Channel Width (μm)	Sensitivity (mW.sr^{-1}.m^{-2}.μm^{-1})	Purpose
1	14.95	0.067	134.	
2	14.71	0.216	31.0	
3	14.49	0.252	23.8	Temperature
4	14.23	0.323	15.3	sounding,
5	13.97	0.313	10.8	cloud cover
6	13.64	0.300	12.9	
7	13.35	0.287	11.2	
8	11.11	0.433	8.1	Window
9	9.71	0.236	15.9	Ozone
10	8.16	0.409	24.0	Water vapor
11	7.33	0.214	37.2	sounding
12	6.72	0.361	42.1	
13	4.57	0.048	2.87	Temperature
14	4.53	0.047	1.47	sounding
15	4.46	0.046	2.01	of warm
16	4.40	0.044	1.03	regions
17	4.24	0.041	1.12	
18	3.98	0.056	1.25	Surface
19	3.76	0.139	0.71	temperature
20	0.69	0.044	0.5	Clouds

Microwave Sounder Unit (MSU)				
Channel	Frequency (GHz)	Channel Width (MHz)	NEΔT (K)	Purpose
1	50.31	220	0.3	Window
2	53.73	220	0.3	Temperature
3	54.96	220	0.3	sounding
4	57.95	220	0.3	

Compiled from data in Werbowetzki (1981).

plates 8, 9). This instrument, described by Smith et al. (1979) and by Werbowetzki (1981), measures temperature and water vapor in the atmosphere and
can be used to estimate the influence of clouds using Chahine's technique.
Despite having a field of view roughly 60 km in diameter, the instrument
seems to be able to measure sea surface temperatures with an accuracy of 1K in
partly cloudy regions (Chahine, personal communication).

Despite the many techniques that have been proposed for reducing errors
introduced by the atmosphere, the problem is not yet solved. Most of the techniques discussed here have not yet been well tested, and the true errors may be
greater than the estimated errors. It is interesting to note that almost all estimates of error, from the earliest to the most recent, promise accuracies near
1K. At this point, only a few general statements can be made. The $3.7\,\mu$m
window is clearer than that at $10.5\,\mu$m. But the $3.7\,\mu$m observations are sensitive to scattered sunlight, whether from aerosols or from the surface (fig. 8.3
indicates that an atmosphere or surface that scatters 1% of the incoming solar
radiance will contribute a 2.4K error to the observation of the temperature of a
300K blackbody at this wavelength). Thus many prefer to use nighttime observations. In this case, visible light cannot be used to detect small clouds, and
the variance of the observed radiance must be used for this purpose.

Of the future developments, the multiwavelength techniques look most
promising. While it may be possible to build instruments with very narrow
fields of view capable of peering through gaps in clouds, the exercise may be
self-defeating. Bright light from nearby clouds will be scattered into the
instrument's field of view, adding yet another complicating factor.

Regardless of the particular techniques used, careful observations at several
wavelengths, when combined with good cloud detection schemes, seem capable
of yielding estimates of the surface temperature with an accuracy of slightly less
than 1K. The precision of the observations is much better, and features with a
contrast of around 0.1K can be detected.

8.4 Errors originating at the sea surface

A perfectly accurate infrared measurement of sea temperature may still be
misleading. The upwelling infrared radiation originates in a surface layer only
0.02 mm thick, and the apparent infrared temperature may not be representative of temperatures at slightly greater depths (fig. 8.7). Several phenomena
are involved (Saunders, 1967; Katsaros, 1980):

(a) Dry air blowing across the sea evaporates water and cools the surface.
This influence is confined to a very thin layer, perhaps a millimeter or so thick,
that cannot be stirred by turbulence but that can be 0.5−1.0K cooler than
deeper layers.

(b) Sunlight shining on the surface on a clear calm day warms a surface layer
a few meters thick by 0.5−1K. This can cool down at night, and under very
calm conditions can become warmer from day to day until the wind stirs the
heat into deeper layers. Satellite observations at the same time each day cannot
detect this diurnal variation, and so these measurements are slightly biased

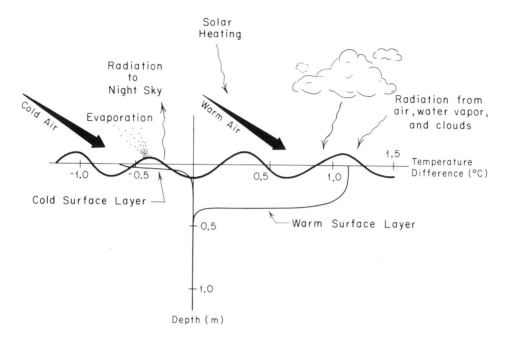

Figure 8.7 Processes influencing the temperature of the skin of the ocean. The skin temperature is measured by infrared radiometers, and it may not be representative of temperatures at slightly greater depths.

depending on the time of day. Of greater concern is the tendency for the best observations of the sea to be made during times when the atmosphere is the clearest and thus when surface heating or cooling tends to be greatest.

(c) Radiation originating from the thin surface layer cools this layer. The effect depends on the radiation balance of the surface with the atmosphere. Very clear, dry nighttime atmospheres allow the radiation to escape into space, producing rapid cooling. Humid, cloudy atmospheres reradiate the energy back toward the surface, keeping it warm.

(d) Surface slicks are an added source of concern, but they are usually not important. The slicks are the result of a thin layer of organic material produced on the sea surface by the decomposition of small animals that die in the water. The layers are usually only one molecule thick, and they strongly damp out small wavelets. This is important for interpreting radio scatter from the sea surface, but the layer is so thin that it does not appreciably change the emissivity of the surface (Jarvis and Kagarise, 1962). Only very thick layers, such as those produced by oil spills, are detectable from space.

(e) In the region of strong horizontal temperature gradients, such as at the northern edge of the Gulf Stream, wind can blow thin layers of warm or cold water horizontally, displacing the apparent position of the current. That is, the position of strong current shear is not always precisely defined by the position of strong temperature contrast.

(f) Finally, cold or warm air blowing over the sea cools or warms the surface layers. The wind also stirs and mixes the layers, however, and the process does not really introduce an error. In this case, the skin temperature is representative of the deeper temperature.

The difference between the surface T_s and the deeper temperature T_b due to various phenomena that warm or cool the surface can be estimated using an equation proposed by Saunders (1967)

$$T_b - T_s = \frac{7\nu\,(H + R + E)}{k\,(\tau/\rho)^{1/2}}$$

where H, R and E are the vertical fluxes of sensible heat, longwave radiation, and latent heat respectively [W.m^{-2}], ν is the kinematic viscosity of water [10^{-6} m^2.s^{-1}], τ is wind stress [Pa], ρ is the density of water [10^3 kg.m^{-3}], and k is the thermal conductivity of water [0.6 W.m^{-1}.K^{-1}]. The constant of proportionality is based on theory and observations, and the value used here is from Simpson and Paulson (1980). Typically $|T_b - T_s| < 1K$ in the open ocean.

8.5 Observations of the ocean

The National Oceanic and Atmospheric Administration produces several types of sea surface temperature measurements. One is the absolute value of sea temperature in designated surface areas, usually 1° on a side (Brower et al., 1976; Strong and Pritchard, 1980). Another is the location of current boundaries, such as the position of oceanic fronts or the edge of the Gulf Stream published in the *Oceanographic Monthly Summary*. The third is imagery of the sea, enhanced to show variations of surface temperature as grey tones in an image. We will consider the first and last of these three measurements, the second being derived from the third.

Absolute measurements of sea surface temperature can be produced by solving the radiative transfer equation using detailed observation of atmospheric structure made by multi-frequency satellite radiometers. In practice, this requires much work and simpler schemes are used. The set of temperature observations is analyzed statistically to determine the presence of clouds and to correct for atmospheric attenuation. Each satellite observation is tested to determine if clouds are present in the field of view, and obviously cloud-contaminated values are rejected. The others are assumed to contain a mixture of cloud and surface observation, and the highest temperature, or sometimes the mode of the distribution of higher temperatures, is used to define an apparent sea temperature in a small area over a short period of time. This is corrected for the influence of the clear atmosphere by adding an empirical correction factor. Finally, the set of all sea surface temperatures from all areas, made in one month, are compared with ship observations of temperatures in the same areas. In general, the comparison between the two sets of observations is very close, typically within 1K for the mean difference. What is overlooked in this assessment is that the empirical correction factors come from similar sets of surface data similar to those used for the comparison. Unless typical worldwide distributions of types and amounts of cloud cover change, the comparison will tend to show close agreement between satellite derived temperatures and those measured by ships.

The close agreement in the global averages of satellite derived and ship measured sea surface temperatures does not necessarily imply that the satellite measurements are useful. We must first ask what measurements are required and to what accuracy. A definite answer, in one specific instance, is provided by the North Pacific Experiment, NORPAX. This continuing study of the North Pacific and its influence on weather and climate has outlined the nature of the variability of surface temperatures in this area (fig. 1.1). At mid-latitudes, sea temperatures typically deviate from their 20-year mean, by $\pm 2K$ over distances of 2000—4000 km. It is against these figures that satellite accuracies must be compared.

The comparison of early satellite observations with regional and seasonal surface observations has been done carefully by Bernstein (1977, unpublished manuscript) for mid-latitude North Pacific areas and by Barnett et al. (1979) for tropical Pacific areas. Bernstein subtracted the 20-year mean temperature for 5° square areas from all ship reports of temperature in the area made over a period of one month and smoothed these maps of anomalous temperature. These were compared with satellite observations treated in the same manner. He found poor agreement between the two fields, and concluded that: (a) satellite measurements had errors of a few degrees Kelvin; (b) these errors had spatial scales of a few thousand kilometers; and (c) the errors were so similar to the signal that they could not be used to define and observe the departure of sea surface temperature from a 20—year mean. The satellite observations were no better, and considerably more expensive, than atlases of surface temperature.

Barnett et al. (1979) came to identical conclusions for observations of the tropical Pacific. They compared sea surface temperature along a line of longitude, measured by calibrated air-dropped expendable bathythermographs (AXBT), with the early Noaa Global Operational Sea Surface Temperature Computations (GOSSTCOMP) derived from the first Noaa satellite data. The comparison was along the 150th and 160th meridians between 20°N and 20°S in the tropical Pacific during a period of several months. Both sets of data were obtained weekly. The investigation showed: (a) The two sets differed by 1— 4K. (b) The difference had a strong latitudinal and longitudinal structure (fig. 8.8). (c) The difference appears to be correlated with both cloud cover and atmospheric water vapor (figure 8.9). (d) The size and spatial structure of the difference are such that satellite observations, as presently analyzed, cannot be used to observe departures of sea surface temperature from their historical mean.

Since these comparisons were made, Noaa has continued to improve the sea surface temperature analyses, primarily by correcting for the influence of atmospheric water vapor using observations from the TOVS and by using the multi-wavelength observations from AVHRR. Estimates of the surface temperature fields are further improved by melding satellite and surface observations, and these are reported in the Noaa *Oceanographic Monthly Summary*.

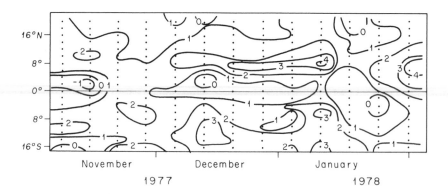

Figure 8.8 The nature of time and space scales of errors due to atmospheric influences on satellite measurements of surface temperature. The plot shows surface temperature (°C) measured by air expendable bathythermographs (AXBT) minus satellite measurements of sea temperature along 150°W (from Barnett et al., 1979). Note that the difference is significant, and that it has spatial and temporal scales similar to naturally occurring variations in sea surface temperature.

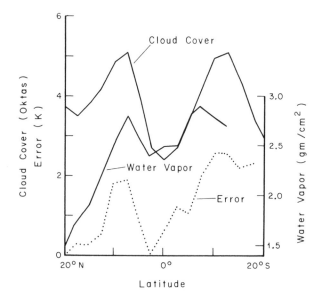

Figure 8.9 Atmospheric contributions to the errors shown in figure 8.8. Water vapor, clouds, and error are all correlated (from Barnett et al., 1979).

In contrast to the accuracy of sea surface temperature, the precision of many observations is very good, allowing oceanic features to be clearly seen. The edge of the Gulf Stream was barely discernible in the earliest images of the sea surface (Warnecke, et al., 1971). Since then the images have become better and better, and now show even the most subtle features in the surface temperature field (fig. 8.10).

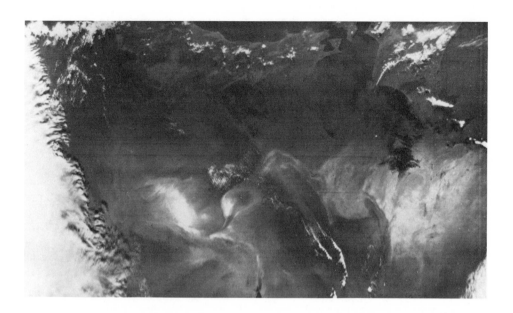

Figure 8.10 Visible and infrared images of the Pacific east of Japan from the Optical Line Scanner on a DMSP satellite on 10 May 1977 showing the subtle variations in surface temperature and a correlation between temperature and the surface roughness which modulates sun glint. Upper: Visible light (0.4−1.1 μm) image showing variations in sun glint. Lower: Infrared (8−13 μm) image showing warm and cold eddies in the Kuroshio. Warm (dark) areas correspond to areas of smooth water; thus warm areas near the subsatellite track produce strong specular reflection and the image is bright, but further from the region of specular reflection the warm areas are dark (from R. L. Bernstein, Scripps Institution of Oceanography).

Numerous studies of ocean thermal features have defined the motion of eddies with scales of 100 or so km (the weather of the sea), and the instabilities of the strong ocean currents such as the Gulf Stream or Kuroshio. For example, Bernstein, Breaker, and Whritner (1977) used both Noaa and Defense Meteorological Satellite observations at a wavelength of 10.5 μm in the infrared to follow the motion of a warm intrusion along the California coast. At the same time, the area was surveyed by a routine California Cooperative Fisheries Investigation cruise and by air expendable bathythermographs. These surface measurement confirmed that the feature seen by the satellite was real and that it represented temperatures typical of the upper several hundred meters directly below the sea surface. It should be noted that the region they observed is characterized by a thin (1 km), cool surface layer of air under a warm, dry upper layer. Thus the total water vapor in the air column could be expected to be relatively small and uniform. This lack of water vapor helps produce very clear satellite images of the sea, but is not essential. Studies of currents at higher latitudes (Legeckis and Gordon, 1982) show the usefulness of the technique for more typical atmosphere conditions. In addition, the latter study confirmed the results of Bernstein, Breaker, and Whritner (1977) and showed that features seen in infrared images of the Brazil and Faulkland currents extend from the surface to the depth of the thermocline, around 100 m, and sometimes to depths of 400–600 m.

Similar satellite observations have been used by others to study the surface motion of the sea and the relationship of currents, temperature, and oceanic productivity (fig. 8.11). Legeckis (1977), using VISSR data, observed long planetary waves in the equatorial region similar to those observed by Harvey and Patzert (1976). These observations were further extended using AVHRR observations (Legeckis, Pichel, and Nesterczuk, 1983) in order to study the development of the waves over a period of years. Legeckis (1978) has also used many different VHRR images of the sea from many different oceanic areas to document the typical location of oceanic fronts (regions of strong temperature or salinity gradients) throughout the world's oceans. Maul et al. (1978) have used VISSR images to monitor continuously the position of the Gulf Stream in the Gulf of Mexico and the Western North Atlantic, and to study the variability of this current.

The strong temperature contrast of the northern edge of the Gulf Stream is particularly prominent from space, and our understanding of the temporal and spatial variability of this western boundary current is known primarily from satellite observations. Legeckis (1979) mapped the variability in the position of the Gulf Stream between Florida and Cape Hattaras and noticed that the meanders propagated northward as waves (fig. 8.12). The frequency-wavenumber spectrum of the motion was then calculated by Halliwell and Mooers (1979) using 104 weekly charts of the Gulf Stream produced from VHRR images supplemented by ship reports.

Taken together, these studies have contributed substantially to our knowledge of oceanic variability and productivity, especially in regions of high temperature contrasts. Such studies constitute perhaps the most widespread use of satellite data in oceanography.

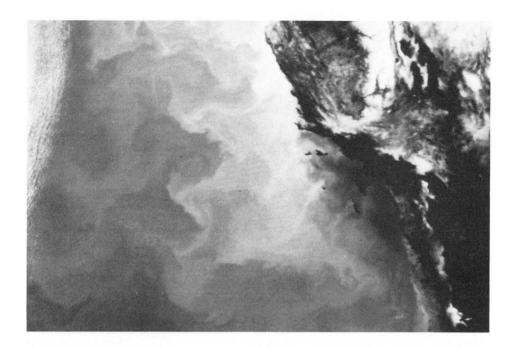

Figure 8.11 The correlation between temperature and chlorophyll concentration in the Pacific west of Southern California on 15 June 1981. Upper: Sea surface temperature measured by an AVHRR on NOAA-6 using 10.5 μm radiation. Lower: Chlorophyll concentration measured by the CZCS on Nimbus-7 using the techniques described in chapter 7 (from Pelaez-Hudlet, 1984).

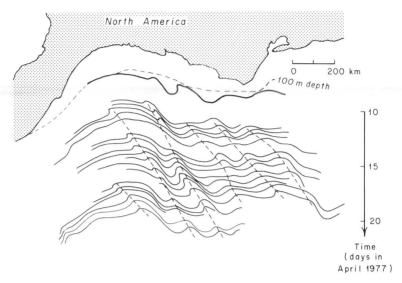

Figure 8.12   The edge of the Gulf Stream offshore of the Carolinas, as seen in infrared images taken by the VHRR on successive days in April 1977 (from Legeckis, 1979).

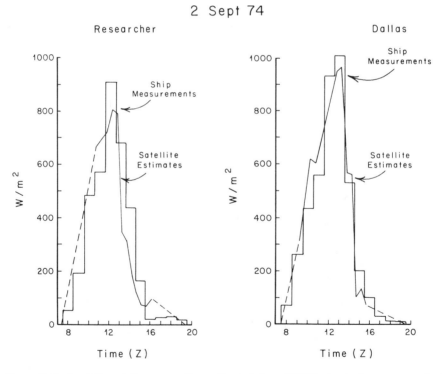

Figure 8.13   Insolation at the sea surface estimated from VISSR observations compared with surface measurements made by instruments on the research ships "Researcher" and "Dallas" during the GATE experiment in 1974 (from Gautier, 1981).

8.6 Radiation budgets

The net balance of incoming and outgoing radiation in the atmosphere and the ocean is important to studies of climate, ocean dynamics, and the interaction of the ocean with the atmosphere. Incoming radiation is primarily solar, and it has a peak wavelength near $0.5\,\mu$m. Outgoing radiation is mostly thermal, with a peak wavelength near $10\,\mu$m. Thus visible and infrared observations can be used to estimate the two quantities and to compute the net balance.

Meteorologists and climatologists have long used a variety of instruments on satellites to observe the balance for Earth as a whole and its geographic and temporal variability. The techniques used for these studies are beyond the scope of this book, although the subject is discussed elsewhere (see Barnett, 1974; Smith, et al., 1977; Gruber and Winston, 1978; Jacobowitz, et, al., 1979; Stephens, Campbell, and Vonder Haar, 1981). Essentially, the various instruments measure the reflected and emitted thermal radiation. If the radiation were observed over all angles, the problem would be solved as long as the radiometers made accurate measurements. Unfortunately, the measurements are usually confined to a narrow range of incidence angles near vertical, and the angular distribution of the radiation must be inferred from other sources. Nonetheless, the various techniques tend to give consistent estimates, and the global heat budget is now fairly well known, although for only a relatively short period of time.

The balance for the Earth as a whole is simple. The incoming and outgoing radiation are accurately calculated. The incoming is the solar constant times the cross-sectional area of the Earth. The outgoing radiation is the same, since the Earth is in thermal equilibrium. Thus satellites need measure only the variability of the budget. Averaged over a year, the tropics absorb heat while the polar regions lose heat. Thus heat must be transported from low to high latitudes, partly by the oceans and partly by the atmosphere. Because the atmospheric winds and temperature are measured, the atmospheric transport can be computed. By comparing this with the amount of heat that must be transported according to the satellite observations, the oceanic transport can be inferred (Vonder Haar and Oort, 1973).

The radiation budget at the surface of the ocean is not as easy to measure from space as the balance at the top of the atmosphere because the atmosphere obscures the outgoing longwave radiation. The incoming solar radiation can be estimated accurately from visible observations, however, based on the fact that almost all radiation not reflected back out to space by clouds is absorbed by the sea. The relatively small amounts of radiation absorbed by clouds and the atmosphere can be estimated from the amount of cloud cover (itself estimated from the visible observations) and from knowledge of the average properties of the atmosphere (Gautier, et al., 1980, Gautier, 1981). Hourly observations of reflected sunlight made by the VISSR can be used to estimate the insolation with an accuracy of around 9% despite the lack of an accurate calibration of the sensor (figure 8.13), and are now being used to study the heat budgets of oceanic areas (figure 8.14).

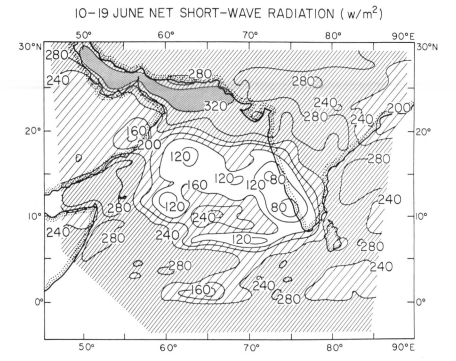

Figure 8.14 Net incoming shortwave radiation at the sea surface in the Indian Ocean during 10–19 June 1979, measured by the VISSR on the Goes-1 satellite; see figure 7.9 for winds measured by the same instrument during this period (data from C. Gautier, Scripps Institution of Oceanography).

Closely related to the radiation fluxes are the fluxes of sensible and latent heat at the sea surface. These are, in general, not observable from space, although there is at least one notable exception. When cold, dry air blows out to sea from a continent, the heat fluxes can be calculated by noting in an image of the area, the distance from the shore to the first formation of clouds, knowing the structure of the boundary layer at the shore and the air-land temperature difference (plate 16; Chou and Atlas, 1982). Similar analyses may apply to the general thickening of the trade wind boundary layer as cool dry winds blow from mid-latitudes toward the equator, but the idea has not yet been tested for accuracy.

9

OBSERVATIONS USING
RADIO-FREQUENCY RADIATION

Super high frequency radio waves, those with wavelengths between roughly 2 and 30 cm, are emitted to a greater or lesser extent by all material. Thus downward-looking microwave radiometers that observe such radiation in space can be used to measure those atmospheric and oceanic variables that influence the emission of radiation. At the longest wavelengths, the atmosphere is clearer than the clearest air for visible light, and radiation from the surface dominates. This radiation is used to map sea surface temperature. At shorter wavelengths, the atmosphere is less transparent, and clouds and water vapor contribute to the signal. Their contribution is used not only to map the distribution of water vapor in the atmosphere but also to make small corrections to the observations of the longer wavelengths. Still other wavelengths are sensitive to changes in surface emissivity due to foam and wind-generated waves. This information is used to map wind speed. But most important of all is the general transparency of even cloudy atmospheres to super high frequency radiation. As a result, all measurements can be made day or night, cloudy or clear, even through persistent clouds that hide the surface from view of visible or infrared radiation.

The usefulness of radiometer observations is further enhanced by the low emissivity of sea water, although this makes more difficult the measurement of sea surface temperature. Sea water has an apparent temperature near 120–130K, while ice, rain, and water vapor radiate at temperatures near 270–290K. Over the open ocean, the latter signals tend to stand out clearly against the very cold background of the sea surface and are easily observed.

The advantages of microwave radiometers must be balanced against one important disadvantage: antennas for observing radio wavelengths have much poorer resolution than lenses used to observe shorter wavelengths. Typically, the area on the surface viewed by spaceborne instruments varies between 10 and 100 km in diameter depending on wavelength; the shorter the wavelength, the smaller the field of view.

9.1 Resolving power of antennas

The *resolving power* of an antenna is its ability to distinguish between two closely spaced points of radiation. This is identical to the resolving power of a microscope or telescope, and for both light and radio waves the theory for the formation of images is handled in similar ways.

The description of the imaging ability of an antenna is simpler if we assume the antenna is a transmitter, and then calculate the pattern of the emitted radiation at a distance that is great compared with either the size of the antenna or the wavelength of the radiation. Furthermore, the algebra is simpler, but the concepts are unchanged, if we consider the antenna in one dimension, assuming infinite extent on the other. This is the problem we will consider here. In the next section we will consider the relationships between transmitting and receiving antennas.

Consider a transmitting antenna with an aperture $2a$ wide and with a distribution of electric field $E(x,t)$ across the aperture (fig. 9.1) given by:

$$E(x,t) = f(x) \exp(i\omega t) \,.$$

By Huygen's principle, the electric field at some distant point p is found by summing the phases of the field from each point in the aperture. After the phase is propagated along r to p, the field at p due to a point in the aperture is

$$E_p \approx E(x,t) \exp(-ikr)$$

using

$$r = R + x \sin \theta$$

$$k = 2\pi/\lambda$$

where (table 9.1) λ is the wavelength of the radiation and r, R, and θ are defined by the figure. Using the relation for $E(x)$ and integrating over all of x yields

$$E_p(s) \approx C \int_{-\infty}^{\infty} f(x) \exp(-kxs) \, dx$$

where

$$s = \sin \theta$$

and

$$C = \exp(i\omega t) \exp(-ikR)$$

is the mean phase delay along the path R. Thus the angular distribution of the electric field in the far field of the antenna ($R \gg \lambda$) is the Fourier transform of the distribution of the electric field in the aperture of the antenna (Bracewell, 1965:275; Born and Wolf, 1970:358).

The intensity of the radiation is proportional to the square of the electric field

$$I(\theta) \sim |E(\theta)|^2$$

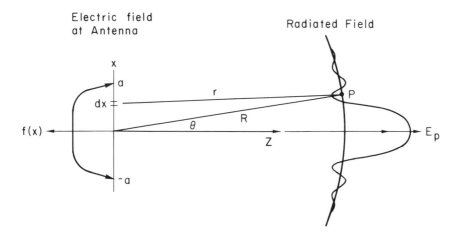

Figure 9.1 Geometry for calculating the electric field E_p at a point p far from an antenna due to an arbitrary distribution of electric field $f(x)$ at the antenna.

Table 9.1 Notation for Describing Antennas

a	Half-width of an antenna [m]
A_e	Antenna effective area [m²]
C	Mean phase delay
$E(x,t)$	Electric field [V.m⁻¹]
$E(\theta,\phi)$	Radiant flux density [W.m⁻²]
E_p	Electric field [V.m⁻¹]
$f(x)$	Electric field [V.m⁻¹]
G	Antenna gain
I	Radiant intensity [W.sr⁻¹]
k	Wavenumber [m⁻¹]
K	Boltzmann's constant [1.38×10^{-23}J.K⁻¹]
L	Spectral radiance [W.m⁻².sr⁻¹.Hz⁻¹]
p	A point
P_n	Power pattern
P_r	Received power [W]
P_T	Transmitted power [W]
r	Radial distance [m]
t	Time [s]
T_A	Antenna temperature [K]
T_B	Blackbody temperature [K]
x	Distance [m]
Z	Electrical impedance [S⁻¹]
$\Delta\theta$	Antenna beamwidth at half power points
λ	Wavelength [m]
η	Absorptance
θ,ϕ	Polar coordinates
ν	Frequency [Hz]
Ω	Solid angle [sr]
ω	Angular frequency

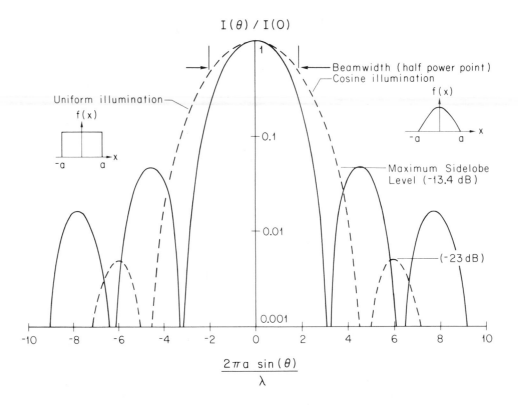

Figure 9.2   The angular distribution of radiant intensity from an antenna with either a uniformly illuminated aperture or a cosine distribution of electric field across the aperture. The beamwidth of the antenna is the width of the distribution at the half-power point. Note that the tapered illumination produces a distribution that has much lower sidelobes at the expense of a slightly broader beamwidth. The scale on the abscissa also reads in degrees if the antenna has a width $(2a)$ of 1 m and radiates at a wavelength (λ) of 5.48 cm. Such an antenna has a beamwidth of 2.8° when uniformly illuminated.

Thus the angular distribution of the radiated field is given by the square of the Fourier transform of the electric field in the aperture of the transmitting antenna.

These ideas can be illustrated by a simple example: an antenna with a uniformly illuminated aperture. The illumination and its Fourier transform are:

$$f(x) = 1 \qquad -a < x < a$$

$$E_p(s) \approx \int_{-a}^{a} \exp(-iksx) \, dx$$

thus

$$E_p(s) \approx \frac{\sin ksa}{ksa}$$

and

$$I(\theta) = \left(\frac{\sin ksa}{ksa}\right)^2 I(0)$$

Satellite Images
The Emerging Global Perspective

Shown on this and the following few pages are samples of the many images of the ocean now being made from space by cameras, scanners, radars, and radiometers. Such images are giving the first global view of the ocean and its interaction with the atmosphere, as well as an overview of important oceanic regions.

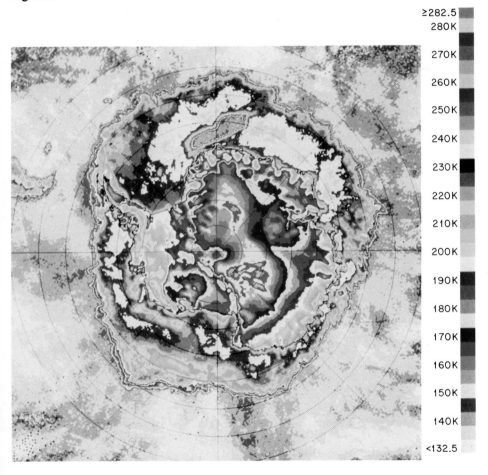

≥282.5
280K
270K
260K
250K
240K
230K
220K
210K
200K
190K
180K
170K
160K
150K
140K
<132.5

Plate 1. The brightness temperature of the Antarctic region observed by the Electronically Scanned Microwave Radiometer on the Nimbus-5 satellite on 26 August 1974. The image shows the extent of ice cover, snow temperature, and an area of open water in the Weddel Sea. Such images are used to study the interannual variations of ice cover and the interaction of ice with currents and winds (from NASA Goddard Space Flight Center).

Plate 2. The boundary between two water masses approximately 300 km south of Bermuda as photographed by a Skylab astronaut using a 70 mm hand-held camera. Similar features are seen in radar and infrared images of the sea (Skylab photo SL3-28-050).

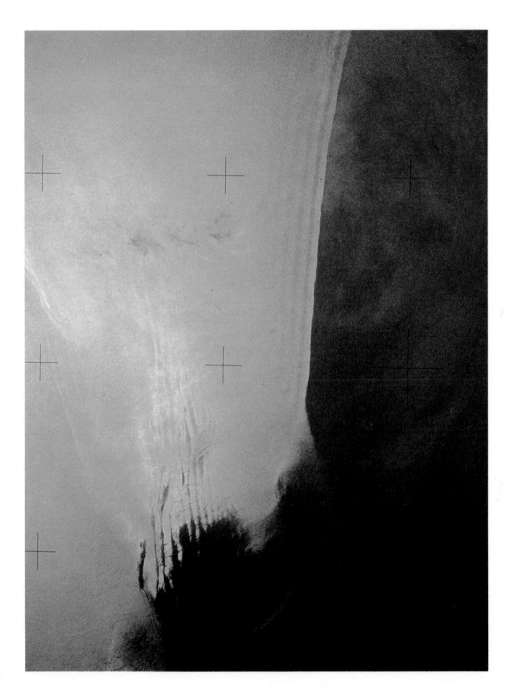

Plate 3. Variations in surface roughness apparently produced by a tidally generated group of internal waves have modulated the sun glint in this photograph of the ocean south of Kangaroo Island offshore of Adelaide, Australia. The photograph was made by a Skylab astronaut using a 70 mm hand-held camera (Skylab photo SL4-137-3564).

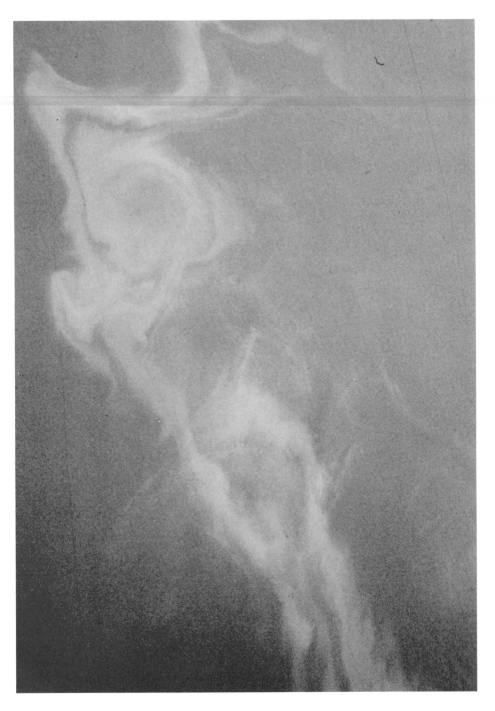

Plate 4. A plankton bloom in the Falkland Current photographed by a Skylab astronaut using a 35 mm hand-held camera. Quantitative measurements of phytoplankton distributions are now being made by the Coastal Zone Color Scanner on Nimbus-7 (Skylab photo SL4-196-7350).

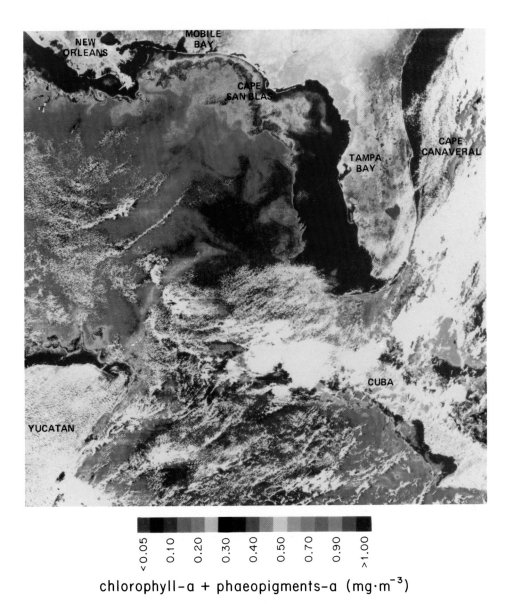

chlorophyll-a + phaeopigments-a (mg·m^{-3})

Plate 5. Chlorophyll and phaeopigment concentration in the Gulf of Mexico observed by the Coastal Zone Color Scanner on the Nimbus-7 satellite on 2 November 1978. The image shows strong concentrations of pigments in coastal waters, and weaker concentrations in the gulf and in the Gulf Stream east of Florida (from Hovis et al., 1980).

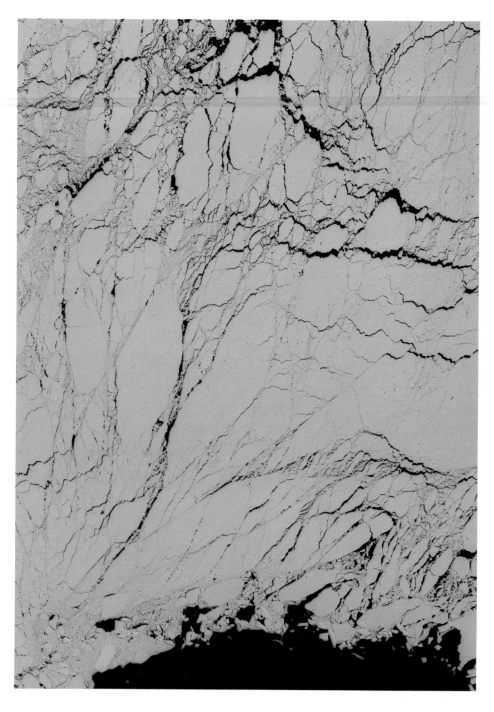

Plate 6. Sea ice in the Amundsen Sea offshore of Antarctica (71°S, 115°W) on 7 February 1975 observed by the Multispectral Scanner on Landsat 2. The image is approximately 100 km on a side, and is constructed from observations of light made in three color bands by the scanner (Landsat image 2016-15062).

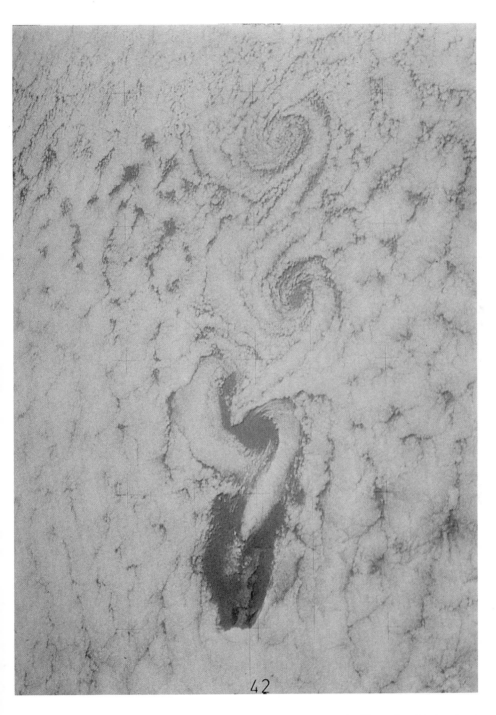

Plate 7. A vortex street downwind of Guadalupe Island in the east Pacific outlined by low-level clouds and photographed by Skylab astronauts using a hand-held 70 mm camera. Motion of low-level clouds seen from space can be used to estimate surface wind velocity or to study atmospheric dynamics (Skylab photo SL3-122-2497).

Mean Cloud Cover for January 1979

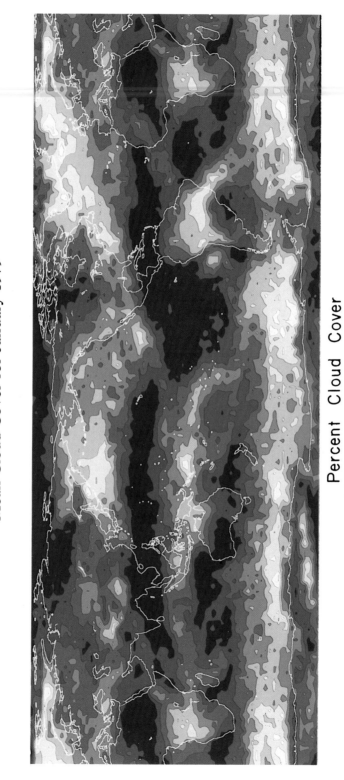

Percent Cloud Cover

10 20 30 40 50 60

Plate 8. Mean cloud cover during January 1979 calculated from data collected by the High Resolution Infrared Sounder and Microwave Sounding Unit on the Noaa meteorological satellites using the technique described by Chahine (1974). The technique also yields maps of surface temperature (plate 9), cloud-top pressures, and cloud-top temperatures (NASA Jet Propulsion Laboratory).

Mean Sea Surface Temperature for January 1979

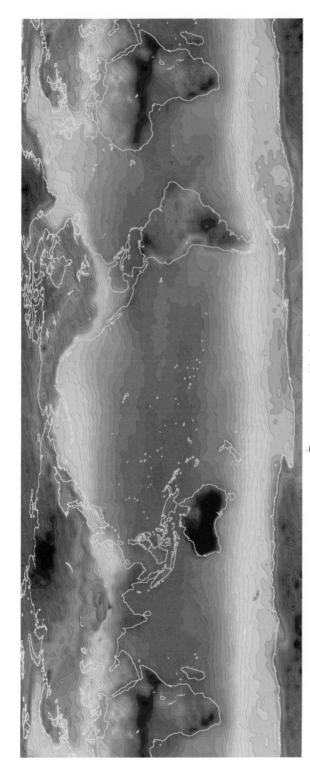

Degrees Kelvin

243 253 263 273 283 293 303 313

Plate 9. Surface temperature of land and oceans in January 1979 calculated from data collected by the High Resolution Infrared Sounder and Microwave Sounding Unit on the Noaa meteorological satellites using the technique described by Chahine (1974) (NASA Jet Propulsion Laboratory).

Mean Water Vapor for July 10 — October 10, 1978

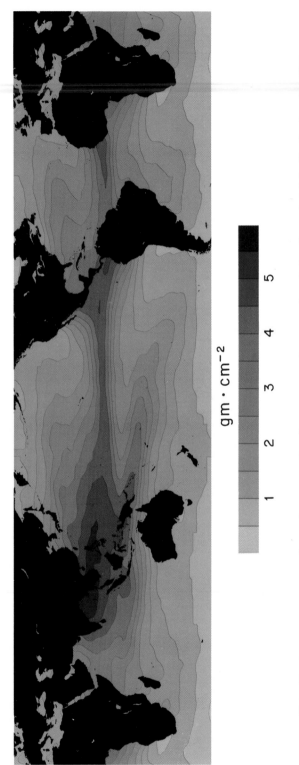

gm · cm^{-2}

1   2   3   4   5

Plate 10. Average water vapor in a column extending from space to the sea surface during the period July 10—October 10, 1978 observed by the Scanning Multichannel Microwave Radiometer on Seasat (NASA Jet Propulsion Laboratory).

Mean Wind Speed for July 10 — October 10, 1978

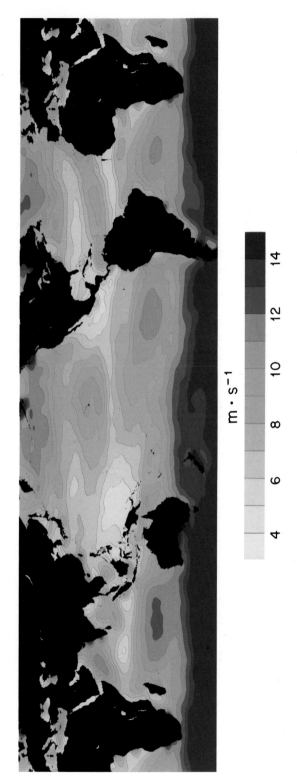

m · s⁻¹

Plate 11. Average surface wind speed during the period July 10—October 10, 1978 observed by the Scanning Multichannel Microwave Radiometer on Seasat (NASA Jet Propulsion Laboratory).

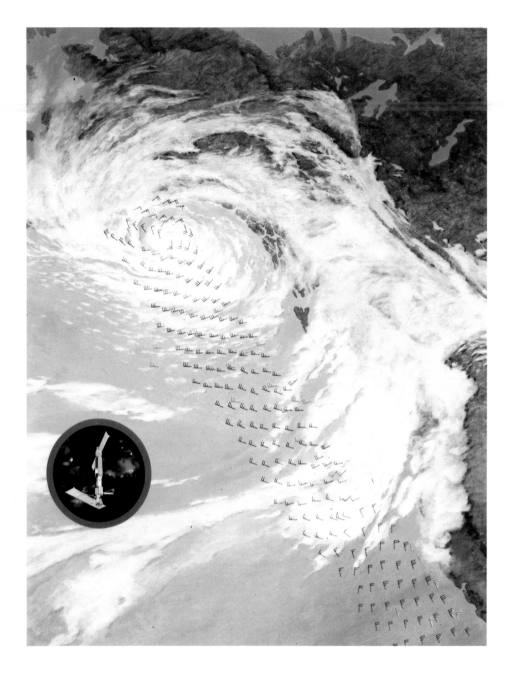

Plate 12. A storm in the Gulf of Alaska on 14 September 1978 observed by satellite techniques. The surface winds marked in black were observed by the Seasat scatterometer; the cloud image is from the Visible and Infrared Spin Scan Radiometer on a Goes geostationary meteorological satellite, and the winds marked in red were measured by surface ships. The Seasat satellite is shown in the inset (NASA Jet Propulsion Laboratory).

Wind Velocity on September 14, 1978

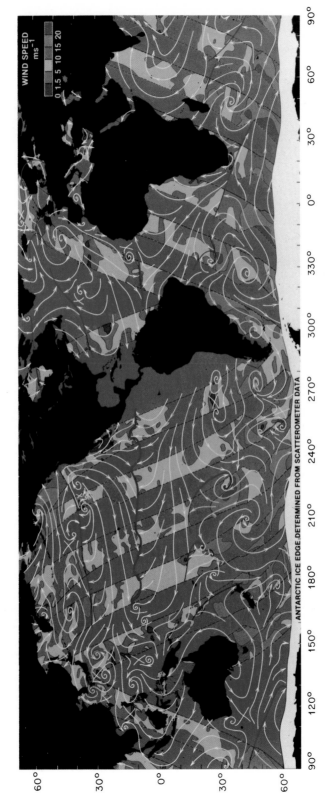

Plate 13. Global wind field at the sea surface observed by the Seasat scatterometer on 14 September 1978 together with the northern extent of the sea-ice pack around Antarctica seen by the same instrument (NASA Jet Propulsion Laboratory).

Gravity Field at the Sea Surface

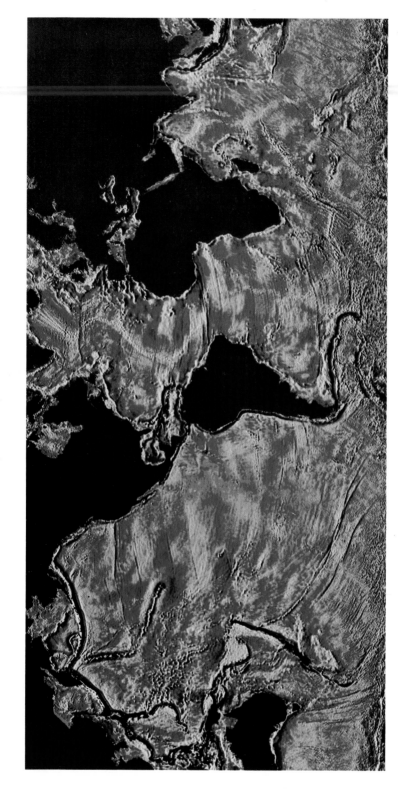

Plate 14. The marine geoid observed by the altimeter on Seasat. The relief is due to variations in Earth's gravity field resulting from the uneven distribution of mass within the Earth. Trenches, seamounts, transform faults, and oceanic ridges are clearly seen. The surface is displayed as though it were illuminated by light shining from the northwest, thus accentuating the features on the surface (Lamont-Doherty Geological Observatory of Columbia University).

Variability of Sea Level due to Surface Currents

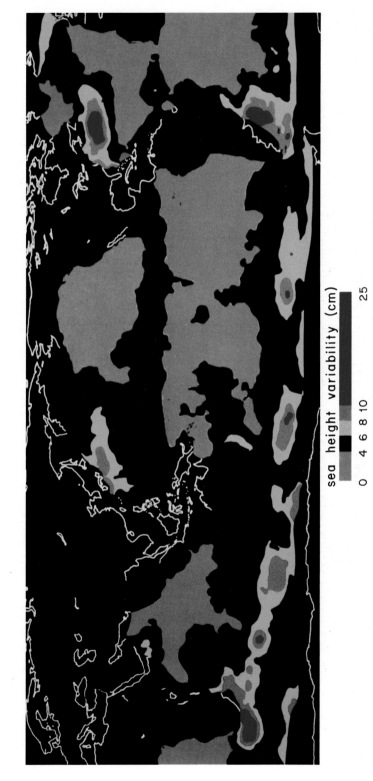

sea height variability (cm)

0   4 6 8 10   25

Plate 15. Variability of sea level due to time-varying surface geostrophic currents during July—October 1978 calculated from repeated observations of sea level made by the Seasat altimeter. The map clearly shows the variability due to the western boundary currents and the circulation in the Southern Ocean (NASA Goddard Space Flight Center).

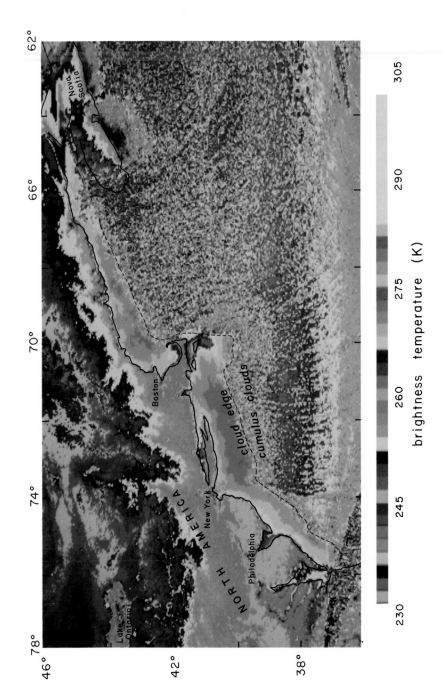

Plate 16. Clear, cold air blowing off the east coast of North America in winter is warmed by the ocean producing convection and cumulus clouds that grow in height with distance from shore. This infrared image of the area made on 17 February 1979 by the Advanced Very High Resolution Radiometer on TIROS-N shows a decrease in cloud-top temperature that is closely related to heat loss from the ocean, a quantity that can be calculated from the image knowing the structure of the atmosphere at the coast and the distance from the shore to the first formation of cumulus clouds (from Chou and Atlas, 1982).

This function occurs frequently in the theory of Fourier transforms and is plotted in figure 9.2.

The *beamwidth* of the antenna is defined to be the angular width $\Delta\theta$ between the two points where

$$I(\theta)/I(0) = \tfrac{1}{2}$$

For this example, the half-power points of a uniformly illuminated aperture are at

$$ksa = \pm 1.391$$

so the total width is $ksa = 2.782$, and

$$\Delta\theta = 0.886\,\lambda/(2a)$$

assuming $\sin\theta \approx \theta$ for a narrow-beam antenna.

If the illumination is varied across the aperture, then other similar relations hold. For example, if

$$f(x) = \cos(\pi x/2a) \qquad -a < x < a$$

then

$$\Delta\theta = 1.19\,\lambda/(2a)$$

In general, tapering the illumination slightly increases the beamwidth.

The primary reason for varying the illumination across the aperture is to control the secondary beams to the side of the main beam, the sidelobes. In the case of the uniformly illuminated aperture, the first sidelobe has a maximum value of 0.047 the value of $I(0)$; that is, it is down only 13.3 dB from the main lobe. By tapering the illumination, the level of the sidelobes can be substantially reduced. Thus the aperture with a cosine illumination has sidelobes down 23 dB from the main lobe.

This discussion demonstrates the two primary characteristics of antenna beams:

(a) Resolving power is determined by size. The larger the antenna, compared with the wavelength of the radiation, the better the antenna can resolve two emitting regions close together in angle.

(b) The distribution of intensity across the aperture of the radiation field determines the shape of the beam. A uniformly illuminated aperture has a narrow main beam but relatively large secondary beams (sidelobes). A tapered distribution produces a beam with much lower sidelobes, at the expense of slightly broadening the main lobe. But such tapering makes little change in the overall resolving power of the antenna. This is fixed by its size.

9.2 Antenna gain and received power

The angular distribution of intensity radiated by an antenna is called either the power pattern $P_n(\theta,\phi)$ or the gain $G(\theta,\phi)$, depending on how the distribution is normalized. The *power pattern* is convenient for describing mathematically the distribution of radiated power, and it is defined to be the radiated intensity divided by the maximum intensity I_{max}:

$$P_n(\Omega) \equiv I(\Omega)/I_{max}(\Omega)$$

where we have switched to a discussion of two-dimensional antennas radiating into a solid angle $\Omega\,(\theta,\phi)$ in spherical coordinates θ,ϕ, and $I\,(\theta,\phi)$ is the radiant intensity of the transmitted signal. P_n is the function plotted in figure 9.2.

The *gain* is useful for discussing the power either radiated or received by an antenna, and it is defined to be

$$G\,(\Omega) \equiv 4\pi I\,(\Omega)/\int_{4\pi} I\,(\Omega)\,d\Omega$$

Thus the gain is normalized so that the integral of gain over all solid angles is 4π.

The total power P_T transmitted by the antenna (in watts) is

$$P_T = \int_{4\pi} I\,(\Omega)\,d\Omega$$

Finally, the power density $E\,(\theta,\phi)$ in the beam at some distance R from an antenna transmitting P_T watts is

$$E\,(\theta,\phi) = P_T G\,(\theta,\phi)/(4\pi R^2)$$

Consider now an antenna used as a collector of radiance. Recall that the power incident on a surface is $dP = L\,(\theta,\phi)\,A\cos\theta\,d\nu\,d\Omega$ watts, where $L\,(\theta,\phi)$ is the radiance from the direction θ,ϕ; $d\nu$ is the bandwidth of the radiance; $A\cos\theta$ is the projected normal area of the surface; and $d\Omega$ is the solid angle of the beam of radiation. Because an antenna can be a wire as well as a collecting surface, the *effective area* of an antenna is defined to be

$$A_e\,(\theta,\phi) \equiv \frac{P_r\,(\theta,\phi)}{L\,(\theta,\phi)d\Omega\,d\nu}$$

This relates power $P_r\,(\theta,\phi)$ received by a lossless antenna to the brightness $L\,(\theta,\phi)$ of a distant scene in the direction θ,ϕ. The total received power is then

$$P_r = \frac{d\nu}{2}\int_0^{2\pi}\int_0^{\pi} L\,(\theta,\phi)\,A_e\,(\theta,\phi)\,\sin\theta\,d\theta\,d\phi$$

where the factor of ½ accounts for an antenna responding to polarized radiation.

The relationship between gain and effective area can be established using thermodynamic relationships. Consider an antenna terminated by a resistor, and both in thermal equilibrium within a large enclosing blackbody having a temperature T_B (fig. 9.3). The antenna both radiates and receives energy, and by reciprocity the two must be equal when the antenna and resistor are in equilibrium with these surroundings. The power transmitted in a particular direction is

$$P_T G\,(\theta,\phi)d\Omega/(4\pi)$$

and the power received from the same direction is

$$L_B\,(\theta,\phi)A_e\,(\theta,\phi)d\Omega\,d\nu$$

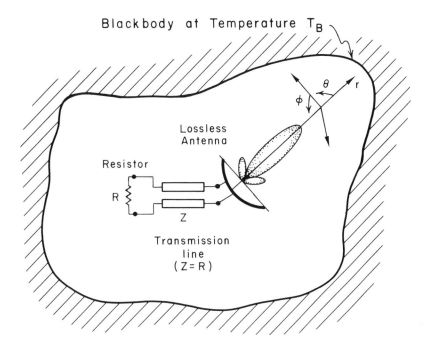

Figure 9.3   Geometry for relating the effective area A_e of an antenna to its gain G.

Equating the two gives

$$P_T G\,(\theta,\phi)/\,(4\pi) = L_B\,(\theta,\phi)A_e\,(\theta,\phi)\ d\nu$$

The power radiated into a single linear polarization by a blackbody is

$$L_B\,(\theta,\phi) = KT_B/\lambda^2$$

and the power produced by a resistor is (Nyquist, 1928)

$$P_T = KT_B\ d\nu$$

where K is Boltzmann's constant, and where we have used the Rayleigh-Jeans approximation for the blackbody radiance. Thus

$$G\,(\theta,\phi) = 4\pi\ A_e\,(\theta,\phi)/\lambda^2$$

These relations can now be used to discuss the power received by a lossless antenna viewing a scene with brightness $L\,(\theta,\phi)$. First, the received power P_r is converted to an *antenna temperature* T_A, defined to be

$$T_A \equiv P_r/\,(Kd\nu)$$

Then antenna temperature is related to the scene temperature T_s through the equation for received power together with the equation relating effective area to gain:

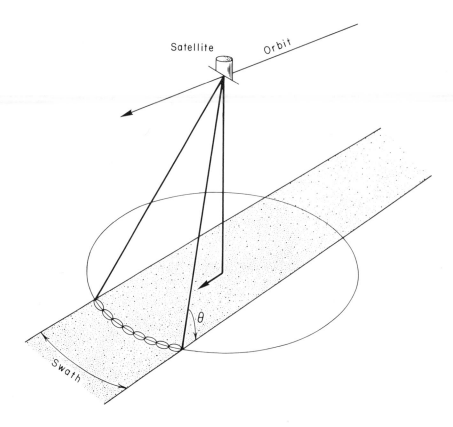

Figure 9.4   The conical scan geometry is often chosen because the incidence angle at the
sea surface is constant along the swath.

$$T_A = \frac{\iint\limits_{4\pi} T_s(\theta,\phi) G(\theta,\phi) \sin\theta \; d\theta d\phi}{\iint\limits_{4\pi} G(\theta,\phi) \sin\theta \; d\theta d\phi}$$

where we have used

$$4\pi = \iint\limits_{4\pi} G(\theta,\phi) \sin\theta \; d\theta d\phi$$

and where the scene radiance has been converted to an apparent blackbody
temperature using the Rayleigh-Jeans approximation

$$T_s = \lambda^2 L / (2K)$$

Thus an antenna views a weighted average of the scene temperature, where the
weighting is determined by the gain of the antenna. If the gain is close to a
delta function, then the antenna temperature is close to the scene temperature
in the direction that the antenna is pointing. If not, then other sources of radiation in the scene, particularly those radiating into the antenna's sidelobes, also
make important contributions to the signal.

It is important to note that the antenna temperature is not the physical temperature of the antenna. Rather, it is the temperature of the scene viewed by the antenna. If the antenna absorbs a fraction η of the power it receives, then the antenna temperature T_A is reduced and the antenna itself must radiate energy at a rate determined by its physical temperature T_0 through Planck's equation. The apparent temperature of the signal at the output of the antenna T_{out} is then

$$T_{out} = (1-\eta)T_A + \eta T_0$$

Thus η is an absorptance, and it influences the signal in the same way as, say, the absorptance of the atmosphere.

The practical calculation of the gain of a multifrequency, polarized antenna is difficult and complex; yet the accuracy of the calculation directly influences the accuracy of observations made by the antenna. The calculation often goes under the general name of *antenna pattern corrections*, and the nature of the problem is well illustrated by the calculation for a particular antenna used by the multifrequency radiometer on Seasat as described by Njoku (1980) and by Njoku, Christiansen, and Cofield (1980).

Antennas for radiometers are designed to have very low sidelobes; thus the sidelobes of a particular radiometer, the SMMR, were roughly 35 dB lower than the main lobe, the actual values depending on frequency. The geometry for viewing the surface is usually chosen to keep the incidence angle constant at the sea surface in order to increase the accuracy of the interpretation of the observations (fig. 9.4), so conical scanning geometries are common.

9.3 Radiometers

The internal noise generated within practical radiometers is much larger than the signal received by the antenna, and would completely dominate if its influence could not be reduced. The reduction is achieved by averaging the output of the receiver over many observations of the surface, and then comparing the averaged output with many observations of a known source, either within the radiometer (a noise source) or outside (cold space). Such a system is called a *switching radiometer*.

The averaging reduces the thermal noise; and the number of measurements that can be averaged depends on the bandwidth of the receiver and the duration of the observation. The bandwidth df of the receiver defines a time Δt:

$$\Delta t = df^{-1}$$

over which a measurement is correlated. Measurements made closer together than this in time are correlated, while more sparsely spaced measurements are not. Over an interval of time t, N independent measurements can be made, where

$$N = t/\Delta t = t \, df$$

Typically $df = 2 \times 10^8$ Hz and $t = 0.1$ sec, so $N \sim 2 \times 10^7$. Because noise is reduced by $\sqrt{N}$, noise within the receiver can be reduced by factors of 5×10^3.

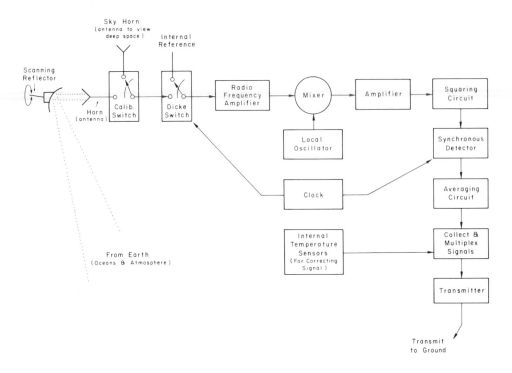

Figure 9.5   Components of an idealized Dicke radiometer.

(Note that many writers use *B* for bandwidth and *T* for time to obtain an expression in the form $\sqrt{BT}$ for the reduction of noise.)

The low frequency, nonthermal noise generated by changes in gain of the receiver is also large, and it is reduced by switching rapidly between the antenna and a source of known power. The switching is done at a rate faster than that at which the gain changes; the variation in the output of the receiver is proportional to the difference between the temperature observed by the antenna and the temperature of the calibration source. The technique was first proposed by Dicke (1946), and radiometers of this kind are called *Dicke-switched radiometers* (fig. 9.5). More elaborate systems are also used, and a description of the more common designs can be found in Reeves (1975:516–527), and in Price (1976:201ff.).

Super high frequency radiometers have been flown in space on Nimbus 5, 6, and 7, on Skylab, on Seasat-1, and on satellites flown by Russia and India (tables 9.2 and 9.3). The early instruments used one or at most two frequencies to observe Earth's surface, and thus were unable to sort out the multiple phenomena that can influence radiometric observations. The multifrequency radiometers NEMS and SCAMS were designed to probe the atmosphere. In theory, observations made at their lowest frequency, when combined with the observations from ESMR on the same satellite, could be used to form a multifrequency radiometer capable of observing the surface; however, this was not done in practice.

Table 9.2 Radiometers for viewing the sea

Satellite	Instrument	Acronym	Swath (km)	Launch Year
Skylab	S-193 Radiometer/ Scatterometer	RADSCAT	11−170	1974
Skylab	S-194 L-band Radiometer	S194	280	1974
Nimbus-5	Electronically Scanned Microwave Radiometer	ESMR	3000	1972
Nimbus-6	Electronically Scanned Microwave Radiometer	ESMR	1270	1975
Nimbus-5	Nimbus E Microwave Spectrometer	NEMS	185	1972
Nimbus-6	Scanning Microwave Spectrometer	SCAMS	2400	1976
Nimbus-7 Seasat	Scanning Multichannel Microwave Radiometer	SMMR	600	1978
Bhaskara-1	Satellite Microwave Radiometer	SAMIR	2000	1979
Bhaskara-2	Satellite Microwave Radiometer	SAMIR	2000	1981
DMSP	Special Sensor Microwave/Imager	SSM/I	1400	1984 (planned)
MOS-1	Microwave Scanning Radiometer	MSR	300	1986 (planned)

It is the multifrequency microwave radiometers on both Seasat-1 and Nimbus-7 (fig. 9.6), identical units, that have made the first simultaneous measurements of rain, water vapor, sea temperature, and surface winds. The radiometers observed two polarizations of the radiation emitted at five frequencies and at an incidence angle of 50°. Each frequency and polarization was sen-

Table 9.3 Characteristics of SHF Radiometers

Satellite	Instrument	Resolution (km)	Frequency (GHz)	Bandwidth (MHz)	$NE\Delta T^*$ (K)
Skylab	S-193	10	13.90	200	1.0
Skylab	S-194	280	1.41	27	1.0
Nimbus-5	ESMR	25	19.35	250	1.5 (47 ms)
Nimbus-6	ESMR	20	37.00	250	
Nimbus-5	NEMS	185	22.24	250	0.24 (2 s)
		185	31.40	250	0.23 (2 s)
		185	53.65	250	0.29 (2 s)
		185	54.90	250	0.29 (2 s)
		185	54.90	250	0.29 (2 s)
		185	58.80	250	0.24 (2 s)
Nimbus-6	SCAMS	145	22.24	220	1.0 (1 s)
		145	31.65	220	1.0 (1 s)
		145	52.85	220	1.5 (1 s)
		145	53.85	220	1.5 (1 s)
		145	55.45	220	1.5 (1 s)
Nimbus-7	SMMR	121	6.63	250	0.9 (126 ms)
Seasat		74	10.69	250	0.9 (62 ms)
		44	18.00	250	1.2 (62 ms)
		38	21.00	250	1.5 (62 ms)
		21	37.00	250	1.5 (30 ms)
Bhaskara-1,2	SAMIR	125	19.35	250	1.0 (0.21 s)
		200	22.24	250	1.0 (0.3 s)
DMSP	SSM/I	41	19.35	500	0.6 (8 ms)
		36	22.24	500	0.6 (8 ms)
		22	37.00	2000	0.4 (8 ms)
		10	85.50	2000	0.6 (4 ms)
MOS-1	MSR	40	23.8	400	0.75 (47 ms)
		30	31.4	500	0.90 (47 ms)

*Includes integration time given in parentheses.

sitive mostly to one or another surface or atmospheric variable, and when used together with all other frequencies, enabled the instrument to measure this one variable with great accuracy free from the influence of the others.

Scanning Multichannel Microwave Radiometer

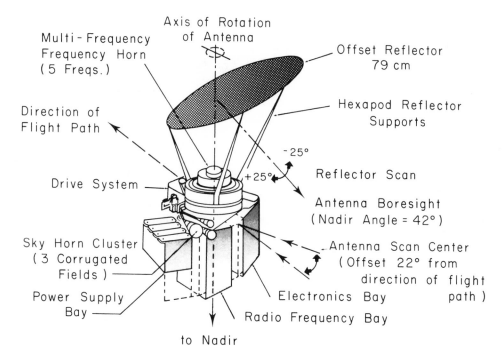

Figure 9.6   The Scanning Multichannel Microwave Radiometer (SMMR) carried on Seasat and
Nimbus 7 (from Stewart, 1982).

9.4 Sources of radio emission

Radiation emitted by both the atmosphere and the surface contribute to the
radiance observed by spaceborne radiometers. In the last chapter we noted that
atmospheric radiance partially obscures the surface, and that the radiance seen
by an instrument in space is the sum of the radiance from the atmosphere,
from the surface, and from the atmospheric radiance reflected by the surface.
For infrared wavelengths, clouds and water vapor are the primary atmospheric
emitters. The same is true for super high frequency radiation. In this section
we investigate the nature of radiation from the surface and from the atmosphere in slightly more detail, beginning with the equations of radiative
transfer. The theory is applicable to both infrared and radio signals, and provides a foundation for describing observations made in either band (except for
cosmic and galactic radiation, which are negligible at infrared wavelengths).

The, equation of radiative transfer for an optically thin atmosphere (section
6.3) is:

$$T(\theta) = (1-e)t^2 \, T_{\text{ext}} + (1-e)tT_d + T_u + etT_s$$

where $T(\theta)$ is the brightness temperature at the top of the atmosphere propagating upward at an angle θ relative to the vertical; t is the transmittance of the atmosphere; $e(\theta)$ is the sea-surface emissivity; $(1-e)$ its reflectance; and T_s its temperature. The other brightness temperatures are due to external downwelling radiation at the top of the atmosphere T_{ext}, downwelling radiation at the sea surface T_d, and upwelling radiation at the top of the atmosphere.

The external downwelling radiation is:

$$T_{ext} = T_{gal} + T_{cos} + T_{sun}$$

where

$$T_{gal} = \text{galactic noise } [< 1\text{K for } f > 3\text{Ghz}]$$
$$T_{cos} = \text{cosmic blackbody radiation } [3\text{K}]$$
$$T_{sun} = \text{solar radiation}$$

The cosmic radiation is isotropic, but the galactic radiation is strongest in the direction of the galactic center and its contribution depends on direction in galactic coordinates. The solar temperature is roughly 6000K for frequencies above 30 GHz, but it can be considerably hotter at lower frequencies. Thus radiometers must avoid reflected solar radiation. Ionospheric noise is negligible for frequencies above 3 GHz.

The atmospheric brightness temperatures are:

$$T_u = \int_0^\infty T(z)\kappa(z) \exp[-\tau(z,\infty) \sec\theta] \sec\theta \ dz$$

$$T_d = -\int_0^\infty T(z)\kappa(z) \exp[-\tau(0,z) \sec\theta] \sec\theta \ dz$$

where

$$\tau(z',z'') = \int_{z'}^{z''} \kappa(z)dz$$

and

$$\tau = \tau(0,\infty)$$
$$t = \exp[-\tau]$$

where $T(z)$ is the thermodynamic temperature of the atmosphere at height z above the surface, τ is the optical depth of the atmosphere, and κ is an absorption coefficient. At SHF radio frequencies, the processes contributing to κ are:

$$\kappa = \kappa_{liq} + \kappa_{oxy} + \kappa_{vap}$$

where

$$\kappa_{liq} = \text{absorption coefficient of water drops}$$
$$\kappa_{oxy} = \text{absorption coefficient of oxygen}$$
$$\kappa_{vap} = \text{absorption coefficient of water vapor}$$

The absorption coefficient for cloud droplets is given by the Rayleigh approximation for absorption by particles:

$$\kappa_{\text{liq}} = \int_0^\infty n(a)\sigma_A(a)\,da$$

$$\kappa_{\text{liq}} = \frac{8\pi^2}{\lambda}\,\text{Im}\left|-\frac{m^2-1}{m^2+2}\right|V$$

where a is the droplet radius, $m(T)$ its index of refraction, λ the wavelength of the radiation, σ_A the Rayleigh absorption cross-section, $n(a)$ the number of drops per unit volume per unit of size, T the thermodynamic temperature of the drop, and V the volume of water per volume of atmosphere observed by the radiometer. If the liquid drops are large enough to fall as rain, the absorption coefficient must be calculated using Mie theory, as discussed in the last section of this chapter.

The absorption coefficients for oxygen and water vapor are calculated from the pressure, temperature, and water vapor density along the path traversed by the radiation, using the analytic expressions for oxygen absorption given by Van Vleck (1947a) and Meeks and Lilly (1963), and those for water vapor given by Van Vleck (1947b) and Staelin (1966).

The brightness temperature of the surface is also due to several processes

$$T_s = T_0 + T_w$$

where

$$T_0 = \text{thermodynamic temperature of the surface}$$

$$T_w = \text{additional brightness due to wind}$$

Note that wind generates waves (which change the local incidence angle at the surface) and foam. Both change the emissivity of the surface.

This completes the description of the transfer of radiation from the surface to space at super high frequencies. The problem in applying radiometry to oceanography is to use observations of brightness temperature at different polarizations, frequencies, and incidence angles in order to calculate variables of interest, such as sea surface temperature or wind speed. This requires that we invert the radiation equations and solve them for the variables of interest.

At this point, the problem looks hopelessly complicated. Observed brightness is a function of three unknown continuous quantities $T(z)$, $\kappa_{\text{liq}}(z)$, and $\kappa_{\text{vap}}(z)$, and two unknown variables T_0 and T_w. Thus we are faced with an inverse problem with no unique solutions. The various possible solutions that satisfy the observations tend to be close together, however, and the solutions provide useful estimates of the variables of interest.

The ability to solve the radiation equations rests primarily on two conditions: (a) the signal from various variables tends to dominate all other signals in some frequency bands (fig. 9.7); and (b) the influence of $\kappa(z)$ and $T(z)$ is small. Thus the problem can be linearized about an expected value of each variable. This leads to a set of linear equations that can be solved if enough observations are available. But remember that the solution is not unique, only a good

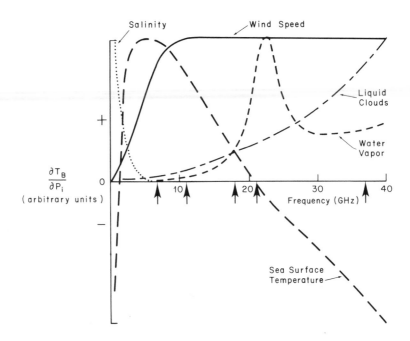

Figure 9.7 Sketch of the variation of brightness temperature T_B due to variation of various phenomena P_i as a function of frequency. The arrows indicate the frequencies observed by the SMMR (from Wilheit, Chang, and Milman, 1980).

approximation to the true solution. The trick is to pick a set of frequencies, polarizations, and incidence angles such that the uncertainty in the calculated solution is acceptably small.

As may be expected, many methods for solving this particular inverse problem have been proposed. Thrane (1978) discusses the general problem; Hofer and Njoku (1981), Wilheit, Chang, and Milman (1980), and Wilheit (1978) discuss the particular problem of interpreting observations from the five-frequency SMMR; and Chang and Wilheit (1979), Grody (1976), Staelin et al. (1976), and Blume, Kendall, and Fedors (1978) discuss the simplified form of the problem applicable to earlier radiometers that observed only a few radio frequencies. Of more general use, however, is the discussion by Parker (1977) of the inverse problem in its widest context.

The general scheme of the solution is roughly the following, although substantial variations have been explored.

(a) A set of frequencies, polarizations, and incidence angles is carefully chosen to reduce the errors in the solution, to enable important variables to be observed, and to avoid radio frequency interference from transmitters on the surface. The SMMR observed two polarizations at each of five frequencies at an incidence angle of 50°.

(b) The mean brightness temperature is calculated for each frequency, polarization, and angle using many different but typical realizations of $T(z)$, $\kappa(z)$, T_0, and T_w.

(c) The deviation from the mean brightness is expressed as a linear (or sometimes a nonlinear) function of the variables of interest. For SMMR, these were: integrated liquid water, integrated water vapor, wind speed, sea surface temperature, and rain rate. (An *integrated value* is the total amount between the surface and the top of the atmosphere. This is also called the *columnar value*.)

(d) The linear matrix relating brightness to the variables of interest is inverted.

(e) Finally, observed values of brightness are used together with the inverted matrix to calculate the variables of interest.

In general, the matrix and its inverse are nearly diagonal, and usually only the most sensitive terms are used in the calculation (Hofer and Njoku, 1981; Pandey and Kniffen, 1982). This not only simplifies the calculation but also increases the accuracy of the solution. Noise in the observations is amplified by the least sensitive terms in the matrix and this overrides the additional information carried by them. Alternatively, the importance of each observation in the calculation of the solution is weighted by the signal-to-noise ratio of the observation.

9.5 Radiometer observations of the sea

The early radiometers, those on Nimbus 5 and 6 and that on Skylab, observed the surface using only one or two frequencies; the other frequencies were used to observe the atmosphere and did not penetrate to the surface. This limited the observations to those surface variables that dominate the brightness temperature to the exclusion of other, less bright phenomena. Fortunately, a few processes produce strong emission and can be studied with little error by single-frequency radiometers:

(a) *Multi-year ice concentration in winter.* In this season, water vapor in the atmosphere is very small, open water is rare, and ice in the Arctic fills the radiometer's view. Under these conditions the different emissivities of ice types allows the radiometer to measure the relative concentration of multi-year ice relative to first-year ice.

(b) *Percent open water in summer.* In summer, thin melt layers on the ice erase the distinction between first-year and multi-year ice, and the variation in brightness temperature is related almost entirely to variation in ice cover. Sea water has low emissivity, ice has high emissivity, and the observed brightness varies linearly between the apparent sea surface temperature and the apparent temperature of the ice, depending on the concentration of ice (Gloersen et al., 1978).

(c) *Ice boundaries.* The strong contrast in emissivity between ice and water also allows the edge of ice sheets to be mapped easily . For example, the boundaries of Arctic and Antarctic ice were regularly mapped using the Nimbus radiometers (Gordon, 1981), and ice around Antarctica was found to be related to the flow of the circumpolar current. Where it is deflected by underwater ridges, so, too, is the edge of the ice sheet (figure 9.8; plate 1).

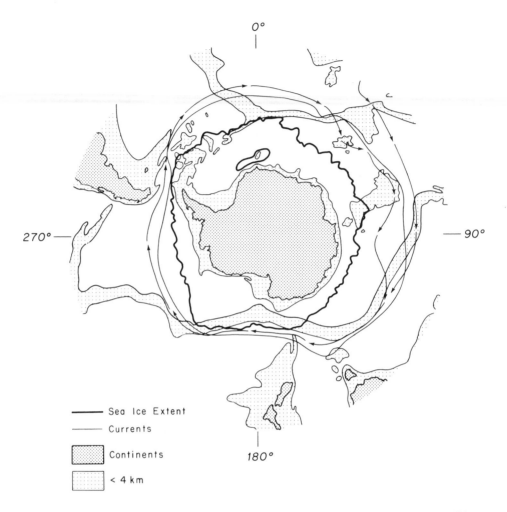

0°

270° —

— 90°

——— Sea Ice Extent

——— Currents

Continents

< 4 km

180°

Figure 9.8 Distribution of Antarctic sea ice on 31 August 1974. The boundary is the 15% sea ice contour derived from brightness temperature observed by the Nimbus-5 ESMR. Note the area of open water near 0°, the Weddel Sea Polynya. Arrows show the approximate position of the Antarctic circumpolar current (from data in Zwally et al., 1976). Underwater ridges appear to deflect the current, and this, in turn, influences the ice boundary.

(d) *Ice grain size on glaciers.* Variation in ice grain sizes result from variation in snowfall rate, and this variation changes the emissivity of glaciers at microwave frequencies. When the snowfields occur over large areas, such as the continental glaciers of Greenland and Antarctica, they can be mapped from space to yield the rate of snow accumulation (Zwally, 1977).

(e) *Rainfall over the oceans.* The large absorption coefficient of liquid water, and the low emissivity of sea water, ensure that liquid water in the form of rain appears in sharp contrast over oceanic areas, allowing satellites to map rain from space. This is discussed in greater detail in the next subsection.

These early satellite observations have been very useful, and the clarity of interpretation of microwave data has resulted in satellites making major contributions to our understanding of the polar seas, the distribution of ice, and ice dynamics. The full potential of microwave radiometry, however, was realized with multifrequency radiometers such as the SMMR, although the ten different combinations of frequency and polarization observed by this instrument may be more than necessary for many purposes. Using the many observations, it is possible to calculate the amount of liquid water and water vapor in the atmosphere, rain rate, sea surface temperature, and surface wind speed, all with good accuracy.

The accuracy of the measurements has been estimated by comparing satellite observations with surface and radiosonde observations. This is not easy. The radiometer observes a large area on the surface, so the surface observations must be extended to provide estimates of the surface variables over a similarly large area. Despite this, careful comparisons have been made in several experiments. These have used surface measurements made during the Gulf of Alaska Experiment (GOASEX); the Joint Air Sea Interaction Experiment (JASIN) in the northeast Atlantic; tropical radiosonde observations, particularly from central Pacific islands; and worldwide ship reports of weather and sea temperature (Wilkerson et al., 1979). The variables commonly studied by multifrequency radiometers include:

(a) *Sea surface temperature.* The measurement of sea surface temperature with an accuracy of 1°C is the most difficult radiometric measurement. Sea water has a low emissivity, typically around 0.35—0.40 at frequencies below 20 GHz, and has a brightness temperature between 120K and 140K depending on incidence angle and polarization. The brightness temperature is converted to thermodynamic temperature by multiplying by the reciprocal of emissivity. This magnifies the influence of small errors produced by atmospheric clouds and water vapor, surface winds, and uncertainties in antenna patterns. All must be accounted for with good accuracy. Present estimates of the accuracy of the observations lead to values around $\pm 1.0°$C globally (fig. 9.9) with somewhat better accuracy in mid-latitudes (Njoku and Hofer, 1981; Hofer, Njoku, and Waters, 1981). This accuracy requires some restrictions: (i) the oceanic areas must be at least 300 km from land, else the influence of the relatively hot land viewed through the sidelobes of the antenna add additional error that is difficult to remove; (ii) sun glint must be avoided, including certain sun angles at the spacecraft for the SMMR on Nimbus 7; and (iii) the radiometer must be out of sight of terrestrial transmitters operating at the same frequency as the radiometer; in some extreme instances this requires that the land be below the horizon seen by the satellite.

(b) *Water vapor.* Radio emissions from water vapor stand out clearly against the cold background of the sea, and the SMMR observations of integrated water vapor are especially accurate. In mid-latitudes they are at least as accurate as radiosonde observations (fig. 9.10). In the tropics, where water vapor is much greater, too few comparisons have been made to assess the quality of the

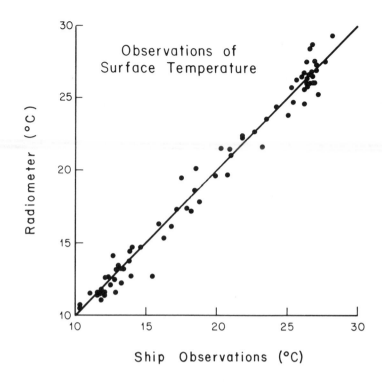

Figure 9.9 Observations of sea surface temperature made by the SMMR on Seasat, compared with sea temperature measured by expandable bathythermographs deployed from ships (from Lipes, 1982).

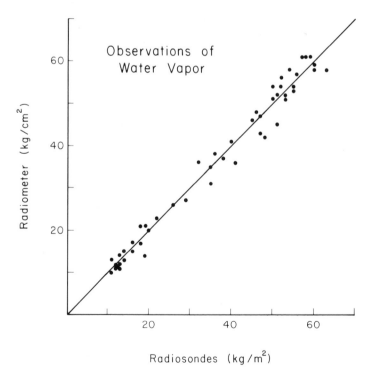

Figure 9.10 Observations of water vapor in the atmosphere made by the SMMR on Seasat, compared with water vapor measured by radiosondes (from data in Katsaros et al., 1981).

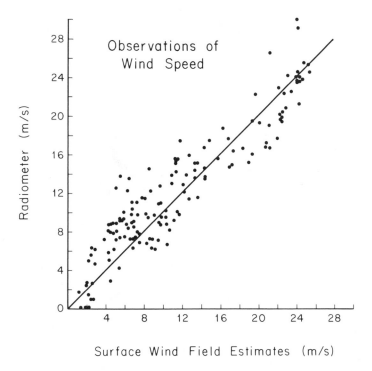

Figure 9.11 Observations of surface wind speed made by the SMMR on Seasat, compared with estimates of the surface wind field compiled from ship and buoy measurements (from T. J. Chester, Jet Propulsion Laboratory).

observations accurately. Overall, the instrument seems to measure water vapor with an accuracy better than 9% (Katsaros et al., 1981; Lipes, 1982); the SMMR on Seasat has been used to map the worldwide distribution of water vapor during July, August, and September 1978 (fig. 4.8; plate 10). Similar observations and maps have been made using the SCAMS on Nimbus 6 (Staelin et al., 1977; Grody, Gruber, and Shen, 1980) and the NEMS on Nimbus 5 (Staelin et al., 1976).

(c) *Wind speed.* The SMMR observations of wind speed are encouraging, but they are perhaps not quite as accurate as observations made by other types of instruments. Preliminary work yields estimates of accuracy that are around ± 2.6 m.s^{-1} (fig. 9.11). More recent work, however, gives slightly better accuracy (Lipes, 1982; Wentz, Cardone, and Fedor, 1982), and data from the Seasat SMMR has been used to produce maps of wind speed (plate 11). Of additional interest is the dependence on fetch. Remembering that surface roughness and foam change the emissivity of the sea surface, it is useful to ask whether or not the known fetch dependence of wave slopes and foam coverage must be accounted for in converting brightness temperature to wind speed. Webster et al. (1976) observed that brightness temperature is independent of fetch at least at an incidence angle of 38° and a wind speed of 20 m.s^{-1} (fig. 9.12), for frequencies between 5 and 37 GHz. The result is surprising. It is attributed to the decrease in shortwave slope with fetch (the influence of the α term in Hasselmann's form of the wave spectrum described in section 2.2), which balances the increase in foam coverage with fetch.

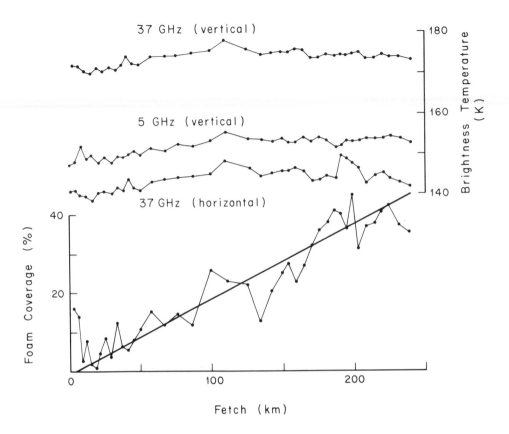

Figure 9.12 Brightness temperature of horizontally and vertically polarized radiation observed at an incidence angle of 38° by a radiometer on an aircraft as a function of fetch, together with percent foam cover estimated from photographs (from data in Webster et al., 1976).

(d) *Oil films.* Thick, new oil films are readily observed by airborne instruments, but old, dispersed oil spills have only a very weak signal, and most films are not sufficiently extensive to be observed from space. Troy and Hollinger (1977) note that films of #2 diesel fuel that are 0.1–2.0 mm thick change the brightness temperature of the sea surface by 5–80K, the temperature change depending on film thickness, when the sea is observed using 20–30 GHz radiation. Such films, however, do not long remain as a continuous, extensive film. The volatile material evaporates, and the film breaks into small globs of tar. For these reasons, SHF radiometers have only a limited usefulness for detecting oil on the sea surface.

9.6 Observations of rain

Liquid water in the atmosphere absorbs radio frequencies and consequently radiates energy at the temperatures typical of the lower atmosphere. In contrast, sea water has very low brightness temperature of around 120K. Against this cold background, rain is easily detected, and when present tends to dominate the brightness temperature at the higher frequencies when seen from space. Frozen water, in the form of snow or ice, has a very low absorption coefficient, 0.01 that of a drop of liquid water with the same water content (Battan, 1973:70), and is nearly invisible compared with liquid water. Thus the signal observed at satellite heights is dominated by water and water vapor between the surface and the freezing level in the atmosphere. The problem is to relate the observed brightness to rain rate, accounting for the brightness due to cloud drops and water vapor, neither of which contributes to rainfall.

Recall from the discussion of scatter of radiation (section 5.1) that the scattering σ_S and absorption σ_A cross sections are

$$\sigma_S = \frac{128\pi^5 a^6}{3\lambda^4} |g(m)|^2$$

$$\sigma_A = \frac{8\pi^2 a^3}{\lambda} \text{Im}[-g(m)]$$

where

$$g(m) = \frac{m^2 - 1}{m^2 + 2}$$

and m is the complex index of refraction, λ is the radio wavelength, and a is the particle radius. The absorption and scattering coefficients are then:

$$\kappa_S = b_S \int_a n(a)\, a^6\, da$$

$$\kappa_A = b_A \int_a n(a)\, a^3\, da$$

where b_A and b_S are constants and $n(a)$ is the particle drop size distribution. Thus both coefficients are expressed as moments of the drop size distribution. The rainfall rate R can be put in similar form. The contribution of a single drop to the rain rate is the drop's volume times its fall velocity $v(a)$, and for an ensemble of drops

$$R = b_R \int_a n(a)\, v(a)\, a^3\, da$$

Battan (1973:86) gives $v(a) \approx \sqrt{a}$, a relation that follows if the drops fall through still air at a rate governed by a constant drag coefficient. Under these conditions,

$$R = b_R \int_a n(a)\, a^{7/2}\, da$$

This expression is only approximately true: the drops do not fall through still air, rather rain falls in cells, and the net influence of the falling drops is to create a downdraft that increases their velocity. Nor is the drag coefficient constant; large drops falling at high velocity are flattened and have an internal circulation, resulting in an increasing drag coefficient (Foote and du Toit, 1969). But these are relatively minor concerns and do not seriously degrade the usefulness of the technique.

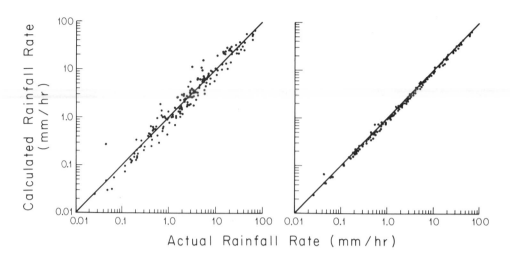

Figure 9.13   Left: rainfall rate calculated from backscattered radio waves. Right: rainfall rate calculated from backscatter plus optical extinction. Backscatter and extinction are different moments of the rain drop size distribution, and rainfall calculated from two moments has considerably less scatter than that calculated from only one moment (from Ulbrich and Atlas, 1977).

The drop size distribution is usually approximated by (Marshall and Palmer, 1948)

$$n(a) = N_0 \exp\left[-2\Lambda\, a\right]$$

with $N_0 = 0.08$ cm^{-4} and $\Lambda = 4.1 R^{-0.21}$ when Λ is expressed in units of per millimeter and R in millimeters per hour. Ulbrich and Atlas (1977, 1978) show that both N_0 and Λ are independent variables, and that if any two moments of the drop size distribution are known, then any other can be calculated with an accuracy of around 10%. If only one is known, then the accuracy is only 50% (fig. 9.13). This surprisingly good accuracy is due in part to the insensitivity of the moments of a function to the particular form of the function, $n(a)$ in this case.

Because it is difficult to devise spaceborne systems to measure two different moments of the drop size distribution, practical systems for observing rain rate measure only one, the brightness temperature due to emission by raindrops. A method for relating brightness to rain rate over the ocean is described by Wilheit et al. (1978). First, they calculated the absorption and scattering coefficients for rain drops from Mie theory using the Marshall-Palmer drop size distribution; then they related the coefficients to rainfall rate through the coefficient Λ. The influence of water vapor and cloud drops was calculated assuming a known surface temperature and a profile of relative humidity that varied linearly from 80% at the surface to 100% at the freezing level. A 0.5 km thick cloud with net density of 25 mg.cm^{-2} was assumed to be below the freezing level plus an additional contribution that varies with rainfall rate. The freezing level was determined from the assumed surface temperature using a 6.5°C.km^{-1} lapse rate.

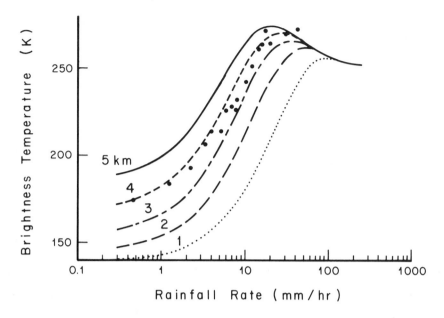

Figure 9.14 Brightness temperature of 19.35 GHz radiation at satellite heights due to rain in the atmosphere, with height of the freezing level as a parameter. The points are temperatures observed by a radiometer on the surface, converted to temperatures that would be seen in space, compared with rain rates measured at the surface (from Wilheit et al., 1978).

This model of the atmosphere is sufficient for solving the radiative transfer equation. A trial solution was obtained numerically by first computing upwelling and downwelling brightness, ignoring scatter; then scatter was included and the brightness was recomputed in a series of iterations until the solutions converged. The calculations indicate that absorption was 2–5 times larger than scatter for all rainfall rates, and that the brightness temperature is determined almost entirely by absorption. These calculations of radiative transfer through a rainy atmosphere give brightness temperature at satellite heights as a function of rainfall rate, with surface temperature (and thus the height to the freezing level) as a parameter (fig. 9.14). Although the brightness is nearly independent of rain, for rain rates less than 1 mm.hr^{-1}, and greater than 20 mm.hr^{-1} the signal can be used to measure rain rate between the two extremes. If lower frequencies are observed, the curves are shifted to the right and heavier rain rates can be measured. Conversely, higher frequencies can be used to measure lighter rains.

Rain rates calculated from ESMR data, when compared with rain rates observed by a weather radar in Florida, indicate agreement within a factor of 2. The scatter in the comparison results from several contributions: (a) The relation between rainfall rate and absorption is inexact, and data in Ulbrich and Atlas (1977) indicate this should contribute ± 30% to the scatter. (b) The relation between rainfall rate and backscatter measured by the surface radar is also inexact by a similar amount, and for the same reason. (c) The particle drop

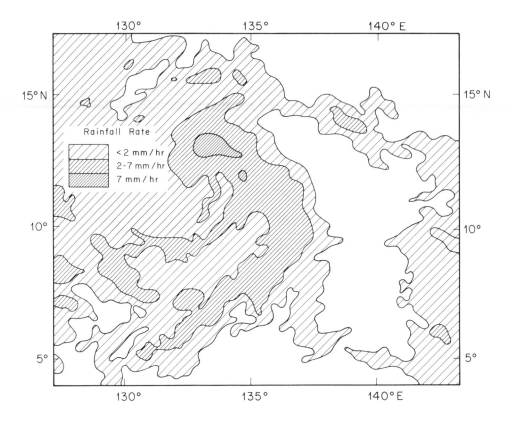

Figure 9.15   Rain cells imbedded in Tropical Depression Nora observed by the ESMR on Nimbus-5 in October 1973 (from Allison et al., 1974).

size distribution is not necessarily that of Marshall and Palmer. Light rain is often the result of a few large drops rather than many small drops (Altshuler and Telford, 1980). (d) In thunderstorms such as those near Florida, frozen water in the form of hail is frequently covered by a thin film of water, and the height of the freezing level is not the same as the maximum height of liquid water. (e) The satellite viewed a large area of sea, over which rain rate was expected to vary. The nonlinear relationship between rain and brightness results in overemphasis of heavy rain in the areal average. Taken together, these errors are no worse than those of shipboard measurements, but the satellite is able to map rain over large areas compared to the very limited number of shipboard observations.

Using the technique developed by Wilheit et al. (1978), Rao, Abbot, and Theon (1976) used ESMR to map the distribution of rain over the world's oceans during 1973 and 1974. Even over this short time period, they saw marked differences in total rain in the North Pacific averaged over a year, indicating the importance of rainfall variability. Most of these maps are unreliable, however, as a result of errors in the computer programs used to generate the maps.

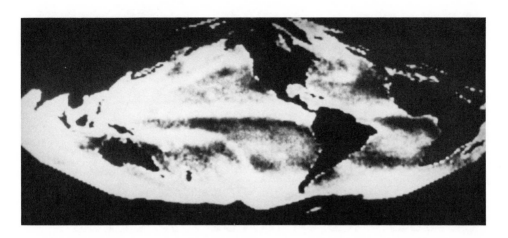

Figure 9.16   Liquid water in the atmosphere calculated from observations at 22.2 and 31.7 GHz made by the Scanning Microwave Spectrometer on Nimbus-6 during 15—29 August 1975. Drier air masses are darker (from Staelin et al., 1977).

The same group has also studied the rain bands in tropical storms (Allison et al., 1974), using the same radiometers but with a somewhat cruder relation between radio brightness and rain rate (fig. 9.15). They could trace the development of the storm and could differentiate between high cirrus and convective rain cells which are normally hidden below the cirrus. A few years later, Staelin et al. (1977) mapped liquid water and water vapor over the oceans using observations at 22.2 and 31.7 GHz made by the Scanning Microwave Spectrometer SCAMS on Nimbus-6 (fig 9.16).

10

THEORY OF RADIO SCATTER
FROM THE SEA

Radio waves transmitted by a downward looking satellite radar are scattered into many directions after striking the sea. A small portion of this scattered energy, carrying with it information about phenomena at the surface, is eventually received by the radar. To use radars to study the sea, it is necessary first to calculate the scattered electromagnetic field, knowing the properties of the surface; and then, conversely, to infer the surface characteristics (but not necessarily its exact shape) knowing the scattered field. We are concerned here with the first problem. The second is the subject of the following chapters, where we will discuss the interpretation of the radar signal.

The theory of radio scatter from the sea is well understood and is in good agreement with observations. Two regimes are recognized: near vertical incidence, scatter is from mirror-like wave facets that reflect energy toward the receiver; away from the vertical, the number of facets rapidly becomes very small, because almost no waves have slope greater than 25°. At greater angles, another mechanism reflects energy toward the receiver, resonant scatter from single sinusoidal components of the spectrum of ocean waves that match the projected wavelength of the radio signal onto the sea surface. The sinusoids act exactly like a diffraction grating, and the scattered power is given by Bragg's equation. As might be expected, the resonant scatter provides a powerful and very selective tool for observing the sea surface and the waves upon it.

The theory of the scatter of electromagnetic waves from rough surfaces is comprehensively described in textbooks (see, for example, Beckmann and Spizzichino, 1963; Ishimaru, 1978; Bass and Fuks, 1979). The scattered field is calculated from the incident field by properly matching the electric and magnetic fields across the boundary, knowing its exact shape. The exact mathematical techniques for effecting such solutions are varied, and often lengthy and difficult, so we will be content merely to state the answer, referring to the literature for the details of the calculation. Usually, the ratio of reflected and incident energy is expressed by a scattering cross section typical of the surface or target illuminated by the radar. It is this we wish to describe.

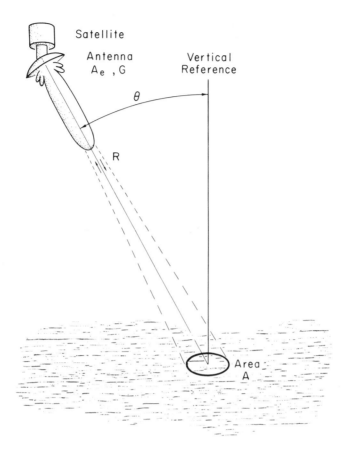

Figure 10.1   Radar geometry for describing scatter from a sea surface having a dielectric constant ϵ_r. θ is the angle of incidence, A is the surface area illuminated by the radar at range R from the surface, A_e is the effective area of the radar antenna, and G is its gain.

10.1 The scattering cross section

Consider a radar illuminating an oceanic area (fig. 10.1). The *scattering cross section* σ of the area is defined to be 4π times the ratio of the radiant intensity I_s scattered back toward the transmitter—to the power density (watts per unit area) in the wave incident on the area (Ridenour, 1947:21):

$$\sigma = 4\pi\ A\ \cos\theta\ \left(\frac{I_s}{\Phi_i}\right)$$

where (table 10.1) we have written the power density in the beam as the radiant flux Φ_i divided by the projected surface area $A\cos\theta$. Note that this definition is equivalent to the definition of extinction cross section given in section 5.1.

The relationship between the power transmitted P_T and power received P by a radar is the *radar equation* (Ridenour, 1947):

Table 10.1 Notation Used to Describe Radio Scatter

a	Amplitude of a sinusoidal ocean wave [m]
A	Area on surface [m^2]
A_e	Effective area of antenna [m^2]
G	Gain of antenna
$g(\phi)$	Directional distribution of ocean waves
$g_{ij}(\theta)$	Angular dependence of scattered radio intensity
h	Wave height [m]
H	Constant
I_s	Radiant intensity [W.sr^{-1}]
k_R	Radar wavenumber [m^{-1}]
$p(\zeta_x, \zeta_y)$	Joint probability distribution of wave slopes
P, P_T	Received, transmitted radar power [W]
r	Radius of curvature of sea surface [m]
r_1, r_2	Two components of radius of curvature [m]
R	Radar range [m]
s_u, s_c	Standard deviation of wave slope in upwind and crosswind directions
x, y, z	Cartesian coordinates
α	Ocean wave spectral constant
α, γ	Sine and cosine of incidence angle
α_i	Sine of local incidence angle
ϵ_r	Dielectric constant
ϕ	Angle of ocean wave propagation
ϕ_m	Mean wind direction
Φ_i	Radiant flux [W]
ζ	Sea surface elevation [m]
ζ_x, ζ_y	Components of sea surface slope
θ	Incidence angle measured from vertical
θ_i	Local incidence angle
λ	Radar wavelength [m]
ρ	Fresnel reflectance
σ_0	Backscatter cross-section per unit area
σ	Backscatter cross-section [m^2]
$\psi(k_x, k_y)$	Wavenumber spectrum of sea surface elevation [m^4]
ψ, δ	Local sea surface slopes

$$P = \frac{P_T G}{4\pi R^2} \ \frac{\sigma}{4\pi} \ \frac{A_e}{R^2}$$
$$(a) \qquad (b) \qquad (c)$$

where the terms in the expression are related to the definition of cross section:

(a) is the power density in the electromagnetic wave at a range R. The transmitted power P_T, if isotropic, will expand outward as a spherical wavefront having an area of $4\pi R^2$. The gain G is the increase in radiance relative to an isotropic radiator, and multiplying by the gain in the direction of the surface gives the power density at the surface.

(b) is the radiant intensity in the direction of the radar, produced by scatter from a surface with a scattering cross section σ.

(c) is the solid angle subtended by a radar antenna with an effective area A_e. Recalling that antenna gain and effective area are related by (section 9.2)

$$A_e = G\lambda^2/(4\pi)$$

we obtain, assuming the antenna gain for receiving is the same as for transmitting:

$$P = \frac{P_T G^2 \lambda^2}{(4\pi)^3 R^4} \sigma$$

This equation is correct for a point target with cross-sectional area σ. Distributed targets, such as the sea, are spread over some area A, and the scatter is described by the *scattering cross section per unit area* σ_0 defined by

$$\sigma = \int_A \sigma_0 \, dA$$

where A is the area observed by the radar. Note that σ_0 is dimensionless and that it is defined relative to a unit area of the surface, not the projected area in the direction of the beam. Thus a Lambert surface, one with uniform brightness, has a cross section that varies as the cosine of the incidence angle.

The problem of calculating the amount of power scattered toward the receiver now devolves to calculating σ_0 knowing the scattering surface.

10.2 Specular-point scatter theory

For angles of incidence near vertical ($\theta \sim 0$), scatter comes from mirror-like facets of the sea surface aligned so as to reflect energy back toward the transmitter (fig. 10.2). In order to calculate the scattered energy, we need to define a facet; to estimate the probability that a facet will be correctly oriented; and to calculate the energy reflected by each facet.

Mathematically, this approach for calculating σ_0 based upon facets is called the *tangent plane* approximation; and two general methods for finding a solution can be used. When the solution is found from the equations for the electromagnetic field, the procedure is called the *physical optics* approach. Alternatively, we could consider the radiation as a bundle of rays and could calculate the scatter of these rays, an approach designated as *geometrical optics*. Both approaches based on the tangent plane approximation give essentially the same results.

Using physical optics, the facet is approximated by an elliptic paraboloid with radii of curvature r_1 in the x,z plane and r_2 in the y,z plane. The scatter from each paraboloid is calculated and the scatter is summed over the illuminated area, using the probability distribution of surface slopes. For a radar looking in the x direction, this procedure gives (Barrick, 1968)

$$\sigma_0 = |\rho(0)|^2 \sec^4\theta \; p(\tan\theta, 0)$$

where $\rho(0)$ is the Fresnel reflection coefficient for normal incidence, θ is the angle of incidence, and $p(\zeta_x, \zeta_y)$ is the joint probability density distribution of slope ζ_x in the x direction and ζ_y in the y direction, and where

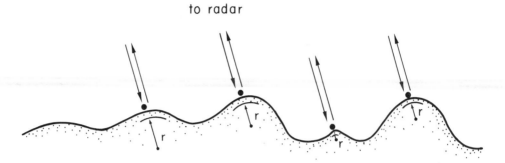

Figure 10.2 Scatter from facets occurs when each facet is oriented so as to reflect radio energy back toward the radar. The strength of the scatter depends in part on the radius of curvature r of the facet.

$$\zeta_x = \frac{\partial \zeta}{\partial x}$$

$$\zeta_y = \frac{\partial \zeta}{\partial y}$$

provided

$$2\pi r \cos^3\theta / \lambda \gg 1$$

$$r = \text{lesser of } r_1, r_2$$

To a good approximation, the sea surface slopes have a Gaussian distribution that is anisotropic about the wind direction. For a radar looking upwind

$$\sigma_0(\theta) = \frac{|\rho(0)|^2}{2 s_u s_c} \sec^4(\theta) \exp[-\tan^2\theta / (2 s_u^2)]$$

with

$$s_u^2 = \overline{\zeta_u^2}$$

$$s_c^2 = \overline{\zeta_c^2}$$

where s_u^2 is the mean square upwind slope and s_c^2 is mean square cross-wind slope. Note that the slopes must be evaluated using only ocean wavelengths greater than the radar wavelength; shorter waves do not contribute to the scatter.

10.3 Resonant scatter

Specular point scatter applies only to that small range of incidence angles near vertical which are perpendicular to the wave slopes, there being few slopes that exceed $20-25°$ and none at great angles. Yet scatter is observed at all angles. The explanation for this phenomenon has been well known for nearly a century, but its application to the ocean was first noticed by Crombie (1955) only a few decades ago.

Surfaces with a regular lattice of roughness act as diffraction gratings, and scatter radiates into angles given by the geometry of the lattice spacing relative to the wavelength of the radiation. Because scatter from the repeating lattice pattern adds coherently in the far field (far from the surface), the scatter is resonant; and because the geometry of the scatter is given by Bragg's equation it is often called *Bragg scatter*. Because the ocean surface can be considered to be a superposition of plane waves, and because radio scatter is a linear process, each wave component can act independently as a diffraction grating. Within the limits set by the spectrum of ocean waves, all wavelengths and angles are found on the sea surface, so all radio frequencies in resonance with these waves are scattered. And the scatter is surprisingly strong.

To calculate the scattering cross section, a physical optics approach is again used, but the boundary conditions are linearized (note that this does not imply that radio scatter is a nonlinear process). Thus the ocean waves are replaced by effective electric currents on the mean surface. This requires that

$$k_R h \ll 1$$

where $k_R = 2\pi/\lambda$ is the radar wavenumber and h the wave height. The condition is easily met by long, dekameter radio waves; but it is almost always violated by short, SHF radio waves. Composite surfaces introduced in the next section must be used to calculate the resonant scatter of SHF signals from the sea.

The theory also requires that

$$k_R a \ll 1$$

where a is the amplitude of the ocean wave causing Bragg scatter. This condition is almost always true (Jordan and Lang, 1979). Water waves become unstable and break long before they become large enough to influence the linearization of the boundary conditions.

The resonant-scatter theory yields a backscatter cross section (Rice, 1951; Wright, 1966; Barrick, 1972; Jordan and Lang, 1979)

$$\sigma_0 = 8\pi k_R^4 |g_{ij}(\theta)|^2 \psi(2k_R \sin\theta, 0)$$

where

$$g_{HH}(\theta) = \frac{(\epsilon_r - 1)\cos^2\theta}{[\cos\theta + (\epsilon_r - \sin^2\theta)^{1/2}]^2}$$

$$g_{VV}(\theta) = \frac{\cos^2\theta(\epsilon_r - 1)[\epsilon_r(1 + \sin^2\theta) - \sin^2\theta]}{[\epsilon_r \cos\theta + (\epsilon_r - \sin^2\theta)^{1/2}]^2}$$

where $\psi(k_x, k_y)$ is the two-dimensional ocean wavenumber spectrum and ϵ_r is the complex dielectric constant of sea water. In these equations the subscripts *VV* and *HH* refer to polarization of the transmitted and received signal: *VV* = vertical transmit/vertical receive, *HH* = horizontal transmit/horizontal receive. The ocean wave spectrum is normalized such that

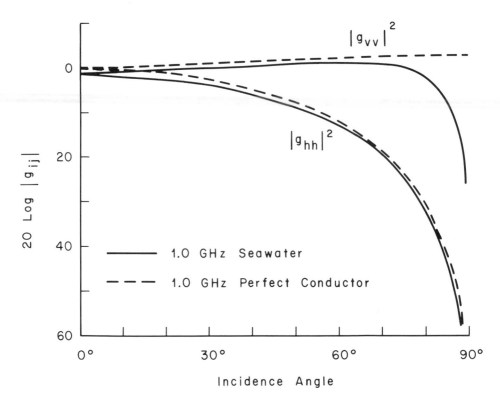

Figure 10.3   Angular dependence of scattering from a slightly rough sea (from Wright, 1968).

$$\overline{\zeta^2} = \int\limits_0^\infty \int\limits_0^\infty \psi\,(k_x\,,k_y\,)dk_x\,dk_y$$

Note, however, that the various spectra used in the literature are normalized in various ways. Electrical engineers usually assume a double-sided spectrum (k has both negative and positive values), while oceanographers use a single-sided spectrum (k is always positive).

The usefulness of radio scatter for the study of ocean waves is immediately apparent. The scattering cross section is directly related to the two-dimensional wavenumber spectrum of ocean-surface waves. Actually, because the spectrum is multiplied by k_R^4, the relationship is to the spectrum of wave slopes; however, the two are linearly related and interchangeable in this case. Further discussion of the use of radar scatter for studying ocean waves is included in the next two chapters.

For high frequency radio waves, the ocean is nearly a perfect conductor. Recalling (section 3.1) that the dielectric constant of a conductor is

$$\epsilon_r = [\epsilon' - i\sigma/(\epsilon_0\omega)]$$

where σ is the conductivity of the surface, ω the radian frequency of the radiation, and ϵ_0 the permittivity of free space, and noting that for a perfect conductor $\epsilon_r \sim -i\infty$, the expressions for $g_{ij}(\theta)$ become

$$g_{HH}(\theta) \sim \cos^2\theta$$

$$g_{VV}(\theta) \sim (1+\sin^2\theta) \; .$$

Thus the scatter of vertically polarized radiation (fig. 10.3) is nearly independent of incidence angle; and the scatter of horizontally polarized radiation varies as $\cos^4\theta$. As θ approaches 90°, horizontally polarized radiation is not scattered, but the cross section for vertically polarized radiation scattered from a good conductor remains constant:

$$\sigma_{VV} = 8\pi k_R^4 \psi(2k_R,0) \quad \theta = 90°$$

Note that at this grazing angle of incidence, the incident electric field is twice that which would occur in the absence of the boundary. This is due to the image of the transmitter being reflected in the surface, so the field in the scattering area is the sum of the direct and reflected fields. The influence can be accounted for by modifying either the radar equation or the cross section. The latter is commonly done, and the above equation includes a factor of 1/4 due to this effect.

In the case of scatter at grazing incidence from a good conductor, only radio waves with twice the wavelength of the ocean wave, and traveling in the same (or opposite) direction as the ocean wave, are scattered. Munk and Nierenberg (1969) noted that the scatter should also be nearly independent of wavelength for suitably short radio waves, and that it should be related to the equilibrium constant α of the ocean waves. To calculate the relationship, they assumed that the directional distribution $g(\phi)$ of the wavenumber spectrum is constant within 90° of the mean wind direction ϕ_m and zero elsewhere.

$$g(\phi) = 1 \quad -\pi/2 < \phi_m < \pi/2$$

Thus $H = 1/\pi$ in the notation of section 2.2. With this directional distribution, the directional spectrum $\psi(k,\phi)$ at sufficiently large k is

$$\psi(k,\phi) = \alpha/(2\pi k^4) \quad -\pi/2 < \phi_m < \pi/2$$

where α is a weak function of fetch. Using $\alpha \approx 0.008$ (typical of oceanic waves) they obtained

$$\sigma_{VV}(k,0) = \alpha/4 = 0.002$$

$$\sigma_{VV}(k,0) = -27\text{dB}$$

a value close to that observed by radio scatter experiments. Conversely, these experiments can be used to obtain better estimates of both α and the directional distribution of waves, $g(\phi)$.

to radar

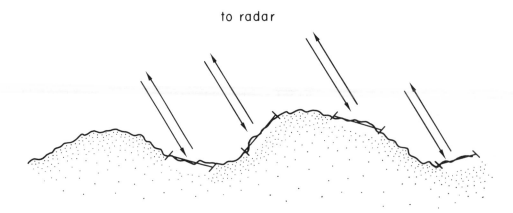

Figure 10.4 Bragg or resonant scatter from short waves tilted by longer waves (composite-surface approximation). Note that the vertical scale is exaggerated. Wave slopes rarely exceed 15°, and spectral reflection contributes little scatter to the indicated incidence angle.

10.4 Scatter from composite surfaces

The scatter of short radio waves can be calculated by assuming the ocean surface has two scales: (a) short wavelength ripples, riding on (b) longer, larger waves (fig. 10.4). The Bragg scatter from each small area is calculated using the local small wave field and the local orientation of the surface. The scatter is then integrated over the entire area, using the probability density function for surface slopes due to longer waves. The local cross section is (Wright, 1968; Valenzuela, 1978; Brown, 1978)

$$\sigma_{HH}(\theta_i) = 8\pi k_R^4 \left\| \left(\frac{\alpha \cos\delta}{\alpha_i} \right)^2 g_{VV}(\theta_i) + \left(\frac{\sin\delta}{\alpha_i} \right)^2 g_{HH}(\theta_i) \right\|^2 \psi(2k\alpha, 2k\gamma \sin\delta)$$

where

$$\alpha_i = \sin\theta_i$$
$$\gamma = \cos(\theta + \psi)$$
$$\alpha = \sin(\theta + \psi)$$

θ_i is the local incidence angle, and ψ, δ are the local sea surface slopes. A similar expression holds for $\sigma_{VV}(\theta_i)$, but σ_{VV} is much less sensitive to local tilts than is σ_{HH}.

Integrating over all local incidence angles gives

$$\sigma_{HH}(\theta) = \int_{-\infty}^{\infty} \int_{-\infty}^{\infty} \sigma_{HH}(\theta_i)p(\tan\psi, \tan\delta)d(\tan\psi)(\tan\delta)$$

The range of angles over which the various approximations apply, together with typical theoretical and experimental values of the scattering cross section, are given in figure 10.5. Specular point theory is applicable at small angles, Bragg or resonant scatter at larger angles. At very large angles, for the short wavelengths of SHF signals, shadowing can become important; but for dekame-

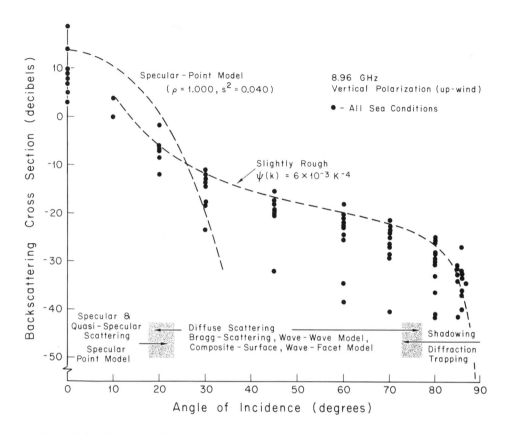

Figure 10.5   Illustration of mechanisms that produce radar backscatter from the ocean at the various angles of incidence (from Valenzuela, 1978).

ter waves at these angles, shadowing is unimportant, and Bragg scatter theory applies out to 90°.

Foam, spray, and breaking waves also occur on the sea surface and influence radio scatter. Scatter from thin foam and spray tends to be small, but breaking waves produce locally strong scatter. Thus the scatter of very short pulses from very small areas tends to be spiky, with spikes coincident with breakers (Lewis and Olin, 1980). Nevertheless, the influence is usually lumped into the wind dependence of scatter from waves and then forgotten.

10.5 Comparison of theory with measurement

To be believable, theories of scatter from the sea must be compared with observations. This is difficult. The two-dimensional wavenumber spectrum of the ocean waves must be measured at the same time the ocean area is observed by a radar; yet conventional techniques cannot measure $\psi(k,\theta)$ well in the open ocean. To avoid the difficulty, microwave radars have been operated above laboratory wave tanks where artificial, and controlled, waves can be made. Experiments by Wright (1966) and Wright and Keller (1971) attest to the accuracy at these frequencies (fig. 10.6), taking into account the difficulty of measuring both the radio scatter and the water waves even under these con-

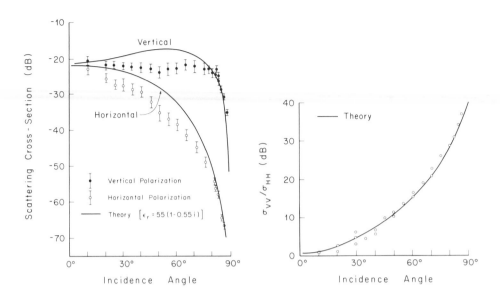

Figure 10.6 Backscattering cross section as a function of incidence angle for capillary waves on tap water. Left: the scattering section calculated from theory compared with observations. Right: the ratio of vertically to horizontally polarized backscatter, also compared with theory (from Wright, 1966).

trolled conditions.

At dekameter wavelengths the experimental difficulties are greatly reduced. Teague, Tyler, and Stewart (1975) compared the observed scatter of 160 m waves with that predicted by measurements of the directional ocean-wave spectrum averaged over all angles, and found good agreement. (They reported a factor of 2 difference between theory and observation, but failed to note a factor of 2 difference between the theoretical definition of the wave number spectrum and the same spectrum measured by a wave buoy.) Similar measurements were made by Barrick et al. (1974) but with less accuracy.

The resonant nature of Bragg scatter of dekameter waves at the intermediate incidence angles is also confirmed by the Doppler spectrum of the scattered signal. For sufficiently long wavelengths (such that $k_R h \ll 1$), the spectrum typically contains two delta functions at plus and minus the ocean wave frequency (see fig. 11.5 in the next chapter). These peaks correspond to approaching and receding ocean waves, respectively.

The theory is also confirmed by observed functional dependence of various variables; for example, the ratio σ_{VV}/σ_{HH} as a function of θ (fig. 10.6), as well as σ_{VV} and σ_{HH} as a function of θ. In particular, observations of this last function by airborne radars observing windy sea surfaces (fig. 10.7) show close agreement with the predicted shape, and indicate the relative importance of the three models for scatter. The uncertainty in the comparison is due primarily to an uncertainty in knowledge of the surface wave spectrum $\psi(k,\phi)$ used in the theory to compute radar scatter, particularly its dependence on local wind speed. Foam from breaking waves is also a good reflection of SHF radio signals, however (Lewis and Olin, 1980), and must also contribute.

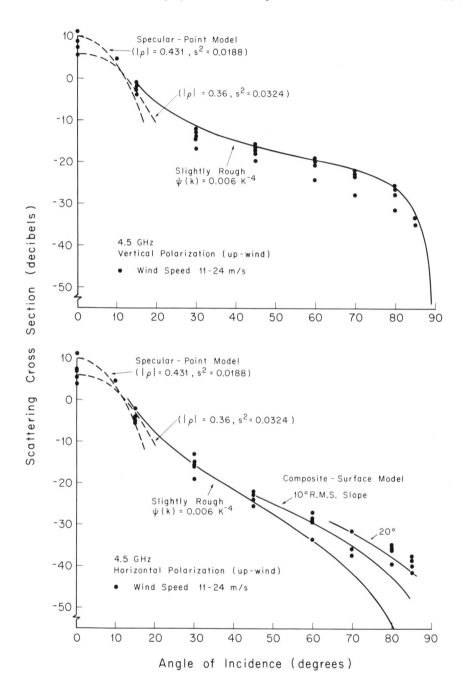

Figure 10.7   Comparison of measured and theoretical cross sections of the ocean for a radar frequency of 4.5 GHz. Top: vertical polarization. Bottom: horizontal polarization. The variation in the observations is due primarily to variation in wind speed at the sea surface (from Valenzuela, 1978).

11

SCATTER OF DEKAMETER
RADIO WAVES FROM THE SEA

The scatter of high frequency radio waves from the sea, although it cannot be observed from space, provides an elegant yet simple example of the usefulness of radio observations. In addition, the description of how these observations are made lays the foundation for understanding the operation of super high frequency radars that do operate from space and for interpreting the scatter of shorter wavelengths from the sea. Simply put, pulse-Doppler radars transmitting dekameter radio waves are used to measure very precisely the direction, frequency, wavelength, and height of typical ocean waves. This information is not only useful in itself but also can be used to deduce surface currents, and wind speed and direction out to several hundred kilometers from shore- or ship-based radars. If the signal propagates via the ionosphere, the distances can exceed 4000 km and bring into view all but the most distant reaches of the ocean.

11.1 Theory of radio scatter

The theory of the resonant scatter of dekameter radio signals, outlined in the last chapter, is identical to that for scatter of light from a grating or even acoustic scatter from a rough surface. The periodic electromagnetic field interacts with that periodic component of the ocean wave field which matches the Bragg equation:

$$\vec{k} = 2\vec{k}_R \sin\theta$$

where $\vec{k}$ is the ocean wave number vector, $\vec{k}_R$ is the radio wave number, and θ is the incident angle measured from the vertical. Usually, backscatter is observed at grazing incidence angles ($\theta = \pi/2$) and the scatter is produced by ocean waves with one-half the radio wavelength travelling radially toward or away from the radar. Furthermore, the scattered signal is shifted in frequency (the Doppler shift) by exactly the frequency of the ocean wave producing the scatter. The amplitude of the scattered signals is related to wave height through the equation for scattering cross section:

$$\sigma_{VV} = 8\pi \ k_R^4 \ \psi[2k_R, 0]$$

where the notation is that of chapter 10 and ψ is the ocean wavenumber direction spectrum.

The accuracy of radio measurements of the wavelength, direction, and frequency of the ocean wave that scatters the radio signal is determined by the geometry and duration of the observation. The size of the scattering area in units of wavelength, and the duration of the observation in units of wave period, determine the accuracy of the wavelength and frequency observation. The size of the receiving antenna determines its angular resolution and fixes the accuracy of the measurement of wave direction. Typically, wavelength and frequency can be measured with an accuracy of 0.5% and direction with an accuracy of better than $\pm 10°$.

Measurements of the amplitude of the scattered signal are less accurate. The transmitter, receiver, and antenna gains are difficult to calibrate, and they tend to change with time, as do propagation conditions. To minimize these influences, most experiments rely on relative measurements of scattered power, usually with radio frequency held constant.

The measurement of scattered power is further complicated by the randomness of the scattered signal, and many independent observations must be averaged together to determine accurately the mean scattered power. Usually, the randomness of individual observations greatly exceeds all other sources of uncertainty in the scattered signal. Each Fourier component of the ocean wave field has an amplitude that is a Gaussian random variable; thus the amplitude of the electric field at the receiver is also a Gaussian random variable. As a consequence, each point in the Doppler spectrum of the received signal is chi-square distributed with two degrees of freedom (before averaging), and accurate spectral estimates require that many independent estimates be averaged together to obtain the final spectrum (Barrick and Snider, 1977).

The validity of the theory of radio scatter depends on the assumptions used to derive the theory. The linear theory assumes that the sea is a good conductor and that the radio wavelength is much greater than the average wave height. The former is well satisfied, but the latter assumption is true only at lower radio frequencies. For frequencies above 20 MHz, second-order terms must usually be included in the theory. These require that all waves on the sea surface influence the scatter, not just the single wave component producing first-order resonant scatter. Thus observations of the entire ocean-wave field can be made using a single radio frequency, albeit with a great increase in the complexity of the processing of the received signal (Lipa, 1977; Lipa and Barrick, 1980). Ultimately, as the radio wavelength becomes short compared to wave height, the composite theory of radio scatter must be applied, and the Doppler shift of the scattered signal carries little useful information about the ocean wave spectrum. However, the intermediate case, where only second-order corrections to the linear theory are required, is of practical importance.

The second-order terms can be described within the general framework of wave-wave interactions (Hasselmann, 1966). Four waves are involved: the incoming radio wave, $\vec{k}_i$; two ocean waves, $\vec{k}_1$ and $\vec{k}_2$; and the radio wave scattered back toward the receiver, $\vec{k}_s$, where $\vec{k}$ refers to the respective wavenumber vectors. The interaction is written as the product of two interactions, each involving three waves. Two combinations are important if the aver-

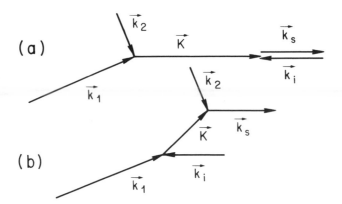

Figure 11.1   The vector wavenumber diagrams of ocean waves ($\vec{\mathbf{k}}_1$ and $\vec{\mathbf{k}}_2$) that scatter an incoming radio wave $\vec{\mathbf{k}}_i$. (a) Two ocean waves interact to produce a forced wave $\vec{\mathbf{K}}$ which scatters the radio wave. (b) The radio wave scatters first from one ocean wave, then from another.

age ocean wave height is sufficiently large (fig. 11.1): (a) Two ocean waves interact to produce a forced wave, a corrugation on the sea surface, that matches the Bragg equation. The radio wave scatters from this forced wave. (b) The incoming radio wave resonantly scatters first from one ocean wave, then from a second, provided the ocean waves are so aligned that after two scatters the radio wave is propagating toward the receiver.

In both cases, two conservation equations must be satisfied:

$$\vec{\mathbf{k}}_s = -\vec{\mathbf{k}}_i \pm \vec{\mathbf{k}}_1 \pm \vec{\mathbf{k}}_2$$

$$\omega_s = \omega_i \pm \omega_1 \pm \omega_2$$

together with the dispersion relations

$$\omega^2 = gk \qquad \text{ocean waves}$$

$$\omega = ck \qquad \text{radio waves}$$

where c is the velocity of light, g is gravity, and ω is the frequency of waves in radians per second. Because the Doppler shift $(\omega_s - \omega_i)$ is small, we can assume that $\vec{\mathbf{k}}_s \approx -\vec{\mathbf{k}}_i$. If, in addition, the ocean wave heights are small, the second-order terms can be ignored and the equations reduce to those for resonant scatter:

$$\vec{\mathbf{k}}_1 = \pm\, 2\vec{\mathbf{k}}_i$$

$$\omega_i = \pm\, (\omega_s - \omega_i)$$

Note that two ocean waves cannot interact to produce a third wave that satisfies the dispersion relation for water waves (Phillips, 1977:82). Instead they produce a forced wave, one that exists only when the original two waves are present.

Of the two interactions, the interactions among ocean waves tends to dominate, but multiple radio scatter cannot be ignored. Multiple scatter is roughly

(a)
Continuous - Wave Transmission

(b)
Pulsed Transmission

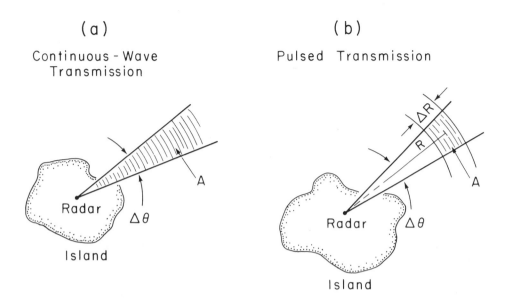

Figure 11.2   Plan view of an oceanic area A illuminated by a radio beam from an antenna with beamwidth $\Delta\theta$. (a) The echo from a continuous wave transmission comes from a long, wedge-shaped area; thus the echo from a particular small area cannot be isolated from the rest of the echo. (b) When a pulsed transmission is used, the echo at a particular instant comes from a particular small area.

proportional to the product of the scattering cross-section for each scatter, and this is small. The ocean wave interactions are proportional to the slope of the sea surface determined by all waves present on the surface, and the forced wave can have nearly the same amplitude as the free wave that produces the first-order scatter.

The four possible sign combinations produce four sidebands to the two peaks in the Doppler spectrum. Because all waves on the sea surface contribute to these processes, all can be measured by properly interpreting the information in these sidebands.

11.2 Observations of scatter using pulse-Doppler radars

The use of radars to study the properties of ocean waves requires measurements of the Doppler spectrum of radio signals backscattered from a particular area of the sea surface. The simplest way to make the measurement would be to place the radar in such a position that it could continuously observe the particular area of the sea using a directional antenna. But this is not possible in many instances. Suppose, for example, the radar is on an island and must observe a particular oceanic area offshore of the island. Such a radar would observe scatter not only from the desired area but also from those areas closer and further away than the desired area (fig. 11.2a). The solution to the problem is to use a pulse-Doppler radar to view the selected ocean area. Because such radars are widely used, it is worthwhile studying their operation.

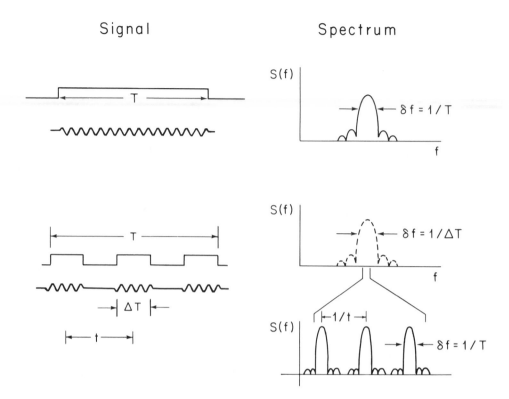

Figure 11.3   The spectrum of a pulsed radar transmission. A continuous wave, turned on and off to form a train of pulses, has a very narrow spectrum (bottom right), much narrower than the spectrum of a single pulse (the envelope at center right). The narrower the spectrum, the better the resolution in the Doppler spectrum of the echo from the sea surface.

The pulse-Doppler radar consists of a directional antenna that transmits and receives short pulses of radio waves with duration ΔT, each pulse containing several hundred wavelengths (fig. 11.2b). The time T for the pulse to travel out and back gives the range R to the area; the length of the pulse ΔR and the antenna resolution $\Delta\theta$ determine the size of the scattering area, A:

$$R = c\ T/2$$

$$\Delta R = c\ \Delta T/2$$

$$A = R\ \Delta R\ \Delta\theta$$

where $c = 3.0 \times 10^8$ m.s^{-1} is the velocity of light. This is the original concept of a radar, and it is the concept that generally comes to mind when we think of the term "radar": an apparatus to determine the range and direction of a particular target. The additional complication we wish to consider here is how to calculate the Doppler shift of the signal reflected from the target using a train of pulses.

Ordinarily, the spectrum of scatter is calculated from a continuous observation of the scatter from a particular area. But this is not essential. A train of

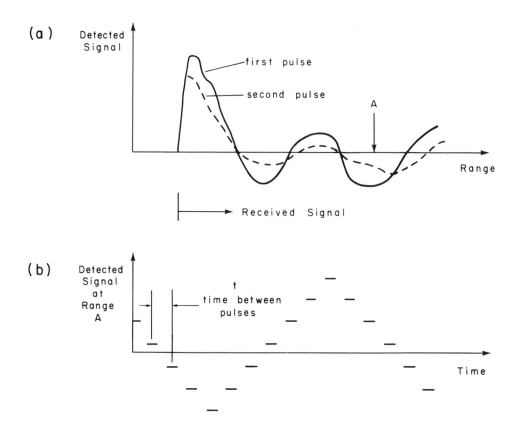

Figure 11.4   (a) Detected signal from a pulse-Doppler radar, as a function of range, showing typical pulse to pulse variation. (b) Variation from pulse to pulse as a function of time at a fixed range. The Doppler spectrum at a fixed range is calculated from the Fourier transform of this signal. Note that the scattered signal is sampled at the pulse repetition rate and is not known continuously.

coherent pulses can be used to sample the phase of the scattered signal and the Doppler spectrum can be calculated from a time series of reflected pulses, provided the pulses occur at a rate that is greater than twice the maximum Doppler shift of the echo from the sea (fig. 11.3). The concept of a pulse-Doppler radar thus comprises an apparatus for transmitting and receiving a train of coherent pulses, and a means of calculating the spectrum of the echo from a particular range.

To retain the information about the phase of the echo without having to handle high radio frequencies, the echo is multiplied by the transmitted signal offset in frequency by a few hertz, say ΔF, then smoothed over times comparable to the pulse length. The product contains the sum and difference of the transmitted and received signals, and smoothing removes the sum frequency, leaving the difference frequency. This is called the *detected signal*. Over the very short duration of the pulse, the sea is frozen; but a short time later, at the time of the next pulse, the sea will have changed slightly, as will the scatter, as a function of range (fig. 11.4). At any particular range, the scatter is sampled

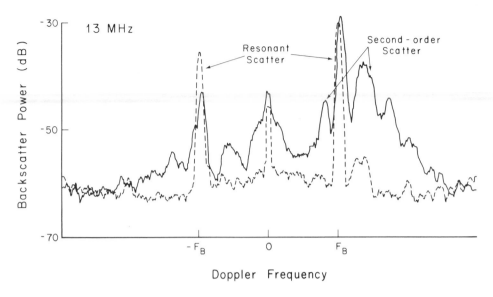

Figure 11.5 Doppler spectrum of 13 MHz (11 m) radio waves backscattered from the sea near San Francisco, showing two sharp peaks at $\pm F_B$ (0.37 Hz), the frequency of ocean waves producing Bragg scatter. *Dashed line*: the radio wavelength is greater than ten times the significant wave height and only first-order scatter is important. *Solid line*: the radio wavelength is less than ten times the significant wave height, and second-order terms produce sidebands to the first-order lines. These sidebands contain information about all components of the ocean-wave spectrum. The scatter at zero frequency is from land. Note that the first-order peaks are not precisely at their expected position, but are slightly displaced due to ocean currents.

at the pulse repetition rate. If this is greater than $2\Delta F$, then the sampled signal is an adequate representation of scatter from a particular area as a function of time. This is the desired signal. The Fourier transform of this signal gives the Doppler spectrum of the scattered signal from the area (fig. 11.5).

Now consider a typical example. The old Loran-A radio stations transmitted 50 μs pulses of 2.0 MHz, 160 m radio waves. Each pulse contained 100 (50 μs × 2 MHz) radio wavelengths, which scattered from 80 m, 0.14 Hz ocean waves. One ms after the pulse is transmitted, a radar receives 2,000,000.00 $\pm$ 0.14 Hz signals from an area 7.5 km (50 μs × 3 × 10^8 m.s^{-1} ÷ 2) deep that is 150 km away (1 ms × 3 × 10^8 m.s^{-1} ÷ 2). When this signal is multiplied by one with frequency of 2,000,001.00 Hz, and smoothed over 50 wavelengths (25 μs), the resulting signal is an estimate of the amplitude of the scattered radio waves as a function of range. The signal from a particular range will vary at a rate of 1.00 $\pm$ 0.14 Hz, the sign depending on whether waves are coming toward or going away from the radar, and this produces two lines or peaks in the Doppler spectrum of the received signal. Roughly 33 pulses are transmitted every second, and the pulses recur every 30 ms. In this short time the sea changes only slightly and the radar adequately samples the time history of the scattered signal. Of course, the amplitude will also vary over periods of minutes to hours as waves in the scattering area slowly change with time.

With its antenna pointed in a suitable direction and receiving scatter from the appropriate range, a pulse-Doppler radar can observe radio scatter from a particular oceanic area. Typically, observations can be made out to ranges of 1000—2000 wavelengths over the 2—30 MHz band using a few hundred watts of transmitted power. By scanning in range and direction (or azimuth in this example), the radar can map the ocean wave field over larger areas.

Note that there are two conflicting factors that must be balanced in the design of pulse-Doppler radars. (a) The pulse rate must be slow enough so that scatter from one pulse dies away before the next pulse is received. (b) The pulse rate must be at least twice the maximum expected Doppler shift of the scattered signal. Doppler radars for observing the sea that operate in the HF band easily meet these design criteria, but Doppler radars operating in the SHF band from rapidly moving satellites are more tightly constrained. Because of the satellite's velocity, the Doppler shift is much larger; yet the surface area to be observed extends over much greater distances, sometimes hundreds of kilometers. As a result, greater care must be taken in the design of these radars. Further information on the tradeoffs among design parameters, and of the inherent ambiguities in pulse-Doppler radars, can be found in Rihaczek (1969).

11.3 Receiving antennas, the concept of synthetic-aperture radar

A fundamental problem at dekameter wavelengths is to construct a sufficiently large antenna to obtain adequate azimuthal resolution for these long radio waves. Using the equation for the resolving power of a uniformly illuminated aperture (section 9.1), we see that at 2.0 MHz an antenna with a resolution of 5° must be 10 wavelengths or 1.6 km (1 statute mile) long. Such an antenna is neither cheap nor portable. To avoid this difficulty, we seek simpler methods for obtaining high angular resolution using much smaller antennas. The solution is to use a simple receiver to sample the radio field over a large aperture. The samples are then combined to synthesize an antenna that is equivalent in size to the aperture. Again, the technique is widely used and worth describing.

The synthetic-aperture principle is explained most simply by considering the Doppler shift of the signal received by a moving radar (fig. 11.6). The Doppler shift f observed by a moving radar is

$$f = (2vF \cos\theta)/c$$

where v is the radar speed, c is the velocity of light, F is the radar frequency, and θ is the angle between radar velocity vector and scattering area. For dekameter wavelengths, the sharp lines in the Doppler spectrum are spread by the radar motion, and scatter as a function of frequency is directly related to scatter as a function of angle. Note that the Doppler shift is symmetric about the radar heading and that there exists a right-left ambiguity in the measurement. This can be resolved by moving along a long, straight shoreline such that there is no ocean on one side of the beam, or by using an asymmetric receiving antenna.

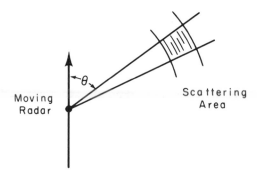

Figure 11.6 Plan view of geometry for viewing a scattering area using an antenna synthesized
by a moving radar.

The maximum angular resolution is related to the frequency resolution in the Doppler spectrum. The derivative of the Doppler shift is

$$df = (2vF \sin\theta \, d\theta)/c = 2v \sin\theta \, d\theta/\lambda$$

and it has a maximum at right angles to the direction of motion

$$df_{max} = 2v \, d\theta/\lambda, \quad \theta = \pi/2$$

where df is the frequency resolution in hertz, and λ is the radio wavelength in meters. The resolution in the Doppler spectrum is inversely proportional to the duration T of the sample used to compute the spectrum:

$$df = 0.886/T$$

thus

$$d\theta = \frac{0.886\lambda}{2vT}$$

During this time, the radar moves a distance

$$D = vt$$

thus

$$d\theta = \frac{0.886\lambda}{2D} \qquad \theta = \pi/2$$

This resolution is twice that of a uniformly illuminated aperture of size D, the distance over which the radar moves, and is independent of radar velocity. The increase in resolution over that of a uniformly illuminated aperture results from allowing both the transmitter and receiver to move. If the transmitter were stationary, the synthesized aperture would be identical in all respects to a real aperture.

The analysis assumes that the Doppler spectrum in the absence of motion contains two sharp lines due to approaching or receding ocean waves. This is another way of saying that the field in the aperture cannot change appreciably while it is being sampled. The technique works well in the lower HF band, but at higher frequencies the Doppler lines tend to be broad, and real antennas are used unless the radar velocity is so large that it dominates the Doppler shift of the signal from the sea.

A simple example can help clarify the discussion. Working with colleagues at Stanford, I have used a simple vertical wire plus loop antenna on a van driven along straight sections of road and along aircraft runways on Wake Island to synthesize 1–3 km long antennas for receiving 2.0 MHz Loran-A signals (Tyler et al., 1974). The simple antenna had a cardioid pattern that was oriented perpendicular to the velocity vector, and the main lobe of the cardioid was switched from right to left to right with alternate pulses. The two simultaneous series of observations were used to remove the right-left ambiguity of the synthesized antenna. The vehicle moved slowly enough, around 20 km.hr^{-1}, that the Doppler bands due to approaching waves did not overlap the band for receding waves. Typically, the synthesized antennas had resolutions of 4–6°, observing scattering areas of roughly 7.5 × 10 km at a range of 100 km.

11.4 Over the horizon radars

Dekameter radio waves, popularly called short waves in the early history of radio communications, were used for many years as the primary means for sending messages around the world. At some times of the day and for some radio frequencies, these wavelengths are reflected by the ionosphere to a greater or lesser extent and propagated to great distances, thus allowing radio broadcasts and messages to be sent between continents. By very careful choice of time and frequency, accurate observations of radio scatter can be made to ranges of 1000–4000 km using the same technique. Ranges greater than 4000 km are also possible, but for these longer distances the signal is scattered more than twice from the ionosphere, with considerable degradation in the signal.

The ionosphere is not a perfect mirror even for a single reflection. Rather, it behaves as a wavy, distorting, absorbing, moving reflector. It can support many ray paths between two points (fig. 11.7); its motion induces additional Doppler shifts, often with frequencies around 0.1–0.5 Hz; and each path has its own particular Doppler shift. These distortions not only limit the usefulness of ionospheric propagation but can also produce features in the Doppler spectrum which closely mimic those produced by the sea. Nevertheless, on some days and at some times, the disturbances are small and can be tolerated.

At best, reflections from some ionospheric layers, especially the E layer, can be very steady. Then scatter from the ocean can be used to survey waves over large areas. For these studies, multifrequency, high power radars with large, narrow-beam antennas are most useful. A number of such installations have been built to study the ionosphere and several are now observing the ocean. Notable examples include the radar operated by Stanford Research Institute in the central valley of California (fig. 11.8), which has a 3-km-long receiving

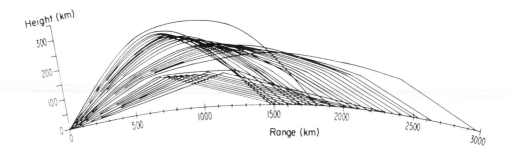

Figure 11.7 Typical ionospherically propagated ray paths showing multiple simultaneous paths between the radar and a particular scattering area, particularly between 1500 and 2300 km in this example (from Barrick, 1973).

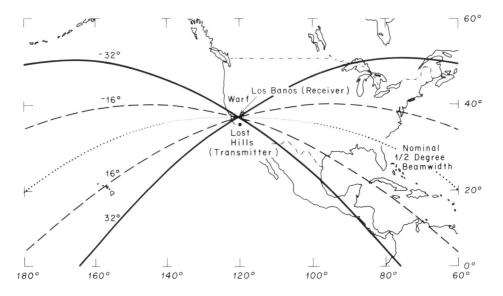

Figure 11.8 Limits of oceanic areas observable by ionospherically propagated radio signals from the Wide Aperture Radar Facility (WARF) operated by Stanford Research Institute. The narrow dotted line in the center is approximately ½° wide, the azimuthal resolution of the receiving antenna (from Barnum, Maresca, and Serebreny, 1977).

antenna, and a new facility in England, operated by the University of Birmingham for studying the North Atlantic. Note that the primary limit in using these radars is the same as that which has always limited shortwave radio communications, the instability of the ionosphere. Anyone who has listened to shortwave radio broadcasts knows the problem. The signal fades in and out due to multipath, and on some days the ionosphere is disrupted, sometimes by solar flares. Finding steady ionospheric layers that permit propagation to particular oceanic areas at particular times is not easy. Finding these conditions routinely is very difficult.

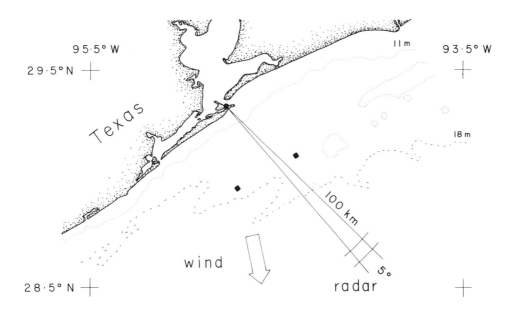

Figure 11.9 Plan view of radar beam, drawn to scale, used to measure directional distribution and growth of waves offshore of the Texas coast. The beam could be directed at any angle to measure the height of seven-second waves travelling in that direction (from Stewart and Teague, 1980).

11.5 Observations of the sea using HF scatter

Dekameter radio waves have provided important new information about ocean waves: measurements of the directional distribution of waves produced by a uniform wind, their growth as a function of angle to the wind, their attenuation by an opposing wind, their reflection by a coast, and the relation between wavelength and frequency of waves in deep water.

The simplest experiment consists of a pulse-Doppler radar on a very small island in the middle of the ocean or on a long, straight coast. A transmitter with an omnidirectional antenna sends out pulses of radio energy, and after being scattered these are received by a very directional, steerable antenna. This measures the scattered energy as a function of azimuth, which in turn is directly related to the directional distribution of waves with one-half the radio wavelength (fig. 11.9).

Just such an experiment was performed by Tyler et al. (1974). We used the synthetic aperture technique to measure the directional distribution of 7 s (0.14 Hz) waves generated by a steady uniform trade wind in the vicinity of Wake Island. The radar beam had an azimuthal resolution of about 5° and was steered in a circle around the island to observe the height of the waves as a function of direction, assuming the wave field was uniform about the island.

We have also used the same technique to measure the growth of waves away from the long, straight Texas shore, as a function of angle to the wind during times the wind was blowing away from the shore (Stewart and Teague, 1980).

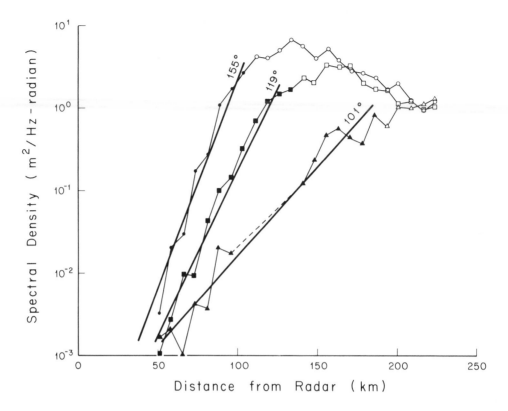

Figure 11.10　Dekameter radar observations of the spectral density of 80 m ocean waves along radial lines as a function of angle θ relative to north, using the geometry shown in figure 11.9. The solid points were used to determine the best-fitting straight lines whose slopes yield growth rates of the waves as a function of angle to the mean wind direction (from Stewart and Teague, 1980).

The radar beam was synthesized by driving along straight beach roads, and was steered in azimuth to observe wave growth as a function of angle along lines radial to the radar out to ranges of 250 km (fig. 11.10). Because the radar distinguished approaching from receding waves, both were observed. The measurements showed that: (a) the waves grow as the cosine of the angle relative to the wind; (b) attenuation of incoming waves going against the wind is very weak compared with growth of waves going with the wind; and (c) approximately 1% of wave energy incident on a very gently sloping natural beach is reflected back to sea.

Dekameter radars, by giving an accurate description of the waves produced by winds, have opened the possibility of applying this information to measure surface winds at great distances using over-the-horizon radars. These transmit relatively short, 10–20 m radio waves that scatter from 5–10 m ocean waves, which are quickly generated by even weak winds. By assuming these waves are in equilibrium with the local wind averaged over the preceding hour or two, the ratio of approaching to receding wave energy, as measured by the ratio of lines in the Doppler spectrum, is related to wind direction but with a right-left ambi-

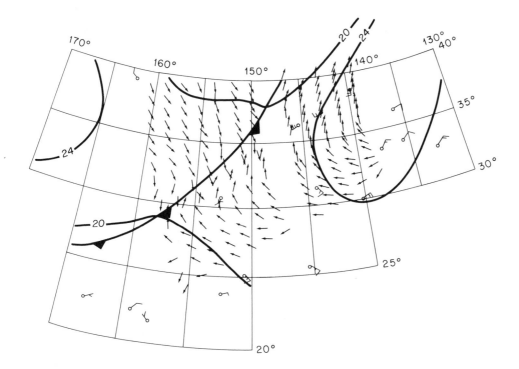

Figure 11.11 Radar map of wind direction for 26 September 1973, in the North Pacific compared with the U.S. National Weather Service surface analysis (from Barnum, Maresca, and Serebreny, 1977).

guity. Winds blowing toward the radar generate predominantly approaching waves and almost no receding waves. Cross winds generate approaching and receding waves of equal energy, but the radar cannot determine if the wind is blowing from right to left or left to right. In addition, strong winds generate relatively large seas, and these produce sidebands to the lines in the Doppler spectrum. The relative strength of the sidebands can be used to estimate wind speed.

Long and Trizna (1973) produced the first maps of wind direction over oceanic areas, and Stewart and Barnum (1975) estimated the accuracy of the radar technique. We used the Stanford Research Institute's Wide Aperture Radar Facility (WARF) to observe areas around ships equipped with anemometers to record the local wind velocity. The standard deviation of the differences between the two sets of measurements was 17° in wind direction and 4 m.s^{-1} in wind speed. Subsequent work by Maresca and Barnum (1977) has improved the estimate of wind speed by accounting for the influence of variable radio frequency and by removing more carefully the distortions produced by the ionosphere. Using these improvements, Barnum, Maresca, and Serebreny (1977) have mapped the wind field in the North Pacific at ranges of 3000—4500 km (fig. 11.11).

The velocity of ocean surface currents has been inferred by dekameter radars

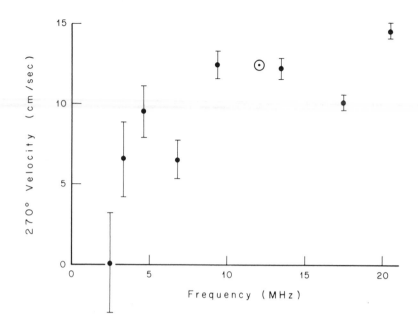

Figure 11.12   Radar measurements of that component of the ocean surface current in the direction of 270° as a function of radar frequency (points), compared with the simultaneous measurement of current made by a drogue (circled point). The depth of the drogue determines the radio frequency with which it must be compared (from Stewart and Joy, 1974).

through their ability to measure accurately the velocity of ocean waves. The radar measures wave length through the Bragg equation and wave frequency from the Doppler spectrum, both to high precision, and their product is the wave phase velocity. If the ocean wave is carried along by a surface current, it will have a velocity that differs from its theoretical velocity, and the difference is the radial component of the surface velocity. The depth over which the current is measured depends on the wavelength of the ocean waves. Short waves are carried along only by thin surface currents, while longer waves are carried by thicker layers. The thickness is determined by the depth of appreciable wave motion, and is about one-eighth of the wavelength of the ocean wave scattering the radio wave.

Stewart and Joy (1974) have compared radar measurements of surface currents with those measured by drogues at the same time and place. We found that 0.20 m.s^{-1} currents could be measured with an accuracy of around 20% (fig. 11.12). Barrick, Evans, and Weber (1977) have developed a radar capable of mapping the surface currents over an area roughly 70 km on a side using 25–35 MHz signals. To test their system, they mapped the flow of the Gulf Stream off the Atlantic coast of Florida while an oceanographic ship measured currents along a line crossing the stream. They, too, found good agreement between the two ways of measuring currents.

Because of the great usefulness of resonantly scattered radio waves for oceanic studies, the limits of the theory have been explored experimentally. In

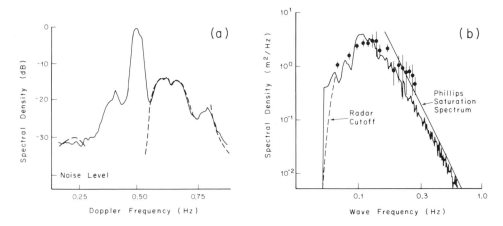

Figure 11.13 The usefulness of the second-order contribution to the Doppler spectrum of 22 MHz radio waves scattered from the sea surface. Left: the positive parts of the Doppler spectrum, with 10 mHz resolution, showing the continuum due to second-order scatter. Right: the ocean wave height spectrum obtained from the continuum in the Doppler spectrum, together with the spectrum measured by a wave buoy. The dashed curve in the Doppler spectrum is the second-order spectrum that exactly corresponds with the calculated ocean wave spectrum within the frequency bands used in the calculation (from Lipa, 1978).

general, little second-order contribution is present in the scatter of signals below 10 MHz (30 m wavelength). At higher frequencies, it becomes increasingly important and is often significant at frequencies above 20 MHz (fig. 11.5; Tyler et al., 1972).

Using the strong second-order components at these frequencies, Lipa (1978) has calculated the ocean-wave spectrum that must have produced these components. She finds that the signals uniquely define the ocean-wave spectrum and that her calculated spectra agree well with simultaneous measurements of the wave spectrum made by a wave buoy, thus demonstrating that at suitably high frequencies a single radio wave can be used to measure the spectrum of waves on the ocean, not just the one ocean wave component in resonance with the radio wave (fig. 11.13).

The technique can even be used by over-the-horizon radars. On some days, the ionosphere is particularly steady and second-order scatter can be observed with little distortion. These observations can be used to estimate the dominant wavelengths and directions of the waves in the scattering region (Lipa, Barrick, and Maresca, 1981).

12

SCATTER OF
CENTIMETER WAVES FROM THE SEA

Observations of the scatter of centimeter-wavelength radio signals from the sea have contributed substantially to our knowledge of short ocean waves (wavelets and capillary waves), their relation to the wind, and their interaction with longer waves. From this knowledge have come radars capable of measuring winds, waves, and currents as well as many other phenomena at the sea surface.

Several classes of radars are useful. *Scatterometers* measure backscatter from the sea at oblique angles and are used to measure surface wind speed. *Synthetic-aperture radars* map the variation or modulation of scatter from point to point along the surface and are used to observe longer surface waves, large internal waves on the thermocline, wind shadows behind islands, ships, and many other surface features. *Altimeters* transmit pulsed signals at vertical incidence to measure the height of a satellite above the sea, and are used to measure surface currents, winds, and waves. Finally, *delta-k* radars measure the coherence between scatter at two closely spaced frequencies and are used to study the spectrum of longer ocean waves.

In this chapter we will be concerned primarily with the nature of microwave scatter at oblique angles and the ability of scatterometers to measure wind speed. Synthetic-aperture and delta-k radars will be described in the next chapter and radar altimeters in chapter 14.

12.1 Observations of the mean scattering cross section and its relation to wind

Early radars, observing the ocean, often saw clutter or noise attributed to rough seas. Later investigations sought to understand the source of the clutter with the goals of (a) improving radar performance; (b) developing techniques to study the sea surface; and (c) testing the new theories that sought to explain scatter from the sea. These investigations comprised for the most part three programs using airborne radars, which provided many new and controlled observations of sea scatter. The first consisted of a series of flights by the Naval Research Laboratory using a four-frequency radar operating at 8.9, 4.5, 1.2 and 0.46 GHz (Guinard and Daley, 1970); the second was a series of flights by NASA aircraft observing scatter at 13.3 and 0.46 GHz (Krishen, 1971); and the third consisted of another series of flights by NASA aircraft using an improved radar (the AAFE RADSCAT) to observe very precisely scatter at 13.9 GHz (Jones, Schroeder, and Mitchell, 1977). All these radars measured very accurately the scattering cross section per unit area of the sea σ_0 as a function of wind speed, wave height, radio frequency, polarization, and incidence angle over North Atlantic areas during the period 1959—1975.

The extensive sets of aircraft data confirmed the then new theories for scatter from rough surfaces. Near vertical, scatter is from specular points; and at angles greater than 20° from vertical, scatter is from small waves tilted by larger waves. In particular, Daley (1973) and Valenzuela, Laing, and Daley (1971) showed that the four-frequency data were reasonably well predicted by Wright's (1968) theory using an ocean wave spectrum of the form (fig. 12.1):

$$\psi(k) = A U^{2\nu} g^{-\nu} k^{-(4-\nu)}$$

where $\psi(k)$ is the ocean wavenumber spectrum, k is the ocean wavenumber, g is gravity, and A and ν are dimensionless parameters determined from the radar data. With $\nu = 0$, the equation reduces to the spectrum described in sections 2.2 and 10.3; and figure 12.1 shows that ν is indeed small and that the spectrum of short ocean waves is approximated by a k^{-4} power law. The data also show, however, that the amplitudes of the shortest waves increase with wind speed. Thus, according to theories of radio scatter from the sea, the effect of these waves is to reduce specular reflection at vertical incidence and to increase resonant scatter at higher incidence angles (figs. 10.7 and 12.2). At some intermediate angle (10° in fig. 12.2), the cross section must be nearly independent of wind speed.

Although the early aircraft measurements indicated that radio scatter is correlated with wind speed, subsequent work has attempted to refine that correlation and to determine the accuracy with which wind can be measured.

Using all available data, Jones and Schroeder (1978) summarized the observed dependence of σ_0 on wind speed. They noted that if only wind speed is allowed to vary, the scattering cross section per unit area varied as the wind speed to some power, and so fitted the data to the equation:

$$\log \sigma_0 = a + \chi \log U$$

where σ_0 is the scattering cross section per unit area and U is the wind speed at some height, usually taken to be 19.5 m for scatterometry. The cross section in the equation can be the cross section for transmitted and received signals that are both vertically polarized, or both horizontally polarized, or one horizontally and one vertically polarized ($\sigma_{VV}, \sigma_{HH}, \sigma_{VH}$, in the notation of chapter 9). They observed that χ varied as the logarithm of ocean wavenumber k (fig. 12.3), and that χ was between 0.5 and 2.0 for the wind conditions investi-

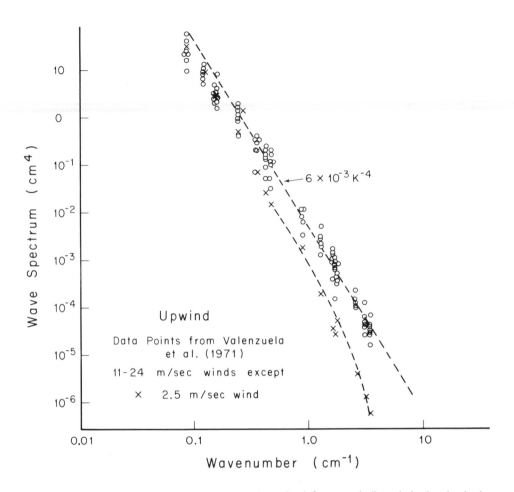

Figure 12.1   Average ocean wave spectrum determined from vertically polarized radar back-scatter at 0.46, 1.2, 4.5, and 8.9 GHz. The spread in observations is due primarily to variation in wind speed (from Valenzuela, 1978).

gated. Thus short waves were more sensitive to wind than were longer waves.

Scattering cross section is also a function of azimuth angle θ relative to the mean wind direction (fig. 12.4). Again, the variation is greatest for short wavelengths ($k = 600/m$), is weak for longer wavelengths ($k = 60/m$), and is nearly independent of polarization and incidence angle. The upwind-crosswind anisotropy is greater than the upwind-downwind variation, and the cross section is symmetric about the mean wind direction. Thus the azimuthal dependence is usually approximated by

$$\sigma_0 = aU^x(1 + b \cos\theta + c \cos 2\theta)$$

where the coefficients b and c are determined empirically.

Because the cross section is symmetric about the mean wind direction, it is difficult to determine wind direction unambiguously using observations of radio scatter made at a small number of different azimuth angles. If an area of the sea is observed from two different angles, up to four possible wind vectors satisfy the observations (fig. 12.5). A third observation can reduce the ambi-

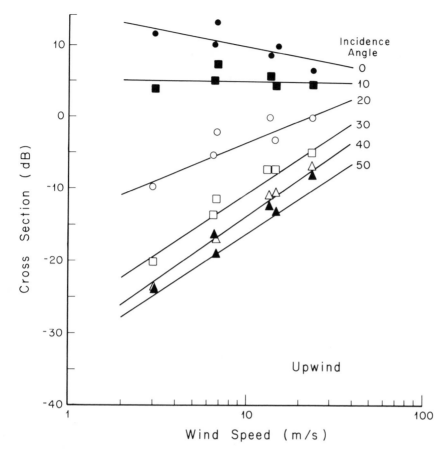

Figure 12.2 Scattering cross section per unit area of the sea for 13.96 GHz vertically polarized signals, as a function of incidence angle and wind speed (from Jones, Wentz, and Schroeder, 1978).

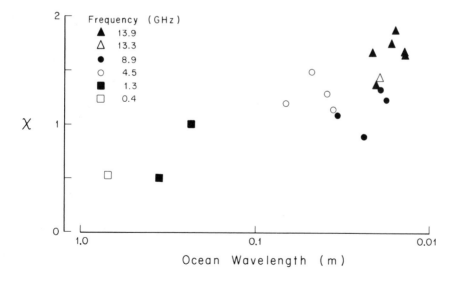

Figure 12.3 The radar scattering cross section varies as wind speed to some power χ. Shown here is χ as a function of the ocean wavelength that scatters the radio signal (from Jones and Schroeder, 1978, with data from Thompson, Weissman, and Gonzales, 1983).

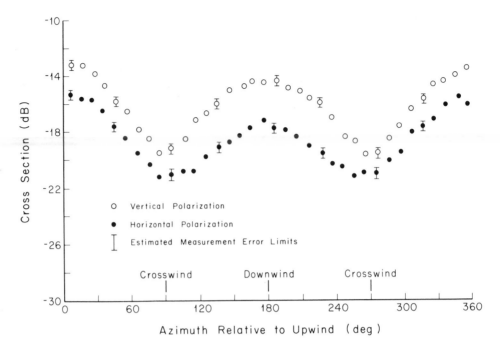

Figure 12.4 The scattering cross section per unit area of the sea as a function of angle relative to the mean wind for 13.9 GHz radio signals at 40° incidence angle (from Jones and Schroeder, 1978).

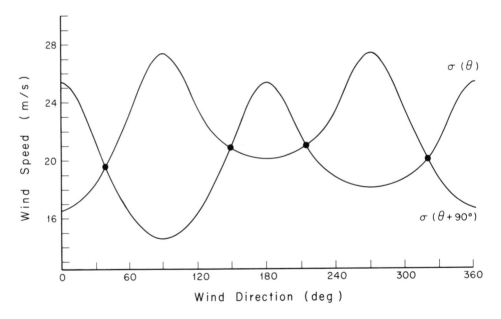

Figure 12.5 If a radar observes a particular cross section $\sigma_0(\theta)$ at an azimuthal angle θ relative to the wind, all points on the curve are possible wind vectors that yield the observed cross section. If the oceanic area is observed from two different directions, θ and $\theta + 90°$ in this example, up to four possible wind vectors satisfy the observations. These are the intersections of the two curves. Thus wind direction is not uniquely determined. If three observations are made, the ambiguity is reduced, but the observations must be very precise because scatter is only weakly anisotropic (from Jones et al., 1982).

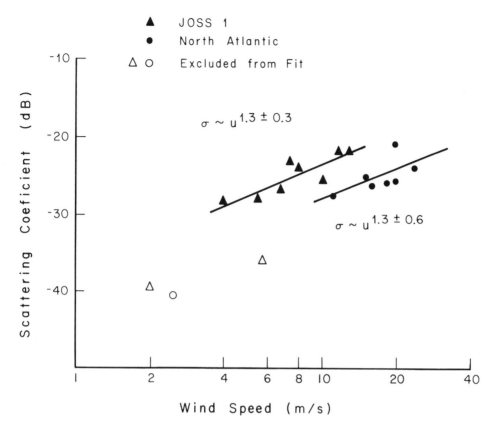

Figure 12.6 Scattering cross section per unit area of the sea as a function of wind speed as analyzed by Claassen et al., (1972), illustrating that the two variables are well correlated, but that the correlation is different for similar experiments on different days or for different areas. Thus other variables must also influence radio scatter.

guity; however, this requires very precise observations because the anisotropy of cross section is weak.

Summarizing their analysis of radar cross section, Jones and Schroeder (1978) conclude that a 14 GHz radar operating at large incidence angles and with horizontal polarization is most sensitive to wind velocity.

12.2 Errors in observations of wind speed

The analysis of data from any one aircraft experiment usually yielded a very good correlation between radio scatter and wind speed, but the correlation was much worse when data from many experiments were compared (fig. 12.6). For example, the various published correlations between σ_0 and U have values of the exponent χ that differ by a factor of 2. This scatter in the correlation is a continuing source of controversy and has not decreased much over the years. The scatter in the correlation has been attributed, at one time or another, to errors in the measurement of either wind speed or radar reflectivity; to the influence of other waves on the sea surface; to surface films; to surface temperature; or to variations in the stability of the atmospheric boundary layer. Of these, the latter three are probably the most important, but it is worthwhile to examine all possibilities.

In one attempt to reduce the scatter in the radar wind speed correlations, Claassen, et al. (1972) reanalyzed the four-frequency radar data. They claimed that the cross section increased rapidly with wind speed, that the sensitivity was greater than that claimed by Daley and colleagues, and that systematic biases existed in the four frequency data (fig. 12.6). The latter claim was strongly refuted by Daley (1974) who pointed out that each day's set of data was carefully and accurately calibrated by observing spherical targets dropped from the aircraft. More recently, de Loor et al. (1981) found systematic differences between backscatter observed from the same area by well-calibrated aircraft and tower-based radars, but could offer no explanation for the difference.

If we assume that both σ_0 and U have been accurately measured, as claimed by those who made the measurements, then other, unmeasured variables must influence radar reflectivity. Recalling that away from vertical the reflectivity depends only on the height of ocean wavelets that match the Bragg relationship, we are led to consider the processes that influence the height of such waves. The height of any wave component results from a balance between input of energy by the wind and other wave components (via wave-wave interactions) and the loss of energy through viscosity, surface films, and to other waves, again through wave-wave interactions. Thus we need to consider all these processes.

The input of wind energy to wavelets is not well understood (see Phillips, 1977; §4.3), but appears to depend on the shape of the velocity profile near the surface and on wave-induced turbulence in the air. Thus it has been hypothesized that wave height, and hence radar reflectivity, is a function of the friction velocity or air drag described in section 2.1, and not wind speed. So in calculating the correlation between wind speed and radar reflectivity, all measurements of wind speed have been corrected to a value the wind would have at a height of 19.5 m under conditions of neutral stability. While this correction improves the correlation, it does not provide a good test of the hypothesis. The correction depends on our understanding of the atmospheric boundary layer, and direct correlations between σ_0 and wind stress are preferred. This correction is difficult because direct measurements of wind stress are themselves difficult to make. A few measurements were made in the North Atlantic at the same time the area was observed by a scatterometer on Seasat (Liu and Large, 1981). Analyses of these data indicate that the correlation between stress and cross section is not statistically different from that between velocity and cross section. The set of data was small, however, and stability was not far from neutral.

Variations in the correlations may also be attributed to variations in significant wave height. Composite-surface scatter theory predicts that longer, larger waves modulate the orientation of the shorter waves that scatter the radio signal, although these long waves contribute little to the total slope. In addition, the theory of water waves predicts that short ocean waves must interact

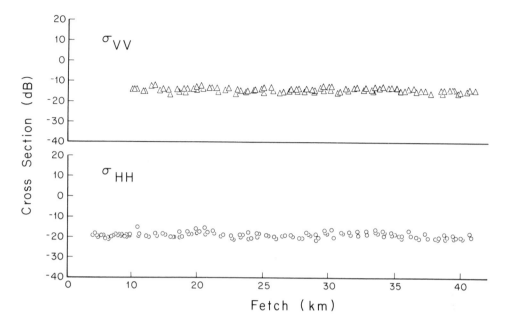

Figure 12.7 Scattering cross section per unit area of the sea as a function of fetch for 13.9 GHz radio signals at 53° incidence angle, showing that cross section is independent of fetch and hence independent of significant wave height. Surface wind speed was 13.0 m.s^{-1} (from Ross and Jones, 1978).

with all other waves on the sea surface. To investigate the influence of long waves, Ross and Jones (1978) measured wave height away from a lee shore while observing the scattering cross section of 13.9 GHz signals as a function of polarization and incidence angle, from the shore out to 40 km. During the experiment the wind blew offshore at a constant speed of 10−13m.s^{-1} while waves grew with fetch. Within the accuracy of their observations they could find no dependence of scattering cross section on fetch and hence no dependence on significant wave height (fig. 12.7).

Viscosity of water must also influence radar reflectivity. If we assume that the spectrum of short waves is that proposed by Lleonart and Blackman (1980), then the spectral energy density of wavelets should vary as the square root of viscosity. Because viscosity of water varies from 1.79×10^{-6} m^2.s^{-1} at 0°C to 0.84×10^{-6} m^2.s^{-1} at 28°C, the spectral density of waves, and hence their radar cross section, should vary by 46% or 1.6 dB over this temperature range, assuming identical wind conditions. The apparent change in measured wind velocity depends on the geometry of the measurement. Using the data in figure 12.6, a change of 1.6 dB in σ_0 should result in a 34% change in measured wind speed. If true, this would be a significant influence, one not accounted for in the relationship between wind speed and cross section; although still not large enough to account for the large differences in the figure. However, the proposed spectrum of short waves results in part from dimensional analysis, and the dependence of wave height on viscosity has not been explicitly tested,

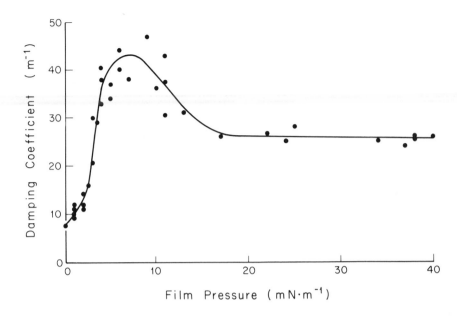

Figure 12.8 Rate of damping of 0.52 cm waves on sea water as a function of film pressure (decrease in surface tension due to a surface film) showing that even tenuous films strongly damp waves (from Garrett, 1967).

although Liu (1983) does find that errors in scatterometer measurements of winds are correlated with water temperature for winds less than 8 m.s^{-1}. Nor has this analysis considered any changes in the growth rate of waves due to concomitant but much smaller changes in the viscosity of air over the waves.

The large, persistent, but apparently random difference between observed and predicted radio scatter suggests that variables other than viscosity, boundary-layer stability, and long waves must influence radar reflectivity. Surface films are thought to be the major unexplored influence. Films are ubiquitous. They alter the surface tension and dynamic elasticity of the sea surface, and through these effects they strongly influence wavelets and capillary waves that cause radio scatter.

Waves shorter than 1 cm result from the restoring force of surface tension and are called capillary waves. Slightly longer waves, those with wavelengths up to 5 cm, are influenced by both surface tension and gravity; longer waves are controlled by gravity. Once generated, capillary and short gravity waves are damped by surface films, and the damping rate is much greater than that of waves on pure water. Because slightly extensible elastic films can be twice as effective as inextensible films, both the dynamic properties of films, such as the dynamic elasticity or surface compressional modulus, and surface tension, a static property, must be considered for wave dynamics (Davies and Vose, 1965; Miles, 1967).

The dynamic properties of films are most important when the surface films are so tenuous that the diminution of surface tension is only about 6 milli-newtons per meter less than the 72.5 mN.m^{-1} of a pure water surface (fig. 12.8; Garrett, 1967; Davies and Vose, 1965). In addition, Scott (1972) showed that similarly weak films decrease the growth rate of capillary waves when wind blows on the water. For these reasons, weak films are perhaps more important for interpreting scatterometer observations than are thicker films.

Naturally occurring organic films are widespread; thus their influence must be ubiquitous. Huhnerfuss, Walter, and Kruspe (1977) observed that weak films were always present in the tropical Atlantic in the summer of 1974, and reported no value of surface tension greater than 70.5 mN.m^{-1}. This widespread occurrence of films can also be inferred from the persistence of bubbles and foam streaks on the sea surface. Scott (1975) states that bubbles do not persist longer than 1.5 s on pure saltwater surfaces. He reports, furthermore, that bubbles rising to the surface through salt water are very efficient in carrying surface-active materials along with them to the surface, materials produced by the death and decay of small marine organisms.

Radar scatter has been correlated with surface films by a number of investigators. Van Kuilenberg (1975) reported that relatively thick ($4-10 \, \mu\text{m}$) films reduce radio scatter by $4-9$ dB for 10 m.s^{-1} winds compared with scatter from relatively clean nearby surfaces (table 12.1). Huhnerfuss, Alpers, and Jones

Table 12.1 Decrease in scatter of 0.76GHz radar signals from oil-covered waters relative to scatter from clean water surfaces (from Van Kuilenberg, 1975).

Oil type	Layer thickness (μm)	Wind speed 5 ms^{-1} Polarization			Layer thickness (μm)	Wind speed 10 ms^{-1} Polarization		
		HH	VH	VV		HH	VH	VV
Crude	5	8dB	8dB	11dB	9	4dB	7dB	9dB
Crude	9	7	5	11	14	9	6	9
Motor oil (SC30)	4	5	7	11	10	7	5	7
Motor oil (SC20W)	4	6	6	11	12	6	6	6
Average	5	7	6	11	11	6	6	8

(1978) reported similar effects for 13.9 GHz signals scattered from monomolecular films of oleyl alcohol. Further work by Huhnerfuss et al. (1981) shows that the reduction of surface wave amplitude by alcohol films is only weakly dependent on wavelength, and that the scatter of radio wavelengths in the band $2-20$ cm ($1.5-15$ GHz) should be similarly attenuated by films.

At this point, all we can say is that the correlation between wind speed and radar reflectivity is not well understood, and that this problem leads to errors of ~ 2 m.s^{-1} in radar measurements of wind speed. The cause of the error is now being investigated and may well result from the nearly universal presence of organic films on the sea surface; however, atmospheric stability, surface temperature, and perhaps other waves on the surface must also influence radar measurements of wind speed.

12.3 Satellite scatterometers

Scatterometers are radars designed to measure accurately the scattering cross section of the sea at particular incidence and azimuth angles. To do this, the

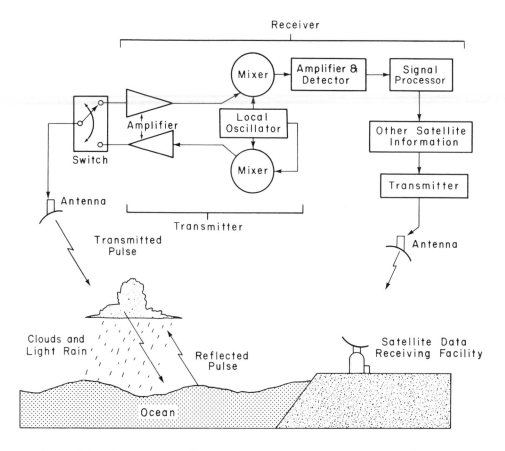

Figure 12.9 The components of a scatterometer, a radar to measure ocean surface winds
(from Stewart, 1982).

radar must consist of a well calibrated, stable transmitter and receiver, antennas
to direct the radiation in the proper directions, and circuitry to process the signal (fig. 12.9).

Two scatterometers have already been flown in space (table 12.2): the S-193
on Skylab and the Sass on Seasat. Further instruments are planned for NASA
satellites and for the European Space Agency's ERS-1. Because the Seasat
configuration has been extensively tested and analyzed, a description of this
instrument illustrates one possible configuration of a scatterometer, its errors
and accuracy.

Table 12.2 Satellite Scatterometers

Satellite	Year	Frequency (GHz)	Resolution (km)	Swath (km)	Incidence Angle
Skylab	1973	13.9	15	800	15−50°
Seasat	1978	14.6	50	750+750	25−65°
Ers-1	1987	5.3	50	400	25−55°

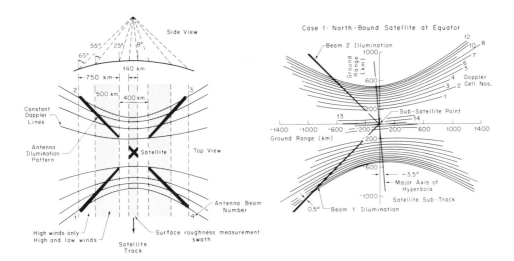

Figure 12.10 Seasat scatterometer geometry. Four antennas viewed two swaths on either side of the spacecraft (left), and the Doppler shift of the scattered signal determines the position of the scattering area in the swath. Because the lines of constant Doppler are skewed by Earth's rotation (right), areas in the forward beam do not exactly correspond with areas in the hind beam (from Johnson et al., 1980 and Bracalente et al., 1980).

The particular configuration of a scatterometer arises from several constraints. First of all, scatterometers must observe the sea at several azimuth angles to determine wind direction; thus they must radiate a signal with a narrow beamwidth in azimuth. At the same time, the radar must map scatter from a wide swath on either side of the spacecraft, so the beamwidth in the vertical must be broad. The two requirements can be met by stick antennas that radiate into a *fan beam* (fig. 12.10).

The number of fan beams determines the ability of the radar to estimate wind direction. Seasat observed the sea at two azimuth angles; but this was insufficient to define wind direction uniquely, and the Seasat estimates of wind velocity had a twofold ambiguity that produced up to four possible vectors (fig. 12.11). Observations at three azimuth angles are expected to reduce the ambiguity by one, producing two possible wind estimates 180° apart. The 180° ambiguity can then be removed automatically with the help of a few independent surface measurements, or even by knowing the pattern of winds at the surface: by knowing, for example, that strong winds circle lows but not highs.

Using a fan beam antenna, a scatterometer illuminates a wide swath on the sea surface, and the range to a particular area in the swath determines the local incidence angle. The range can be calculated either from the Doppler shift of the scattered signal (fig. 12.10) or from the time delay between the transmitted and received signal. Seasat used the Doppler shift because it was more efficient to transmit a few long pulses rather than the many short pulses required to measure range accurately from time delays. However, using the Doppler shifts introduces a problem, because of Earth's rotation, lines of constant Doppler are not symmetric about the satellite's velocity vector, and their position changes

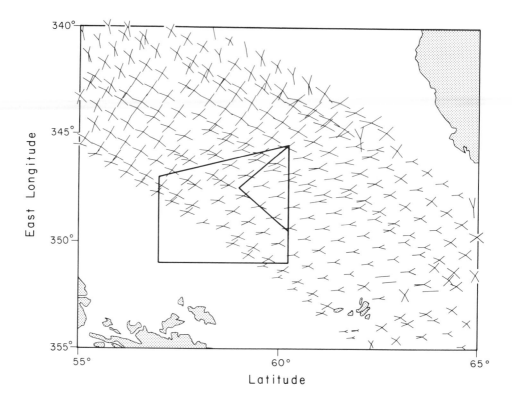

Figure 12.11 Seasat observations of wind direction are ambiguous. In general, four directions satisfy the radar observations, but in some areas the ambiguity degenerates to three or two directions. The boxed area is the region of the JASIN experiment, where surface winds were measured for comparison with the radar observations (from Wurtele et al., 1982).

with latitude. If the Doppler filters are fixed, oceanic areas viewed by one fan beam will not exactly correspond with areas viewed by another beam. Fixed filters can be used with care, but variable digital filters would eliminate the problem.

The last constraint on a scatterometer is that it must measure the cross-section of the sea with very good accuracy. Several sources of error limit this accuracy. For any measurement, the error in cross-section σ_E is given by

$$\sigma_E = (\sigma_I^2 + \sigma_C^2 + \sigma_A^2)^{1/2}$$

The *instrument noise* σ_I is internally generated and is small, on order of 0.1 dB. The major source of error is *communication noise* σ_C due to the random superposition of scatter from many individual areas of the sea. It is approximately

$$\sigma_C = \frac{1}{\sqrt{BT}} \left[1 + \frac{2}{r} \right]^{1/2}$$

where r is the signal-to-noise ratio of the observations, defined to be the signal from the radar when it views the sea divided by the signal when nothing is transmitted; B is the radar bandwidth; and T the time over which data are col-

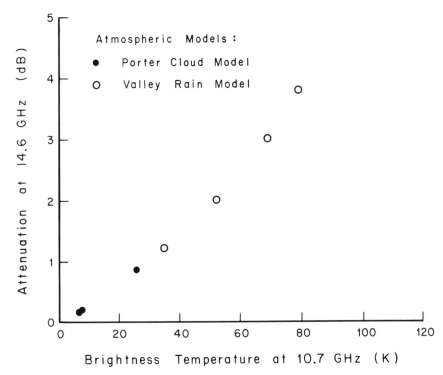

Figure 12.12   The attenuation of 13.9 GHz radar signals is correlated, for a wide variety of rainy or cloudy atmospheres, with brightness temperatures measured in the same direction by radiometers observing at frequencies close to that of the radar. Thus observations of brightness at 10.7 GHz can be used to correct scatterometer observations (from Moore et al., 1982).

lected. Typically, σ_C varied between ± 0.04 and ± 0.4 dB depending on incidence angle and wind speed, the signal being largest for high winds and small incidence angles. Finally, the uncertainty in the attitude of the spacecraft influences the calculation of the area viewed by the radar and thus the calculation of cross section per unit area. The *attitude error* σ_A of Seasat was between ± 0.2 dB and ± 0.4 dB on average. For large signal-to-noise ratios, σ_A dominates σ_C, while the reverse is true for weaker ratios (Bracalente et al., 1980). Overall, the uncertainty in radar cross section σ_E was less than ± 0.47 dB (11%) for high signal-to-noise ratios. Constant errors can be reduced only by calibrating the instrument. For Seasat, this was done with an accuracy of a few tenths of a dB by repeatedly observing the Amazonian rain forest.

Thick clouds and rainy areas also influence the measurements by scatterometers and lead to errors. Both absorb the radar signal, causing the radar to underestimate wind speed. In addition, rain smooths the sea, leading to even lower estimates of wind speed (Guymer et al., 1981). Very heavy rain, however, also scatters the radar signal, further complicating the measurement. Fortunately, clouds and rain emit radiation, and Moore et al. (1982) show that attenuation is closely correlated with brightness temperature for a wide variety of rainy or cloudy atmospheres (fig. 12.12). Thus SHF radiometer observations were used to calculate radar attenuation, and this method provided accurate

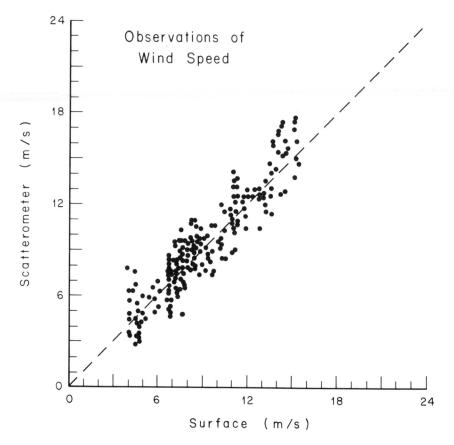

Figure 12.13   Wind speed measured by the scatterometer on Seasat, compared with wind speed measured by carefully calibrated anemometers on ships and buoys during the Jasin experiments (from Pierson, 1981).

corrections to scatterometer observations made through light rain. Observations made through heavy rain must be ignored.

The accuracy of satellite measurements of wind has been assessed by comparing them with winds measured by buoys and ships. This is not easy. Scatterometers observe a region roughly 50 km in diameter, yet surface measurements of wind are usually made at points. Two solutions are possible. A network of surface observations of wind and pressure can be smoothed to produce a surface weather analysis; alternatively, surface observations at a point can be averaged for a period comparable to the time required for a parcel of air to move 50 km, or for about one hour.

The comparison is further complicated by the stability of the surface boundary layer. The correlation of cross section and wind speed assumes neutral stability, but this assumption is only approximately true over most of the ocean. Thus air-sea temperature difference must also be measured and the surface wind observations converted to the values they would have at a height of 19.5 m if the atmosphere were neutrally stable.

The difficulty of assessing the accuracy of scatterometer measurements of wind was first appreciated by those who analyzed the Skylab data. Beaufort-scale estimates of wind reported by merchant ships, surface weather analyses, and winds measured by weather ships, ships with anemometers at known heights, ships with anemometers at unknown heights, low-flying research aircraft, and satellites all disagreed with one another (Young and Moore, 1977). The comparisons between the best surface observations and the Skylab radar measurements agreed within ±40%, but it was not certain which set of data had the greater error.

With these difficulties in mind, the Seasat comparisons were planned to be more accurate. Surface winds were measured during the Gulf of Alaska Experiment (Goasex) and the Joint Air Sea Interaction (Jasin) experiment in the Northeast Atlantic. The latter was the more extensive, with accurate observations being made by carefully calibrated anemometers on ships and buoys. Because two independent experiments could be used to calibrate the Seasat scatterometer, one was used to refine the correlation between wind velocity and radar cross-section and the other was used to test the accuracy of the wind measurements. Thus 3000 observations of the Goasex area were combined with 5000 observations from early aircraft flights in order to define the relationship between scattering cross section and wind speed, azimuth, incidence angle, and polarization (Jones et al., 1981, 1982; Schroeder et al, 1982). This relationship is called the scatterometer *model function*. Then scatterometer measurements of wind velocity calculated using this model function were tested using the Jasin data. The comparisons indicated that the accuracy of the scatterometer measurements was ±1.6 m.s^{-1} in speed and ±18° in direction for winds between 3 and 16 m.s^{-1} at mid-latitudes (fig. 12.13). But global comparisons, particularly in tropical areas, have not yet been completed, and will probably produce slightly different model functions and estimates of accuracy.

Despite the uncertainty about the ultimate accuracy of the scatterometer observations, the data have proven to be very useful. Data from Skylab and Seasat have provided the first detailed maps of surface wind fields (plates 12, 13), particularly the winds in severe storms, both tropical and mid-latitude, and have demonstrated the poor quality of standard weather analyses over large regions of the oceans. However, further applications of the data to oceanography requires unambiguous wind directions, so ways are being sought to remove the ambiguity and to produce maps of velocity and of the wind-stress curl.

In addition to measuring surface winds, SHF radars can also observe and differentiate among various types of sea ice (fig. 12.14). The theory of radar scatter from ice is based primarily on airborne radar measurements closely tied to surface observations of ice type within the field of view of the radar, as well as on surface observations of radar reflectivity and emissivity. As an understanding of radar signals increases, spaceborne instruments will begin to provide useful observations of Arctic and Antarctic ice, especially when scatterometer and radiometer measurements of the same area are combined with detailed images of the ice made by synthetic-aperture radars described in the next chapter.

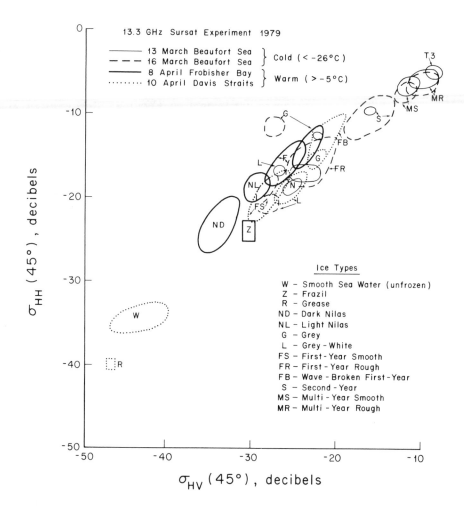

Figure 12.14   Airborne scatterometer observations of Arctic ice types. The radar reflectivity is a function not only of ice type but also of radar frequency, the distinction among various ice types decreasing with radio frequency (from Ramseir and Lapp, 1981).

13

SYNTHETIC-APERTURE RADARS

Synthetic-aperture radars accurately map the radar reflectivity of the sea over areas 50 to 100 km on a side, with a resolution of 10−40 m. Surface slicks swirled around by eddying currents, storm waves, internal waves, and the play of the wind on the water all contribute to the scene; and all, in turn, can be studied by the radar. Likewise, in polar regions, ice floes, leads, ice ridges, and polynias have different reflectivities, and they too can be mapped and studied, even through the thick and persistent clouds typical of the regions.

Synthetic-aperture radars are not without disadvantages. Broad swaths with fine detail require great quantities of data; and the conversion of radar data to images requires elaborate processing of the radar signal. Thus the data links and data processing are at least as important as the radar that produces the data, and just as costly. Along with the elaborate processing comes lack of clarity in the interpretation of the image. The radar uses its motion to produce an image, but the process is distorted by the motion of the water and waves at the sea surface, in ways not easily understood. Yet these problems are surmountable, and radars on Seasat and the Space Shuttle have produced wonderful pictures of many oceanic phenomena.

13.1 Theory of operation

A synthetic-aperture radar, a SAR, is a pulse-Doppler radar that transmits short radio pulses to obtain resolution in range, and that uses the Doppler shift of the scattered signal to obtain resolution in azimuth. In §11.3 we showed how moving radars can synthesize a narrow-beam antenna. Here we will examine the attributes of radars of optimal resolution, and the constraints on spaceborne radars capable of synthesizing antennas with a resolution of a few millidegrees using real antennas with a much coarser resolution of around one degree.

The range resolution δy of a SAR is determined by the length Δr of the

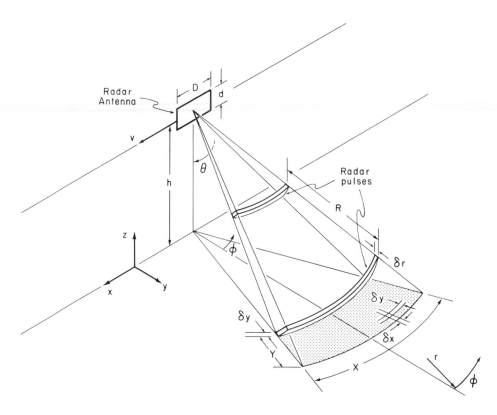

Figure 13.1   Geometry for describing synthetic-aperture radar observations of the sea.

transmitted radar pulse projected on the surface. This is

$$\delta y = \Delta r / (2\sin\theta) = c\tau / (2\sin\theta)$$

where c is the velocity of light, τ is the pulse duration, and θ is the incidence angle (fig. 13.1). Typically δy is around $25-100$ m for satellite radars, but somewhat better resolution is possible, particularly from airborne radars.

The azimuthal resolution depends on the precision with which the Doppler shift of the reflected signal can be determined. The radar transmits a train of coherent pulses, and then samples the reflected wavefront at a series of points along the radar trajectory, the synthesized aperture. This results in the azimuthal resolution being independent of range and dependent only on the size of the real antenna used by the radar.

To understand these properties of a synthetic-aperture radar, and the signal processing required to produce an image, consider a radar with a real antenna beam limited to a few degrees around the perpendicular to the satellite velocity vector. Then the Doppler shift of an area, or target, viewed by the radar is nearly zero, and varies linearly with azimuth. This allows us to write down quickly the approximate equations describing the operation of a SAR, especially those relating Doppler resolution to spatial resolution. More exact equations and a discussion of the operation of real radars is found in an excellent book by

Harger (1970), while simplified discussions can be found in Brown and Porcello (1969) and Develet (1964).

Referring to figure 13.1, the Doppler shift Ω of a target (scattering area) with coordinates (r, ϕ) relative to the radar is

$$\Omega = (2Fv \sin\phi)/c = (2v \sin\phi/\lambda)$$

where F is the radar frequency, $c = F\lambda$ is the velocity of light, λ is the radar wavelength, and v the radar velocity (table 13.1). Assuming that $\phi \approx 0°$, then $\sin\phi \approx \phi$ and

$$\Omega \approx 2v\phi/\lambda$$

Table 13.1 Notation Used to Describe Synthetic-Aperture Radars

c	Velocity of light [m.s^{-1}]
d	Cross-track dimension of antenna [m]
df_0	Shift in focus in image [m]
D	Along-track dimension of antenna [m]
Df_0	Depth of focus [m]
F	Radar frequency [Hz]
f_0	Focal length of optical processor [m]
h	Height of satellite above the sea [m]
r	Range from satellite to the imaged area [m]
R	Interval between radar pulses [m]
T	Duration of radar illumination at the target [s]
v	Satellite velocity [m.s^{-1}]
v_t	Radial velocity of the sea surface in Earth-fixed coordinates [m.s^{-1}]
x, y, z	Cartesian coordinates at sea surface [m]
X	Width of area of surface illuminated by radar [m]
Y	Depth of area on surface illuminated by radar [m]
θ	Incidence angle measured from vertical
θ_0	Mean incidence angle
$\delta\theta$	Spread of incidence angles
δr	Slant range resolution [m]
δx	Range resolution on the surface [m]
δy	Azimuthal resolution on the surface [m]
λ	Radar wavelength [m]
ϕ	Azimuthal angle
$\delta\phi$	Small increment in azimuth
$\delta\Omega$	Small increment in Doppler frequency [Hz]
τ	Radar pulse duration [s]
Ω	Doppler frequency of scattered radar signal [Hz]
Ω_{max}	Maximum Doppler frequency [Hz]

The spread in Doppler frequency $\delta\Omega$ from a small range of angle $\delta\phi$ is

$$\delta\Omega = \frac{2v}{\lambda} \delta\phi$$

and the azimuthal resolution δx is related to angular resolution $\delta\phi$ through

$$\delta x = r\delta\phi$$

where r is the distance from the satellite to the surface area. Thus spatial resolution is related to Doppler resolution by

$$\delta x = \frac{\lambda r}{2v}\,\delta\Omega$$

We now need to calculate the Doppler resolution $\delta\Omega$ achievable with a radar viewing a target on the surface using an antenna of width D. The Doppler resolution is roughly

$$\delta\Omega = T^{-1}$$

where T is the length of time the target is illuminated by the antenna beam as the satellite passes by. The time is just

$$T = X/v$$

where X is the width of the real antenna beam at the sea surface. This width was calculated in §9.1 and is

$$X \approx r\lambda/D$$

Thus the Doppler resolution is

$$\delta\Omega = vD/(r\lambda)$$

Combining this with the relation for spatial resolution gives the relation we seek

$$\delta x = D/2$$

Of course, the derivation is approximate, but it does demonstrate the essential point: the theoretical resolution of a SAR is independent of range and is one-half the size of the real SAR antenna. The greater the range (or the smaller the antenna), the greater the azimuthal swath illuminated, the longer the target remains in view of the radar, and thus the greater the Doppler and azimuthal resolutions. Thus it appears that SAR antennas should be as small as possible. But as we now show, azimuthal resolution is related to swath width, and this relation constrains the minimum size of the antenna.

The maximum swath width that can be mapped by the SAR is determined by the interval between radar pulses: two pulses must not be scattering from the surface at the same time. Wide swaths require long distances between pulses, and this constraint requires a low pulse repetition rate if the surface is to be sampled unambiguously. In contrast, maximum azimuthal resolution requires large Doppler shifts in the scattered signal; thus high pulse repetition rates are necessary if the radar signal is to be sampled unambiguously.

The compromise between azimuthal resolution and swath width can be made more precise. The maximum range of Doppler shift Ω_{max} that must be sampled unambiguously is twice that of the signal scattered from the edges of the area illuminated by the radar. The edge is at a value of $x = X$, so

$$\Omega_{max} = \frac{2v}{r\lambda}\,X$$

But $X = r\lambda/D$, thus

$$\Omega_{max} = 2v/D$$

The criterion for adequately sampling the Doppler signal is that the sampling frequency must be greater than twice the maximum Doppler frequency (Nyquist, 1928). This criterion requires

$$\delta t < 1/(2\,\Omega_{max})$$

so

$$\delta t < D/(4v)$$

where δt is the time between radar pulses.

Because pulses travel at the velocity of light, the spatial interval between pulses is

$$R = c\delta t$$

and the spacing on the ground is (fig. 13.1)

$$Y = R/(2\sin\theta_0) = c\delta t/(2\sin\theta_0)$$

where θ_0 is the mean incidence angle. The factor of 2 results from the reflection at the surface. The pulse from the far edge of the swath must return at least as close as the near edge of the swath before the next outgoing pulse reaches this point. Thus

$$\delta t > 2Y\sin\theta_0/c$$

The two inequalities bounding δt determine the relationship between swath width and azimuthal resolution:

$$\frac{D}{4v} > \delta t > \frac{2Y\sin\theta_0}{c}$$

$$D > \frac{8vY\sin\theta_0}{c}$$

Thus azimuthal resolution, $D/2$, is constrained by the swath width Y mapped by the radar.

Once the swath width Y is determined, so too is the antenna beamwidth in the direction perpendicular to the radar velocity. This must be less than the projected angle $\delta\theta$ between two pulses on the surface:

$$\delta\theta < (Y\cos\theta_0)/R$$

If the antenna is uniformly illuminated

$$\delta\theta \approx \lambda/d$$

where d is the cross-track dimension of the real radar antenna. Thus

$$d > R\lambda/(Y\cos\theta_0)$$

In practice, λ, θ_0, r, v, and Y are determined by the atmospheric windows available for spaceborne radars; the satellite's orbit; the size of the scene to be viewed; and the processes to be observed at the sea surface. Together, these parameters determine the size of the antenna. For example, for Seasat,

$$\lambda = 0.24 \text{ m}$$

$$r = 850 \text{ km}$$

$$\theta_0 = 20°$$

$$c = 3 \times 10^8 \text{ m.s}^{-1}$$

$$v = 7.5 \text{ km.s}^{-1}$$

$$Y = 100 \text{ km}$$

These yield

$$d > 2 \text{ m}$$

$$D > 7 \text{ m}$$

Although these are only rough calculations used to estimate the performance of a radar, we note that the values are close to the 2 m × 10 m antenna used on Seasat (table 13.2).

Table 13.2 Spaceborne Synthetic-Aperture Radars

	Seasat	Shuttle Imaging Radar		ERS-1	JERS-1
	SAR	SIR-A	SIR-B	AMI	SAR
Date	1978	1982	1984	1988	1991
Frequency (GHz)	1.27	1.27	1.27	5.3	1.27
Resolution (m)	25	40	15–50	30–100	25
Swath width (km)	100	50	40–60	75	74
Polarization	HH	HH	HH	HH	HH
Antenna size (m)	2.1× 10.7	2.1× 9.4	2.1× 10.7	10× 10	2.4× 12
Mean incidence angle	22°	50°	15–65°	22°	33°
Data type	analog	analog	digital	digital	digital

13.2 Processing SAR data

Consider now the processing of radar data necessary to produce images of the sea surface. The output of a SAR consists of the phase and amplitude of a received pulse as a function of range; and repeated observations produce a time history of scatter at each range. Thus the processing consists of using the observations at a particular range to produce high spatial resolution in the azimuth direction. The data as a function of range already have high spatial resolution, provided that sufficiently short pulses were transmitted.

If only a single target were illuminated at some particular range, then the scattered signal at that range would have a time history as shown in figure 13.2. When the target first enters the radar beam, the Doppler shift of the scattered signal is large, but it gradually decreases to zero as the radar moves broadside

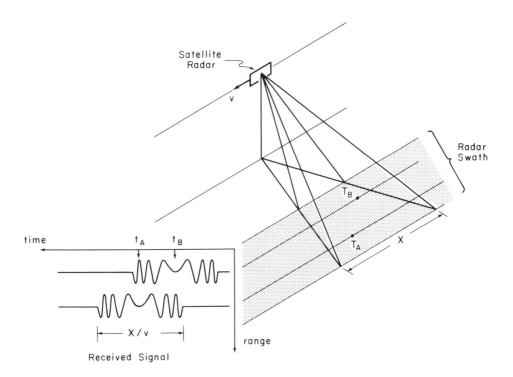

Figure 13.2 Time history of the signal received from two point targets in the radar swath. The received signal at each range is convolved with this point target response to yield the radar reflectivity at that range as a function of time (hence as a function of position in the x direction).

to the target, then increases again as the target passes by the radar. This signal is the point-target response of the radar. The essence of the processing of the radar signal consists of taking the time history of the scatter at each range and convolving it with the point-target response to produce a line in the image. For the example, the output of the convolution is small, or zero, except for a spike at the time when the scattered signal is centered on the point-target response.

Of course, the actual processing of a satellite radar signal is a little more complicated. The curvature of the wavefronts of both the transmitted and the scattered radio signal limits the length of the aperture that can be synthesized without accumulating excessive phase error. Because the nature of the phase error introduced by wavefront curvature is known for a fixed surface, a correction can be introduced and a much longer aperture synthesized. Systems that use this correction, as do all spaceborne SARs, are called *focused* SARs. In addition, the rotation of the Earth alters the Doppler shift of the scattered radar signal, but this too can be accounted for in the processing.

The output of the SAR processor is an image that looks much like a very grainy photograph, with a dense irregular pattern of bright and dark spots. This graininess is called *speckle*, and results from the coherent superposition of radar reflections from different areas of the sea surface within one pixel (Butman and Lipes, 1975). In theory and in practice, the brightness at each point in the image is exponentially distributed, with an expected mean value equal to the radar reflectivity of the surface. Since a process with an exponential distribution has a standard deviation equal to the mean (Hastings and Peacock, 1975:56), then the importance of the speckle can be appreciated.

Better-looking SAR images can be obtained by averaging together a number of pixels (Moore, 1979). For Seasat, four pixels in the azimuth direction, each pixel 6.25 m across, were averaged to produce one pixel every 25 m in the averaged image. This is called a *four-look* image. Because the range resolution was also 25 m, the Seasat images have a theoretical resolution of 25 m × 25 m, with speckle that is half of the brightness of the original speckle. Other radars may use even more averaging to reduce speckle, but little is gained in practice by averaging more than eight pixels together. Many users of radar images prefer high resolution rather than small speckle.

An important practical consideration in the processing of SAR images is the prodigious rate at which data are produced by such a radar, and the great quantities of data that must then be processed into images. The rate at which data are produced is

$$\text{Data rate} = \left(\frac{\text{bits}}{\text{sample}}\right)\left(\frac{\text{samples}}{\text{area}}\right)\left(\frac{\text{area}}{\text{time}}\right)$$

For Seasat,

$$\left(\frac{\text{area}}{\text{sample}}\right) = 25 \text{ m} \times 6.25 \text{ m} = 156 \text{ m}^2$$

$$\left(\frac{\text{area}}{\text{time}}\right) = \text{swath-width} \times \text{radar velocity} = 100 \text{ km} \times 7 \text{ km.s}^{-1} = 0.7 \times 10^9 \text{ m}^2.\text{s}^{-1}$$

$$\left(\frac{\text{bits}}{\text{sample}}\right) \approx 8 \text{ bits/sample}$$

therefore

$$\text{Data rate} \approx 36 \text{ megabits/s}$$

These 36 million bits of data per second must be transmitted to the ground, stored, fed into a processing system, correlated, averaged, displayed, and stored. Clearly, only very large, very fast, special-purpose computers can process data at the rate they are produced. Fortunately, the analysis can be performed in large measure by optical techniques that are relatively simple, fast, and cheap. Only a relatively few images need be processed in digital form.

The optical technique was first developed to avoid the expense and difficulty incurred in processing SAR data using either analog electronic circuitry or digital computers. The technique comprises a system to record the phase history of the received pulse on photographic film (fig. 13.3), a system to reconstruct the original phases of the received signal using laser light; and a set of conical and cylindrical lenses to bring these to a focus (Brown and Porcello, 1969; Cutrona

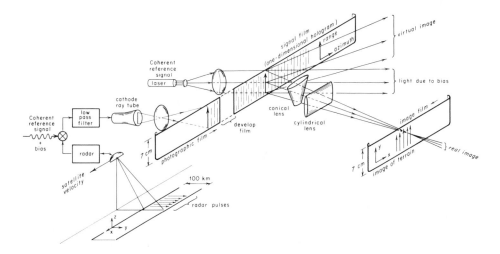

Figure 13.3 The optical processing of SAR images. The phase and amplitude of the scattered radar pulses are recorded on moving photographic film, the *signal film*, to produce a hologram in the azimuthal direction. Illuminating the hologram with laser light produces an image that is then brought to a focus by a combination of conical and cylindrical lenses.

et al., 1966). In essence, the technique replaces the radio waves with light waves. This reduces the geometry by a ratio of the two wavelengths, a factor of 371,000 for Seasat data illuminated with light from a helium-neon laser. Thus a 25 km-wide synthesized aperture can be reduced to a real aperture of only 6.7 cm, an aperture easily formed using lenses of convenient size. Alternatively, the photographically recorded SAR data can be considered to be a one-dimensional hologram from which the SAR image can be reconstructed.

Although optical processing is fast and efficient, some tasks require a more precise analysis than that available from optical systems, and the image must be processed using digital computers. This is the preferred method for testing processing techniques, for achieving the best possible resolution and signal-to-noise ratio in the final image, and for achieving precisely reproducible results. The digital calculations tend to be slow, however, with 4—8 hours of minicomputer time required to process 10^4 km^2 of Seasat data.

The digital processing becomes even more extensive if the images are Fourier transformed to obtain the spectrum of the wavelike features seen in many of them. Hasselmann (1980) noted, however, that the wave spectrum can be calculated directly from the received radar signal without first producing an image. Later, Martin (1981) showed that the calculation can be further reduced to an autocorrelation of the filtered signal from the SAR. In practice, this technique should enable signal processors on a satellite to produce estimates of the ocean wave spectra, directly, thus greatly reducing the amount of data that must be transmitted to the ground. The goal of present work is to reduce this time to a few minutes, and ultimately to less than 14 s, the time required to collect the data.

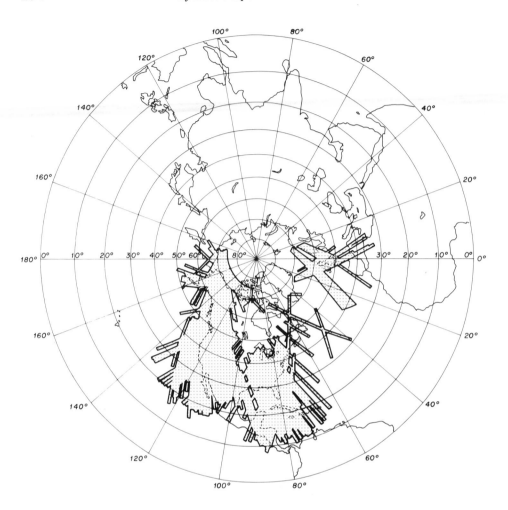

Figure 13.4 Areas mapped by the synthetic aperture radar on Seasat. Each swath is 100 km wide, and because data from the swath could be recorded only while the satellite was in view of one of five receiving stations, the swaths are clustered about the stations. (Map courtesy of Ben Holt, NASA Jet Propulsion Laboratory.)

13.3 Spaceborne synthetic-aperture radars

Several synthetic-aperture radars have flown in space, and more are planned for the future (table 13.2). By far the greatest amount of data was collected by the Seasat SAR, and these data form the basis for much of the discussion in this chapter. A radar nearly identical to that on Seasat, the Shuttle Imaging Radar (SIR-A), has flown on the Shuttle, and an improved version of this radar (SIR-B) is planned for future shuttle flights. Further in the future, the European Space Agency plans to fly an Advanced Microwave Instrument (AMI) on the ERS-1. This radar will operate as a SAR, as an ocean-wave spectrometer, and as a scatterometer. Finally, the Japanese are considering flying a SAR a few years later on their own Earth Resources Satellite, JERS-1.

The Seasat radar produced images 100 km wide and thousands of kilometers long (Jordan, 1980) with a resolution of 25 m. But the radar produced far too much data to be recorded on the satellite, so data were collected only when the satellite was in view of a receiving station on the ground. Five stations, located at Goldstone, California; Merritt Island, Florida; Fairbanks, Alaska; Shoe Cove, Newfoundland; and Oakhanger, England, collected images of 10^8 km^2 of the Earth's surface, mostly oceanic areas, although nearly all of North America and western Europe were mapped (fig. 13.4).

The amount of data produced by Seasat could be processed efficiently and cheaply only using optical techniques. Roughly 1% of the data was processed digitally, but these images were very useful for estimating the accuracy of SAR observations. The optically processed images routinely processed by the Jet Propulsion Laboratory were of swaths 25 km wide and had a resolution of about 40 m. A better resolution of 25 m was obtained using special optical processing, but few images were processed this way. The digital images are of 100 km × 100 km areas, and have a resolution of 25 m. These images are much superior to the optical images but are also much more costly. Work has now begun to develop digital processors that are faster and cheaper, and future SAR data will probably be processed primarily by computers.

The details of the Seasat data, the accuracy of the images, and the calibration of the radar are discussed in Pravdo et al. (1982), while the design of a digital processor is described in Wu et al. (1981).

13.4 Influence of target motion

A synthetic-aperture radar assumes that targets are stationary and that the Doppler shift of the received signal is due entirely to the radar motion. But the ocean surface is always in motion, and interpretation of SAR images of the sea requires some understanding of the influence of this motion on the images. Because most data are processed optically, we will discuss the influence of target motion in optical terms, although identical effects are seen in digitally processed images.

In optically processed images, target motion has three influences: (a) the target is displaced to a new position in the image; (b) the target has a new focal point (either in front of or behind stationary targets in the image); and (c) the depth of focus is altered.

The primary influence of target motion is a displacement of the target from its true position in the image to a new position appropriate to its velocity, provided that its velocity is not greater than the apparent velocity of stationary targets at the edge of the antenna beam. If its velocity is too large, it will not be imaged at all. Targets moving toward the radar are displaced in the direction toward which the radar is moving, and those moving away are displaced in the opposite direction (fig. 13.5).

The displacement x as a function of the radial component of the target velocity v_t is calculated using the relationship between position and Doppler shift Ω stated in §13.1:

$$x = \frac{\lambda r}{2v} \Omega$$

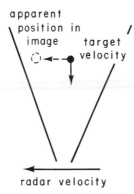

Figure 13.5 Influence of target velocity on its position in the radar image. Radial velocity toward the radar displaces the target's apparent position in the direction of the radar velocity.

and the expression for the Doppler shift of a wave reflected from a moving target:

$$\Omega = \frac{2v_t}{\lambda}$$

to obtain the apparent displacement of the target in the image:

$$x = \frac{v_t}{v} r$$

Using Seasat values of $r = 850$ km and $v = 7.5$ km.s^{-1}, a target moving 1 m.s^{-1} toward the radar is displaced 113 m in the azimuthal direction in the image. Thus target motion is very important for spaceborne synthetic-aperture radars.

Target motion also changes the focal point of an optically processed image of the target. If $v_t \ll v$, the focus is shifted by an amount (Shuchman and Zelenka, 1978)

$$df_0 = 2f_0\, v_t / v$$

where f_0 is the original focal length. If the motion is to produce an observable effect in the image, the change in focus must exceed the depth of focus Df_0:

$$Df_0 = 4f_0(\delta x)^2 / (r\lambda)$$

where δx is the azimuthal resolution. Using

$$Df_0 < df$$

yields

$$v_t > \frac{2v\,(\delta x)^2}{r\lambda}$$

Using Seasat values of $\delta x = 6.25$ m and $\lambda = 0.235$ m, a target will be out of focus if its radial velocity exceeds

$$v_t > 2.7 \text{ m.s}^{-1}$$

This is a relatively slow velocity, and moving targets will usually be out of focus.

13.5 Images of ocean waves

Synthetic-aperture radar images of the sea frequently show wavelike patterns, and this leads to the expectation that Fourier transforms of SAR images can be used to obtain the spectra of longer waves on the sea. For this reason, the images have been examined to determine the conditions under which waves can be imaged, the correspondence between the wavelike patterns in the image and waves on the sea, and the accuracy with which the ocean-wave spectrum can be calculated.

A SAR does not directly measure wave height, but only the variations in the reflectivity of the surface. Thus it is necessary to relate this reflectivity to the properties of longer waves. The physical mechanisms involved were first outlined by J. W. Wright, but were never published. They were subsequently investigated by Stewart (1975), Brown, Elachi, and Thompson (1976), Jain (1977, 1978), Elachi and Brown (1977), Elachi (1978), Alpers and Rufenach (1979), Valenzuela (1980), Alpers, Ross, and Rufenach (1981), Phillips (1981), and Harger (1981), among others. Here we will outline the mechanisms in their simplest form and estimate their relative importance. In the next section, we will compare radar observations of waves with surface observations to estimate the accuracy and usefulness of the technique.

To relate variations in reflectivity to the properties of long waves, we use the two-scale approximation to the sea surface discussed in §10.4, and consider a sea surface covered by short waves of uniform height superposed on a long wave of the form

$$\zeta = A \cos (K_x x + K_y y - \Omega t)$$

The long wave modulates the reflectivity of the sea surface, and the radar cross section can be written as

$$\sigma_0 = \langle \sigma_0 \rangle + \delta \sigma_0$$

where $\langle \sigma_0 \rangle$ is the spatially-averaged cross section per unit area of the sea, and $\delta \sigma_0$ is the deviation induced by the long wave. Assuming further that the radar reflectivity is linearly related to the properties of the long wave, then

$$\delta \sigma_0 = \langle \sigma_0 \rangle |R| A \cos (K_x x + K_y y - \Omega t + \Delta)$$

$$\Delta = \tan^{-1} [\mathrm{Im}(R) / \mathrm{Re}(R)]$$

where the complex quantity R is the *modulation transfer function*. The problem now reduces to determining the real $\mathrm{Re}(R)$ and imaginary $\mathrm{Im}(R)$ parts of R, or alternatively, the magnitude $|R|$ and phase Δ of R relative to the long wave.

The variability in radar cross section is usually attributed to three processes (fig. 13.6):

(a) *Surface tilting.* The scattering cross section per unit area of a small area a few meters on a side, for horizontally polarized radiation, is (§10.3)

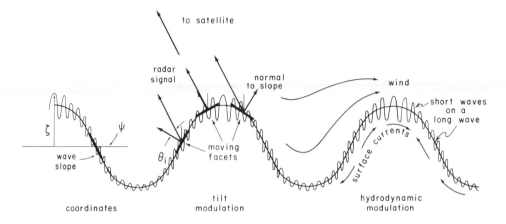

Figure 13.6 Some processes influencing SAR observations of ocean waves. Large ocean waves (a) tilt the surface, and (b) interact with short waves and distort the airflow over the surface. Both processes modulate the heights of the short waves that reflect the radar signal. And (c), large waves move the surface, thus distorting the SAR image of the surface and producing wavelike patterns in the image.

$$\sigma_{HH}(\theta_i) = 8\pi k_R^4 |g_{HH}(\theta_i)|^2 \Psi(0, 2k_R \sin\theta_i)$$

where (table 13.3) k_R is the radar wavenumber, θ_i is the local angle of incidence, $\Psi(k_x, k_y)$ is the ocean wavenumber spectrum, $g_{HH}(\theta_i)$ is a function of surface dielectric properties and incidence angle (fig. 10.3), and the coordinates are defined by figure 13.1. The cross section is a strong function of local incidence angle, and for a two-scale approximation to the sea surface, this is

$$\cos\theta_i = \cos\delta \cos(\theta+\psi)$$

where θ is the mean incidence angle and

$$\psi = \frac{\partial\zeta}{\partial y} \quad \delta = \frac{\partial\zeta}{\partial x}$$

are the slopes averaged over a small area a few meters on a side due to the long wave.

The reflectivity is then

$$\sigma_{HH}(\theta) = \langle\sigma_{HH}\rangle + \frac{\partial\sigma_{HH}}{\partial\psi}\psi + \frac{\partial\sigma_{HH}}{\partial\delta}\delta$$

Recalling that $\psi = iK_y A$, and $\delta = iK_x A$ this can be written

$$\sigma_{HH}(\theta) = \langle\sigma_{HH}\rangle\left[1 + \frac{iK_x}{\langle\sigma_{HH}\rangle}\frac{\partial\sigma_{HH}}{\partial\delta} + \frac{iK_y}{\langle\sigma_{HH}\rangle}\frac{\partial\sigma_{HH}}{\partial\psi}\right]$$

so the modulation transfer function for tilting of small waves by long waves is:

$$R_{\text{tilt}} = R_x + R_y$$

where

Table 13.3 Notation for Describing Radar Images of Ocean Waves

A	Long wave amplitude [m]
a	Maximum acceleration of wave surface [m.s^{-2}]
$g_{HH}(\theta)$	Angular dependence of scattered radar intensity
$g(\theta,\Phi)$	Projected angle of surface velocity
$H_{1/3}$	Significant wave height [m]
$I(k)$	Spectrum of image intensity
k	Short wave number [m^{-1}]
k_R	Radar wave number [m^{-1}]
K_x, K_y	Components of long wavenumber [m^{-1}]
$N(k)$	Wave action [m^4.s]
R	Modulation transfer function [m^{-1}]
r	Distance from satellite to ocean surface [m]
t	Time [s]
U_{10}	Wind speed 10 m above the sea surface [m.s^{-1}]
v	Satellite velocity [m.s^{-1}]
v_t	Surface velocity [m.s^{-1}]
x, y	Horizontal Cartesian coordinates [m]
Δ	Phase of R
δ, ψ	Local sea surface slopes
$\delta\sigma_0$	Deviation from $\langle\sigma_0\rangle$
ζ	Sea surface elevation [m]
ϵ	A small number
θ_i	Local angle of incidence
Λ	Long wavelength [m]
μ	Relaxation time [s]
ρ	Azimuthal resolution of a fixed surface [m]
ρ'	Effective azimuthal resolution [m]
σ_0	Cross section per unit area
$\langle\sigma_0\rangle$	Spatially averaged cross section per unit area
Φ	Azimuthal angle
ϕ	Small wave direction of propagation
$\Psi(k)$	Ocean wavenumber spectrum [m^4]
Ω	Long wave frequency [s^{-1}]

$$R_x = \frac{iK_x}{\langle\sigma_{HH}\rangle}\frac{\partial\sigma_{HH}}{\partial\delta}, \quad R_y = \frac{iK_y}{\langle\sigma_{HH}\rangle}\frac{\partial\sigma_{HH}}{\partial\psi}$$

Using the equations for σ_{HH} derived by Wright (1968), together with the dielectric constant of sea water at 1.2 GHz, Alpers, Ross, and Rufenach (1981) calculated the value of R for both horizontally and vertically polarized radiation (fig. 13.7). As expected, horizontally polarized radiation is more sensitive to wave slopes than is vertically polarized radiation; and *range travelling waves* (those with wave crests parallel to the radar velocity vector) produce a variation in reflectivity that is about ten times larger than the variation produced by *azimuth travelling waves* (those with crests parallel to the radar beam).

To estimate the change in reflectivity that might be expected in a Seasat

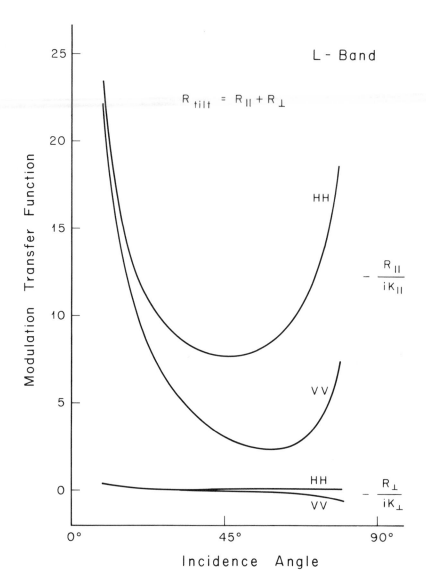

Figure 13.7   The modulation transfer function due to tilting of short waves by a long wave as a function of incidence angle, with polarization as a parameter. $R_{\parallel}$ and $R_{\perp}$ refer to ocean waves whose crests are parallel and perpendicular to the radar velocity vector. The curves apply to a 1.2 GHz radar, such as that on Seasat (from Alpers, Ross, and Rufenach, 1981).

image, consider a SAR viewing the sea at an incidence angle of 23°. If the long wave travels toward the range direction with an amplitude $A = 3$ m and a wavelength of 200 m so that $K_y = 2\pi/200$ m^{-1}, then from figure 13.7, $R/(ik) \approx 10$, so $R \approx 0.3$, and the modulation should be $\delta\sigma_{HH} = \langle\sigma_{HH}\rangle$ $[1 + \sin K_y y]$ or roughly ± 3 dB about the mean value. Such a variation in reflectivity should be clearly visible in a Seasat SAR image. If this same wave were travelling in the azimuth direction, the modulation would be much less, around ± 0.3 dB, and would probably not be detected. Because waves travelling close to the azimuth direction have been clearly imaged by Seasat (Gonzales et al., 1981; Alpers, Ross, and Rufenach, 1981) and by aircraft radars (Elachi and Brown, 1977), other mechanisms must also be important.

(b) *Hydrodynamic interactions.* Contrary to our initial assumption, the sea surface is not covered by short waves of uniform amplitude riding on long waves. Instead, the long wave modulates the amplitudes of the short waves. Several processes contribute: the long wave strains the surface, producing areas of convergence and divergence, and the long wave distorts the airflow over the surface, causing short waves to grow at various rates depending on the phase of the long wave and the mean wind velocity.

The influence of straining on the short wave field is calculated using either the conservation of the density of *wave action* $N(k)$ defined by

$$N(k) \equiv \Psi(K)/\omega$$

or the energy balance of the wave field using the radiation stress tensor (defined by Longuet-Higgins and Stewart, 1964). The former approach has been used by Alpers and Hasselmann (1978) and Alpers, Ross, and Rufenach (1981); the latter by Keller and Wright (1975) and Phillips (1981). Both approaches assume that the short waves are propagating through a medium that changes slowly over distances and times that are long compared with the short wave's length or period, and both produce similar results.

The conservation of wave action leads to a modulation transfer function of the form (Alpers and Hasselmann, 1978)

$$R_{\text{hydro}} = \frac{\Omega - i\mu}{\Omega^2 + \mu^2} \frac{\Omega}{|K|} \, (\mathbf{k} \cdot \mathbf{K}) \left[\frac{1}{\Psi(\mathbf{k})} \cdot \frac{\partial \Psi(\mathbf{k})}{\partial \mathbf{k}} - \tfrac{1}{2} \frac{\mathbf{k} \cdot \mathbf{K}}{|\mathbf{k}|^2} \right]$$

where the *relaxation time* μ incorporates the influence of wind generation and wave attenuation. In the absence of a long wave, the short waves are in equilibrium with the wind. The presence of a long wave continuously disturbs the equilibrium, and the short waves tend to return at a rate determined by μ. If the amplitudes of the short waves are too small, they will increase due to the action of the wind. If too large, they will decrease due to wave breaking. Thus μ depends on several processes.

Keller and Wright (1975) determined μ in the laboratory and could predict the hydrodynamic modulation using a constant value of μ, so long as the wind speed was not too large. But application of the theory to oceanic conditions was less successful (Plant, Keller, and Wright, 1978). It appears that μ depends on wind speed, short wavelength, and perhaps on other waves on the sea surface, but it is not yet sufficiently well known to predict the modulation transfer coefficient for typical oceanic conditions. Thus direct determinations of R_{hydro} such as those by Wright et al. (1980) and by Plant, Keller, and Cross (1983) are still required.

If the spectrum of short waves varies as $\Psi \sim |\mathbf{k}|^{-4}$ (see fig. 12.1), then the modulation transfer function reduces to

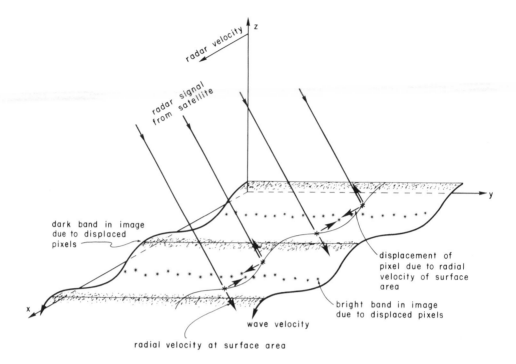

Figure 13.8   The influence of wave velocity on a SAR image. The radial component of the velocity of the scattering areas on the sea surface causes the SAR to displace the position of these areas in the image of the surface. If the wave is not travelling in the range direction, then the periodic velocity field due to a long wave can produce periodic bands of light and dark in the image, bands that look like an image of a wave. This enables a SAR to image waves travelling in the azimuth direction.

$$R_{hydro} = -\frac{9}{2}\,|\mathbf{K}|\Omega\,\frac{\Omega - i\mu}{\Omega^2 + \mu^2}\,\sin^2\Phi$$

where Φ is the angle between the radar velocity and the direction of propagation of the long wave, so $\Phi = 0°$, $180°$ for azimuth travelling waves, and $\Phi = 90°$ for range travelling waves. If $\mu = 0$

$$R_{hydro} = -\frac{9}{2}\,K\,\sin^2\Phi$$

and the peak-to-trough variation of σ_0 due to hydrodynamic interactions is

$$\sigma = \langle\sigma_0\rangle\,[1 \pm 9/2\,KA\,\sin^2\Phi]$$

where KA is the long wave slope. If we assume, as before, that $K = 2\pi/(200\text{ m})$ and that $A = 3$ m, then σ varies by ±0.42 or ±1.5 dB. This is about one-half the modulation due to tilting of the short waves by the long waves. Also, the maximum of R_{hydro} is for waves travelling in the range direction ($\Phi = 90°$), the same as for the tilt mechanism.

c) *Velocity bunching.* A moving sea surface distorts the formation of a SAR image; and a long wave on the sea surface, by producing sinusoidal surface velocities, causes wavelike bands in the image. If the crests of the long wave are perpendicular to the radar velocity vector (fig. 13.8), then surface areas ahead of the wave crest are displaced in the azimuth direction in the SAR image, while areas behind the crest are displaced in the opposite direction. If the long wave amplitude is not too large, the displacement will be a fraction of the ocean wavelength, and the image of the wave will tend to be dark near the wave crests and bright near the troughs even if tilting and hydrodynamic interactions are negligible. If the wave amplitude is too large, however, the displacement can be more than a wavelength, resulting in a very smeared image. Note that the velocity bunching in the wave image results from a nearly uniform distribution of radar reflectivity. If the radar reflectivity were confined to a particular phase of the long waves, say, to an area just ahead of the wave crest, then this area would only be displaced in the image and the wave would still be correctly imaged.

If the wave amplitude is small, or if the wave is travelling close to the range direction, then the velocity bunching is linearly related to wave properties. For these conditions, the modulation transfer function is of the form (Alpers and Rufenach, 1979; Alpers, Ross, and Rufenach, 1981)

$$|R_{vel}| = \frac{r}{v} K \Omega g(\theta, \Phi)$$

where

$$g(\theta, \Phi) = \cos\Phi [(\sin\theta \sin\Phi)^2 + \cos^2\theta]^{1/2}$$

Thus R_{vel} vanishes for waves travelling toward the range direction ($\Phi = 90°$), and is largest for waves travelling in the azimuth direction ($\Phi = 0°$, 180°).

The modulation transfer function is linear in wave amplitude as long as (Alpers and Rufenach, 1979)

$$A |R_{vel}| < 0.3$$

Recalling that the displacement of an element in the SAR image due to surface motion is

$$x = \frac{v_t}{v} r$$

and that the vertical velocity of the ocean surface due to a long wave is

$$v_t = \Omega A$$

it is easy to show that the criterion for linearity applied to azimuth waves is equivalent to

$$x < 0.05 \Lambda$$

where Λ is the wavelength of the long wave.

Assuming, as before, a long wave with $K = 2\pi/(200 \text{ m})$ and the Seasat values of $r = 850$ km, $\theta = 23°$, $v = 7.5$ km.s^{-1}, the linearity condition requires that $A \leqslant 0.3$ m for azimuth waves. Alternatively, the condition requires that

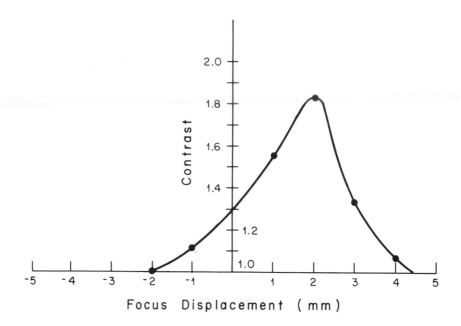

Figure 13.9   Long waves on the sea surface are most clearly imaged when the focal plane of the optical processor is shifted slightly from the position used to image land. Here, a 2 mm shift produces the best contrast in the image of 175 m waves viewed by an airborne SAR. This shift is equivalent to that necessary to image a target moving at a velocity of 20 m.s^{-1} along the surface (from Jain, 1978).

$\Phi > 84°$ for $A = 3$ m. In either case, $A|R_{\mathrm{vel}}| = 0.3$; and σ varies by $\delta\sigma = \langle\sigma\rangle[1\pm0.3]$ or by ±1.2 dB, a value only slightly less than that due to hydrodynamic modulation. More importantly, R_{vel} is the only component of the modulation transfer function that is large for azimuth waves, and it is probably the mechanism that enables a SAR to view these waves.

In addition to velocity bunching, wave motion produces surface oscillations, which tend to defocus and smear out the SAR image of the waves. By varying the focal plane of the SAR processor, it is possible to produce clearer images, where clarity is determined by the contrast or modulation in intensity of the wave image (fig. 13.9). Alpers, Ross, and Rufenach (1981) point out that by adjusting the focus of the processor, the position of maximum azimuthal resolution can be made to coincide with the position of maximum radar reflectivity, and that this adjustment should produce maximum contrast in the radar image. The shift in azimuth focus necessary to produce maximum resolution lies in a range of surface velocities δv:

$$-\left|\frac{ra}{2v}\right| < \delta v < \left|\frac{ra}{2v}\right|$$

where

$$a = A\,\Omega^2 g\,(\theta,\Phi)/\cos\Phi$$

is the maximum acceleration of the surface.

Unfortunately, this range of velocities happens to be nearly the same as the range of phase velocities of the long waves on the sea surface; and this fact has led to the suggestion that the point of maximum contrast coincides with the focal point needed to image a target moving at the phase velocity of the surface wave (Jain, 1978; Shuchman and Zelenka, 1978; Shuchman, 1981). Because it is difficult to explain theoretically how the phase velocity can influence SAR images of ocean waves, the explanation based on wave acceleration is preferred. More work is required to understand the processes that influence the imaging of a moving random sea surface.

At this point it should be clear that the SAR *image* contains only indirect information about wave height; however, the SAR *signal* contains direct information (Jain et al., 1982) that can be extracted from observations of the speckle of images having pixel sizes much larger than typical ocean wavelengths. To understand the technique, we must first understand the cause of speckle.

The radar signal reflected from a surface area corresponding to a particular pixel results from a superposition of signals from a number N of scattering elements on the surface. If the incidence angle is near vertical, these are the facets producing specular reflection; otherwise they are groups of short waves satisfying the Bragg scattering equation. In either case, the electric field E at the satellite is proportional to

$$E \sim \frac{1}{N} \sum_{i=1}^{N} \sigma_i(\mathbf{x}) \sin(\mathbf{k}_r \cdot \mathbf{r} + k_r \cdot \delta \mathbf{r}_i)$$

where σ_i is the cross section of each scattering element, $\mathbf{r}$ is the mean range to the surface, k_r is the radar wavenumber, and $\delta \mathbf{r}_i$ is the additional range to each scattering element. Thus $\delta \mathbf{r}_i$ is directly related to wave amplitude for angles of incidence near vertical.

The received power is proportional to $\langle EE^* \rangle$, where the brackets indicate the expected value of the product and the asterisk denotes the complex conjugate of E. Thus the power is exponentially distributed about the mean cross section of the pixel area. As a result, adjacent pixels in an image will have different brightnesses even if σ_i is constant over the image. If the radar wavelength is changed by a small increment δk, then the phase of the signal from each scattering element will change by $\delta \mathbf{k} \cdot \delta \mathbf{r}_i$, and this phase change will be larger for large waves than for small.

The small increments in wavelength are already contained in the SAR signal, and can be used. Recall that a SAR transmits a pulse of duration $\tau = (2\delta y \sin\theta)/c$ where δy is the range resolution and c the velocity of light; and that this has a bandwidth τ^{-1} in frequency or $\delta k_r = 2\pi/(c\tau)$ in wavenumber. For Seasat, τ^{-1} was 19 MHz, and it is possible to decompose the transmitted pulse into several longer pulses of narrower bandwidth. Images made from these simultaneous pulses will then have less spatial resolution, but more importantly, they will have slightly different speckle depending on the height of the waves on the sea surface. The difference in speckle is obtained by superposing the images and calculating the correlation coefficient

$$\frac{\langle i_j(k_r) i_j(k_r + \delta k_r) \rangle}{\left[\langle i_j^2(k_r) \rangle \langle i_j^2(k_r + \delta k_r) \rangle \right]^{1/2}}$$

where $i_j(k_r)$ is the brightness of pixel j in an image calculated using a narrow band of wavelengths near k_r. If $\Delta k_r = 0$, then the correlation coefficient is unity, and for large Δk_r it tends to zero. Using Seasat data, Jain, Medlin, and Wu (1982) found that the value of the correlation coefficient for $\delta k_r = 0.01$ m^{-1} (corresponding to a change of frequency of 0.5 MHz) gives the best indication of ocean wave height.

13.6 Satellite images of the sea

Synthetic-aperture radar images of the ocean show many interesting features, including surface waves, internal waves, the influence of currents and bathymetry, ice floes, and ships and their wakes (Fu and Holt, 1982). However the radar on Seasat was flown primarily to observe ocean waves, and so much of the analysis of images from this radar has been concerned with the ability of spaceborne SAR to observe accurately the spectrum of waves on the sea surface. The other features in the image spectrum, features that are even more interesting than ocean waves, were examined only at a somewhat later date.

The ability of the Seasat SAR to image waves was investigated by comparing the images with measurements of ocean waves made by buoys and wave gauges in the area viewed by the radar. The primary observations were made during the Gulf of Alaska Experiment (Goasex), the Joint Air-Sea Interaction (Jasin) experiment, and the Duck, North Carolina experiment (Duck-x). These form the basis for the following discussion.

A glance at many images showed that the SAR did not always image waves, even when they were present on the surface. Thus it was necessary to determine the criteria for wave visibility. Several factors are important:

(a) The wind at the sea surface U_{10} must exceed some critical value

$$U_{10} > 2\text{--}3 \text{ m.s}^{-1}$$

so that short waves are generated on the surface. In the absence of short waves, the radar signal would not be reflected and the image would be dark.

(b) The wavelength Λ of the long waves on the sea surface must be at least twice the effective resolution ρ' of the radar. This is just the Nyquist sampling criterion, and it requires that

$$n_{aj} = \frac{\Lambda}{\rho' \cos\Phi} \geqslant 2$$

where the effective azimuthal resolution of a SAR viewing a moving surface (Alpers, Ross, and Rufenach, 1981) is

$$\rho' = \rho (1 + \epsilon^2)^{1/2}$$

where

$$\epsilon = \frac{\pi^2}{2 k_R \rho^2} \left[\frac{r}{v} \right]^2 \frac{A \Omega^2 g(\theta,\Phi)}{\cos\Phi} \cos(K_x x + K_y y - \Omega t + \Delta)$$

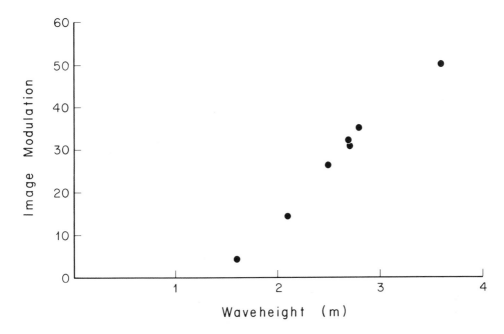

Figure 13.10 The wave height criterion for visibility of waves in a SAR image. If the significant wave height exceeds roughly 1.4 m, then ocean waves can be imaged by the Seasat SAR, provided other criteria for visibility are met. The intensity of the waves in carefully focused images, as measured by a dimensionless modulation index, increases with wave height (from Jain, 1981).

where ρ is the theoretical azimuthal resolution of a SAR viewing a fixed surface. The other symbols are those used in the last section, and are summarized in table 13.3.

(c) The significant wave height $H_{1/3}$ of the long wave must exceed some threshold, which is approximately

$$H_{1/3} > 1.4 \text{ m}$$

Above this threshold the waves become increasingly visible (fig. 13.10), provided the other criteria are met.

The combined influence of the azimuthal and wave height criteria was tested using the Jasin data (fig. 13.11) and were found to apply, provided the wind speed was sufficiently high to generate short waves.

Once the criteria for wave visibility were established, it was possible to estimate the accuracy of SAR observations of the ocean wave spectrum. To do this, optically processed SAR images were digitized and Fourier transformed to produce a wavenumber spectrum of image intensity $I(k,\phi)$, although some images were optically transformed. The centroid of the spectrum gives the dominant wavelength and direction of the waves seen in the image. These agreed well with the surface measurements, although the wavelength tended to be roughly 10% longer than the surface measurements (fig. 13.12). Overall, wavelength could be determined with an uncertainty of $\pm12\%$, and wave

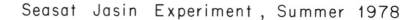

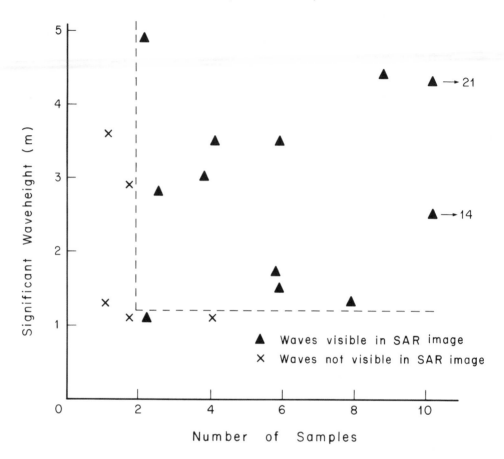

Figure 13.11 The combined wave height—wavelength criteria for visibility of waves in a SAR image. The significant wave height must exceed roughly 1.4 m, and the number of independent pixels between wave crests in the azimuth direction must exceed 2 if waves are to be seen in SAR images of the Jasin experiment (from Vesecky and Stewart, 1982).

direction with an uncertainty of $\pm 15°$ (Vesecky and Stewart, 1982).

Note that the SAR observations of wave direction are ambiguous by 180°, but that the ambiguity can usually be resolved by assuming the waves are travelling downwind, or away from a storm, or toward land. Alternatively, the ambiguity can be resolved by noting that the image of approaching waves comes to a focus at a different plane than does the image of receding waves (Shuchman and Zelenka, 1978; Jain, 1978; Vesecky et al., 1983).

The comparison of the image spectrum with the wave spectrum is less straightforward. In theory, the SAR spectrum should be multiplied by the

Observation   of   Wavelength

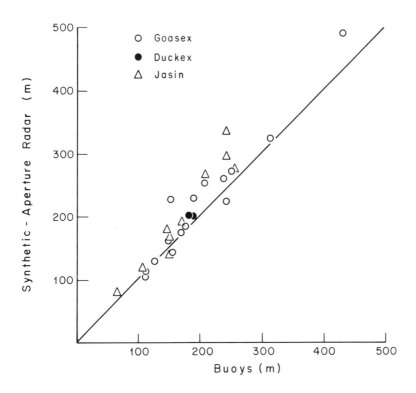

Figure 13.12   Comparison of the dominant ocean wavelength as measured
by the Seasat SAR and by wave buoys (from Vesecky and Stewart, 1982).

inverse of the modulation transfer function relating wave height to image intensity; but in practice this function is not known.   Instead, it is usually assumed
that the fluctuations in the SAR image are proportional to wave height
(Vesecky, Assal, and Stewart, 1981), so that the SAR spectrum can be directly
compared with the wave spectrum.   To facilitate the comparison, the ocean
wavenumber spectrum $\Psi(K,\phi)$ is usually integrated over angle to obtain
$\Psi(K)$, and similarly for the image spectrum.   However, the signal in the image
spectrum is usually confined to a small elliptical area in K,ϕ space; so the
integration must be confined to this region.

A comparison of the two spectra (fig. 13.13) shows that the image spectrum
$I(k)$ is more like the wave height spectrum $\Psi(K)$ and less like the spectrum of
wave slopes $K^2\Psi(K)$, a result that seems to contradict the theory for the
modulation transfer function, which indicates that $I(k)$ should be proportional
to a function that lies between $K^2\Psi(K)$ and $K^3\Psi(K)$ (Alpers, Ross, and
Rufenach, 1981).   However, the image spectrum barely stands out above the

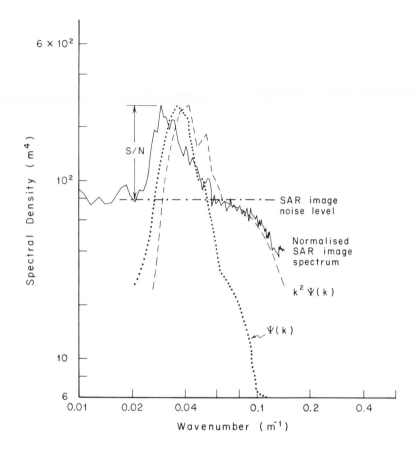

Figure 13.13 The spectrum of ocean surface elevation $\Psi(K)$ and normalized surface slope $K^2\Psi(K)$, compared with the spectrum of SAR image intensity. The spectra of slope and image intensity are normalized along the ordinate so that the peaks have the same values. Data were recorded on 4 August 1978 during the Jasin experiment (from Vesecky and Stewart, 1982).

noise, and more exact comparisons must await spectra calculated digitally from digital images. The optical processing of SAR images is not sufficiently exact or repeatable for such detailed comparisons.

The amount by which the spectrum exceeds the background noise is a function of wave height, and this relation can be used to estimate heights. Large, long waves are clearly visible in SAR images; that is, they produce large modulations in the grey scale of the image. This, in turn, results in spectra of image intensity with large signal-to-noise ratio as indicated in figure 13.13. Jain (1981), using carefully focused optical images, found a close correlation between wave height and image modulation (figure 13.10). Vesecky et al. (1983), using the spectrum of routinely processed optical images, found a similar but poorer correlation between signal-to-noise ratio in the spectrum and wave height, and could estimate wave height with an accuracy of ±0.8 m over

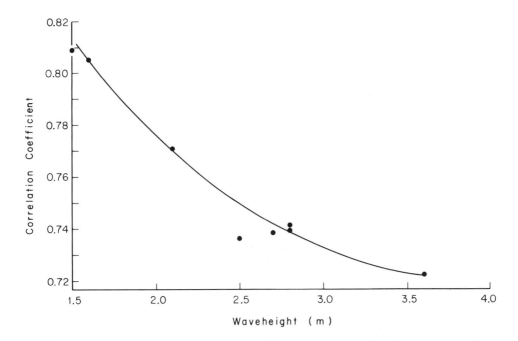

Figure 13.14 SAR observations of significant wave height. If two simultaneous images of the sea are produced using a pair of radar frequencies separated by 0.5 MHz, the correlation coefficient of the image intensities of coregistered images is a function of wave height (from Jain, Medlin, and Wu, 1982).

a range of 0–5 m. The difference between the two calculations appears to be due primarily to the care taken in Jain's study to produce accurately focused images of the waves.

Finally, wave height was also calculated from correlations between speckle in images by using images made from two slightly different radio frequencies present in the Seasat SAR signal (fig. 13.14). This technique, too, produced useful measurements of wave height.

The ability of a SAR to measure surface waves over large areas has facilitated studies of the propagation of waves across the sea and on into shallow water (fig. 13.15). Beal (1980, 1981), using data from the Seasat SAR observed the evolution of low-energy swell ($\Lambda \sim 200$ m, $H_{1/3} \sim 0.7$ m) as it propagated across the Gulf Stream and onto the continental shelf off the East Coast of North America. Wave refraction by shallow water and concomitant changes in wavelength were measured. Shuchman and Kasischke (1981), using similar radar observations offshore of Cape Hatteras, used the measurements of wave refraction to calculate the bathymetry in the area. Studies such as these are particularly important for estimating wave climate in coastal regions, producing information necessary for the design of harbors and coastal installations.

In addition to waves, synthetic-aperture radars also observe variations of wind speed at the sea surface and are used to map the variability of winds over dis-

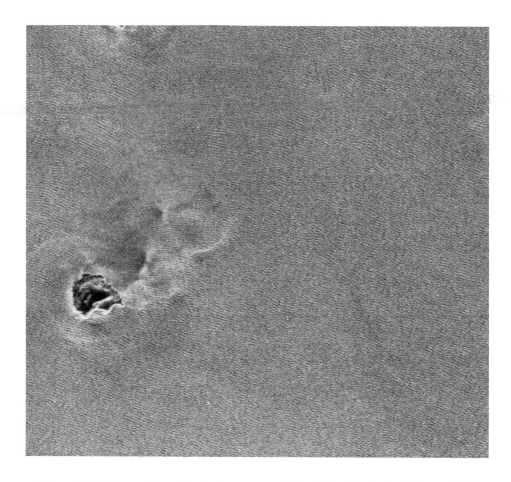

Figure 13.15 Waves refracted by bathymetry near Fair Isle in the Shetland Islands (59.5°N, 1.6°W) as observed by the Synthetic-Aperture Radar SAR on Seasat on 15 September 1978 and shown in this digital image processed by the Jet Propulsion Laboratory.

tances of 0.1 to 100 km. For these observations, the SAR operates as a high resolution scatterometer. Referring to §12.1, the radar reflectivity σ_{SAR} can be related to wind speed U_{10} at a height of 10 m, using

$$\sigma_{SAR} = a \; U_{10}^{\chi}(1 + b \; \cos 2\theta)$$

where χ, a, and b are empirical coefficients and θ is the angle relative to the wind. For the Seasat SAR, the coefficients were determined by comparing the reflectivity measured by the SAR with winds in the same area measured by the scatterometer on the satellite and by low flying aircraft. Jones, Delnore, and Bracalente (1981), using winds measured by aircraft, found that $\chi = 0.4$ and $b = 0.02$, a value of χ that agrees well with the data in figure 12.3. These coefficients were then used to compute winds from SAR data, and produced values that agreed well with winds measured in the same area by the Seasat

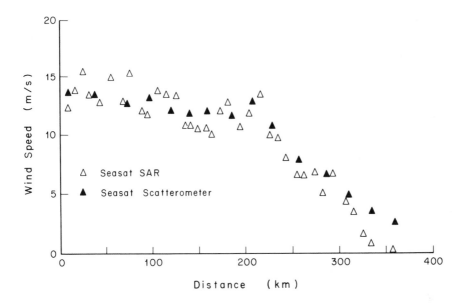

Figure 13.16 Estimates of wind speed at a height of 19.5 m, assuming neutral stability, as measured by the synthetic-aperture radar and scatterometer on Seasat (from Jones, Delnore, and Bracalente, 1981).

scatterometer (fig. 13.16). However, the small value of b precluded any measurement of wind direction, only speeds could be compared.

Maps of the spatial variability of the surface wind field are beginning to show the fine-scale structure of the wind. For example, on September 5, 1978 an image of the Jasin area in the Northeast Atlantic clearly showed wind streaks spaced about 2—4 km apart and aligned in the direction of the wind during a time of steady $12 \mathrm{m.s}^{-1}$ winds. Other images, when Fourier transformed, give the wavenumber spectrum of surface wind speed (fig. 13.17). Of course, these observations must be interpreted with care because other processes can also cause variations of radar reflectivity.

The range of possible phenomena influencing a SAR image of surface winds can be estimated by considering what happens as a homogeneous air mass moves at constant velocity across the boundary of a strong ocean current such as the Gulf Stream. Images taken during these conditions often show a dark band that appears to delineate the edge of the current, as well as other features that appear to be related to the current. These may be due to a number of processes:

(a) The discontinuity of water velocity at the current boundary will produce a discontinuity in the surface wind velocity relative to the surface. Because short waves respond to the relative wind, the radar reflectivity will vary across the boundary.

(b) A discontinuity of water temperature frequently occurs at the edge of currents. If the air mass is in equilibrium with the water on the upwind side, then the boundary layer will be either stable or unstable on the downwind side.

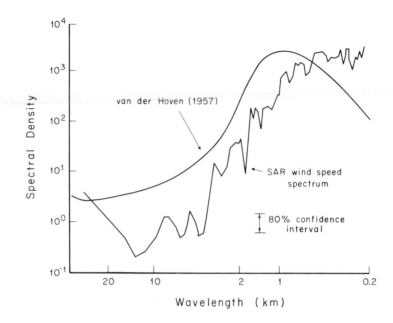

Figure 13.17   The wavenumber spectrum of wind speed at the sea surface, as measured by the Seasat SAR off Cape Hattaras, compared with the spectrum estimated from wind speed measured by a fixed anemometer, assuming that the spatial fluctuations are fixed and moving at a mean wind velocity of 10 m.s^{-1} (from Jones, Delnore, and Bracalente, 1981).

The change in stability will influence the amplitude of the short waves that reflect the radar signal.

(c)   Short waves crossing a current boundary are refracted and sometimes totally reflected; this changes the amplitude of the waves.   Again, this influences radar reflectivity.

(d)   The variations in surface temperature produce corresponding changes in viscosity of the water.   This, in turn, should influence the amplitude of short waves according to the theory of Lleonart and Blackman (1980) stated in §2.2.

(e)   The discontinuity of surface velocity at a current boundary influences the formation of a SAR image, producing a light or dark band in the region of strong current shear.

(f)   Surface films damp short waves, producing dark regions in an image. Films are produced by plankton, which tend to be more concentrated in cool waters on the coastal side of strong western boundary currents such as the Gulf Stream; thus films may be expected to be stronger on one side of a current than on the other.   Furthermore, any convergence of the surface velocity field at a current boundary will concentrate films in this region, producing slicks and a dark band in the image.

Finally, the atmosphere itself may have variations that produce features in a SAR image that resemble those produced by current boundaries.

(g)   Atmospheric fronts associated with larger-scale flow in the atmosphere, or with thunderstorms and squall lines, produce long linear or curvilinear features in SAR images.

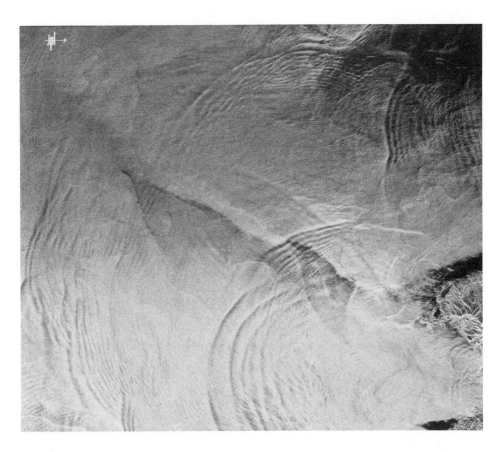

Figure 13.18 Tidally generated internal waves off the west coast of Baja California as observed by the Seasat SAR on 17 September 1970. This digitally processed image from the Jet Propulsion Laboratory shows a region roughly 100 km wide.

(h) Rain absorbs the radar signal, flattens small waves on the sea surface, and produces dark areas in a SAR image.

The relative importance of these phenomena is not yet well understood, primarily because all pertinent variables were not measured at the time the images were obtained. It is worth noting that large features seen in SAR images of the Jasin area were not repeated, leading to the conclusion either that they were ephemeral atmospheric phenomena, or that they were oceanic but that conditions necessary to see them did not repeat (see also Beal, DeLeonibus, and Katz, 1981:184). However large features seen in SAR images of the Gulf Stream do seem to be correlated with the current, and some features are seen in successive images (Lichy, Mattie, and Mancini, 1981). Again, this emphasizes the point that until SAR images of the sea are better understood, they must be interpreted with caution. Furthermore, because a SAR and a scatterometer both measure the amplitude of ocean wavelets, phenomena that influence one must influence the other. Thus analyses of data taken simultaneously by both instruments should lead to improved confidence in interpreting data from either.

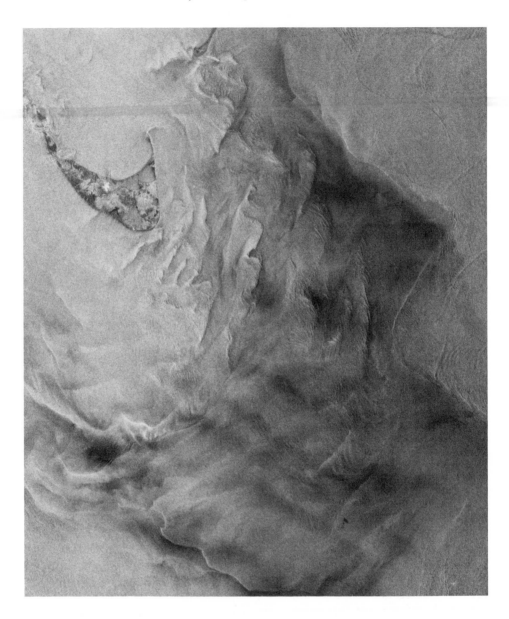

Figure 13.19 Bathymetry in the region of Nantucket Island produced this pattern at the sea surface as seen by the Seasat SAR on 27 August 1978 during flood tide. Shoals produce variations in
the tidal velocity (it must speed up over shallow water) and this modulates the amplitude of short
waves at the sea surface. This digitally processed image from the Jet Propulsion Laboratory shows
a region roughly 100 km wide.

Finally, we note that a wide variety of other surface phenomena are clearly
seen by synthetic-aperture radars:

(a) *Internal waves* are frequently seen over seamounts, at the edge of the
continental shelf, and in regions of current shear (fig. 13.18).

(b) *Eddies* have been seen near strong currents and downstream of points of
land (Fu and Holt, 1983).

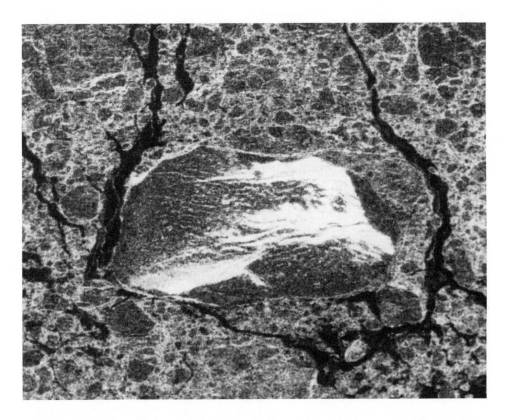

Figure 13.20 Fletcher's Ice Island, commonly known as T-3, in the Arctic as seen by the Seasat SAR on 6 October 1978. The island is about 12 km wide in this digitally processed image from the Jet Propulsion Laboratory. Dark areas are either open or recently frozen water.

(c) *Tidal flows* at coastal inlets are prominent, and show the pattern of dispersal of near shore waters into the offshore waters.

(d) *Bathymetry* is clearly seen in some shoal regions with strong tidal currents, particularly in the English Channel and off the New England coast (fig. 13.19). Tidal currents flowing over shallow ridges speed up, thus distorting the surface wave field. The distortion appears to reflect the bathymetry accurately, and should allow a SAR to map changing positions of sand dunes on the sea floor and other shoals, provide the images are made at flood tide.

(e) *Ships* and *wakes* are seen sometimes, but not always. Slowly moving ships tend not to be seen on windy days; but large ships with strong wakes are easily seen, as are slowly moving ships on calm seas. The Doppler shift induced by the ship's motion causes the image of the ship to be displaced from the ship's wake, and this displacement can be used to estimate the ship's speed.

(f) *Ice* of various types can be distinguished in radar images of the Arctic (fig. 13.20). The SAR responds primarily to surface roughness and can distinguish new from first-year and older ice, but cannot distinguish first-year from multiyear ice using frequencies around 1.36 Hz. Much higher frequencies must be used to distinguish first year from multiyear ice (fig. 12.14). Repeated images of the same area, however, show the displacement of the ice field, its response to wind, and the flow around coastal features (Leberl et al., 1983).

The wide variety of features seen by the Seasat radar have been displayed in an atlas by Fu and Holt (1982), and the phenomena producing these features have been discussed more fully in a book by Beal, DeLeonibus, and Katz (1981). Both make fascinating reading.

13.7 Wideband radar observations of waves

The synthetic-aperture radar is one of a class of wideband, narrow-beam radars capable of making precise measurements of ocean waves. Various other types of these radars have been used from shore and aircraft, and some have been proposed for spacecraft; thus it is useful to discuss these radars briefly here.

Consider a radar transmitting a short pulse into a narrow fan beam such as is shown in figure 13.2. If a synthesized aperture is not used, the reflected power at any particular instant will come from a long, narrow strip on the surface, with azimuth dimension proportional to the antenna beamwidth and with a range dimension proportional to pulse length. If the incidence angle θ is small, around $10° - 15°$, the reflected power will be proportional to the average wave slope in the strip, so the radar will pick out those long waves on the surface with wavenumber K propagating at an azimuth angle Φ parallel to the direction of radar propagation. As this strip propagates along the surface at the radio velocity c, the reflected power will be modulated at a frequency ω

$$\omega = \frac{cK}{2\sin\theta}$$

Thus the spectrum of radio reflectivity is proportional to the slope spectrum of ocean waves $K^2\Psi(K,\Phi)$, where Φ is fixed by the direction of the radar antenna and K is calculated from ω (Jackson, 1981). If successive pulses are averaged together in coordinates fixed relative to the ocean surface, the signal-to-noise ratio of the spectrum can be improved, just as similar averaging reduces speckle in a SAR image. And if the antenna is scanned in azimuth, $K^2\Psi(K,\Phi)$ can be measured for all angles Φ and for all wavelengths longer than the radio pulse length. Such a radar could be used on an aircraft or satellite to measure the spectrum of ocean waves.

If the incidence angle is large, the spectrum of radar reflectivity must be related to the ocean wave spectrum through the hydrodynamic and surface tilting components of the modulation transfer function described in §13.5. This indirect relationship considerably complicates the interpretation of the radar spectrum, and small angles of incidence are preferred.

Yet another type of radar can measure the spectrum of radar reflectivity of the sea surface (Weissman, 1973; Alpers and Hasselmann, 1978). The radar transmits two frequencies of slightly different wavelengths, say λ_1 and λ_2. These have wavenumber $k_1 = 2\pi/\lambda_1$ and $k_2 = 2\pi/\lambda_2$, with a difference

$$\Delta k = k_1 - k_2 = 2\pi \left(\frac{1}{\lambda_1} - \frac{1}{\lambda_2} \right)$$

Hence such instruments are commonly called Δk *radars* or *two-frequency* radars.

The radar observes the amplitudes E_1 and E_2 of each radio wavelength reflected from an area of sea typically several kilometers on a side. The spectrum of the complex product $E_1 E_2^*$ contains a sharp peak at a Doppler frequency $\Delta \omega$ given by (Alpers and Hasselmann, 1978; Alpers et al., 1981)

$$\omega = \pm \sqrt{2g\Delta k \sin\theta} + 2\Delta k v_r \sin\theta$$

where E_2^* is the complex conjugate of E_2, g is the gravity, and v_r is the radial component of the velocity of currents at the sea surface. Thus a Δk radar observes modulations of radar reflectivity of the sea surface having a wavenumber

$$\mathbf{k} = 2\Delta \mathbf{k} \sin\theta$$

travelling toward or away from the radar; so such a radar observes modulations of reflectivity in exactly the same way HF radars observe long waves on the sea surface (see §11.1). But because the modulations of reflectivity can be related to long waves only through the modulation transfer function, wave amplitude is measured indirectly.

The Δk radar is particularly useful for measuring ocean surface currents v using a stationary radar on the shore looking at the sea at angles near grazing ($\theta \approx 70° - 80°$). The phase velocity of the modulations is $\omega / |\mathbf{k}|$, and when this is subtracted from the theoretical phase velocity $(g/|\mathbf{k}|)^{1/2}$ of a wave of wavenumber $\mathbf{k}$, the difference is the component of the surface current going radially toward or away from the radar. Again, this is identical to the HF radar observations of currents discussed in §11.5, with two notable advantages. First, antennas are much smaller at SHF than at HF, so narrow-beam Δk radars are much more compact than similar HF radars. Second, it is easier to measure a wide range of ocean wavelengths at SHF than at HF. The observed ocean wavelength depends on Δk, and observations of ocean wavelengths in the range of $2 - 200$ m require only a small fractional change of radio wavelength of $\Delta k / k_r$ that is less than 1%. In contrast, a wide range of radio wavelengths, from 4 to 400 m, must be used at HF to observe the same range of ocean waves.

Shore-based Δk radars have been used to measure the spectrum of modulations of radio reflectivity, and hence the frequency spectrum of dominant ocean waves (Plant, 1977; Plant and Schuler, 1980) and the velocity of tidal currents in the North Sea (Alpers et al., 1981). Similar radars may also be used in space to measure the spectrum of surface waves (Alpers and Hasselmann, 1978).

Although the short-pulse and Δk radars have been described in quite different terms, they are essentially the same. Both measure the modulation of radar reflectivity as a function of range, and both measure the same oceanic phenomena (Jackson, 1981). The primary difference is practical. The short-pulse radar has a wide bandwidth better matched to the wide bandwidth of ocean waves. As a result, spectra of ocean wave height or modulation calculated from data from this radar should have $20 - 30$ dB better signal-to-noise ratio than should similar spectra from dual-frequency radars.

14

SATELLITE ALTIMETRY

Precise altimeter measurements of the height of a satellite above the sea, when coupled with an accurate determination of the satellite's orbit, allow the topography of Earth's seas, ice fields, and flat lands to be mapped. With increasing accuracy, increasingly useful information can be extracted. Observations with an accuracy of $1-2$ m provide useful determinations of Earth's geoid. But as better orbits and geoids are calculated, altimeters begin to see deviations from the geoid due to the mean (time-averaged) surface geostrophic currents at the sea surface. Repeated observations with precision of a few centimeters, an easier measurement, allow observation of the variable surface geostrophic currents, such as those due to eddies. These observations, when combined with conventional measurements of the internal density structure of the ocean, will allow calculations of the changes in mass transport of major ocean currents, thus improving global calculations of the transport of heat and momentum by the oceans.

Altimeter measurements of the topography of ice sheets to 1 m accuracy allow accurate calculations of the ice flow field. Those with greater accuracy, when repeated from year to year, will determine whether ice sheets are increasing or decreasing in size, a result of great importance for glaciology and climatology (Etkins and Epstein, 1982).

The usefulness of altimeters is not limited to observations of satellite height; the form of the scattered signal and the amount of scattered power also provide useful information. The former is used to measure significant wave height with an accuracy of around $10-20$ cm. The latter is used to observe the sea-ice extent, the boundaries of currents, oceanic wind speed, and the rate of accumulation of snow on glaciers.

The operation of altimeters is straightforward and easily understood. They are radars that transmit very sharp pulses toward Earth. The height of the satellite is found by measuring the time required for the pulse to travel from the altimeter to the surface and back. Observations of pulse amplitude and shape after reflection give additional information about the surface. With care, the satellite height can be measured with an accuracy of 5−10 cm, and the major source of error in the observations of surface topography is the error in determining the satellite's position. The position of even the best tracked satellites is uncertain by about 30−50 cm, (table 15.2, chapter 15), but accuracy should be much better for future satellites.

The interpretation of the altimeter measurements is more complicated. The extraordinary precision of the satellite observations requires equally precise definitions of the geodetic coordinate system (Mather, 1974, 1978), as well as precise definitions of the terms used to describe the measurements. Ultimately, satellite altimetry merges into geodesy, and relies on other very precise geodetic measurements provided by very long baseline interferometry or geodynamics programs.

14.1 The oceanic topography and the geoid

In order to discuss the usefulness of altimeter measurements, it is necessary to examine first a few concepts and definitions. To begin with, consider a rotating, non-homogeneous Earth covered with water, but with no atmosphere; and assume that the water is still, that is, it rotates at the same speed as the solid earth. This water surface must be an equipotential surface. If it were not, water would flow until it were. The surface will be relatively smooth, but will not be an ellipse. Rather, mass concentrations in the solid earth will dimple the surface. Now add an atmosphere that also corotates with the solid earth. Its mass distribution slightly alters the gravitational field (Rummel and Rapp, 1976), and again the sea surface will adjust to remain an equipotential surface.

The potential at any point is

$$W = W_g + W_a$$

where (table 14.1) W_g is the potential of gravity due to the solid earth and water plus the centrifugal acceleration due to Earth's rotation and W_a is the potential of gravity due to Earth's atmosphere. Note that the potential should also include a time varying tidal component, but this can be computed accurately and is usually not retained explicitly. Following Rapp (1974), a *geop* is any equipotential surface on which W is constant. The *geoid* is the geop that corresponds to mean sea level (fig. 14.1). To a very good approximation, the geoid is a biaxial, rotational ellipsoid determined by the mean mass of Earth and its rotation. The difference between this reference ellipse and the geoid is the *geoid undulation*. These undulations are on order of ± 60 m, and are large compared with other causes of variation in the height of the sea surface. The displacement of the sea surface from the geoid is the *sea-surface topography*.

A number of processes contribute to the topography. These are small, with amplitude of around 1 m, but contain all the information about ocean dynamics. Tides distort both the solid earth and the sea. The former very closely correspond to surfaces of constant tidal potential, are accurately computed, and are around 20 cm in amplitude. Ocean tides are a nearly resonant sloshing of water in the ocean basins, and the sea surface deviates greatly from the tidal

Table 14.1 Notation Used for Altimetry

A	Constant [$K.Pa^{-1}$]
B	Constant [K]
c	Velocity of light [3×10^8 $m.s^{-1}$]
d	Diameter of altimeter footprint [m]
E	Columnar value of free electrons [$electrons.m^{-2}$]
e	Partial pressure of water vapor [Pa]
f	Radio frequency [Hz]
F_N	Fourier transform of geoid height [m^2]
F_g	Fourier transform of vertical gravity [$m^2.s^{-2}$]
g	Gravity [9.8 $m.s^{-2}$]
h	Height of satellite above the sea [m]
H	Standard deviation of ocean wave height [m]
k	Wavenumber [m^{-1}]
m_a, m_w	Mean molecular weight of air and water vapor [$kg.mol^{-1}$]
n	Index of refraction
N	Number density of free electrons in the ionosphere [$electrons.m^{-3}$]
P	Pressure [Pa]
P_s	Surface atmospheric pressure [Pa]
r	Geocentric height of satellite [m]
R	Universal gas constant [8.317 $J.mol^{-1}.K^{-1}$]
T	Temperature [K]
T_a	Temperature of lower atmosphere [K]
w	Columnar value of water vapor [$Kg.m^{-2}$]
W	Gravitational potential [$m^2.s^{-2}$]
W_g, W_a	Gravitational potential of solid earth, and atmosphere [$m^2.s^{-2}$]
x, y, z, t	Cartesian coordinates [m,m,m,s]
α	Constant [80.5 $m^3.s^{-2}$]
Δh	Error in height of the altimeter [m]
ρ_{air}	Density of air [$kg.m^{-3}$]
ρ_{water}	Density of water vapor [$kg.m^{-3}$]
τ_p	Altimeter pulse duration [s]

potential. Ocean tides are known with an accuracy of 10–20 cm over most of the deep ocean, where they have an amplitude of around 1 m (Schwiderski, 1980). But in shallow water the tides can have an amplitude of many meters and are less well known.

Ocean currents and winds blowing on the sea surface also displace the surface away from the geoid. Strong currents produce changes in height of around 1 m over distances of 100 km, and weaker currents produce correspondingly weaker displacements. Strong winds blowing across shallow seas can pile water against the coast to a height of more than several meters above mean sea level, but the influence is weak in deeper water.

Changes in atmospheric pressure cause the sea to respond almost exactly as an inverted barometer, producing changes of a few centimeters.

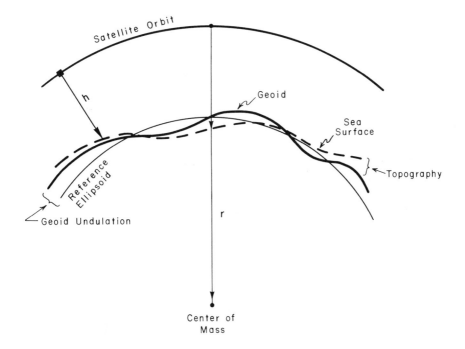

Figure 14.1 A satellite altimeter measures the height of the satellite above the sea surface. When this is subtracted from the height r of the satellite's orbit, the difference is the height of the sea surface relative to the center of the Earth. The shape of the surface is due to variations in gravity, which produce geoid undulations, and to ocean currents, which produce the oceanic topography, the departure of the sea surface from the geoid. The reference ellipsoid is the best smooth approximation to the geoid.

14.2 Satellite altimeters

A satellite altimeter consists of a transmitter that sends out very sharp pulses, a sensitive receiver to record the pulse after it is reflected from the sea surface, and an accurate clock to note the time interval between transmission and reception. Because radio signals travel at the velocity of 3×10^{10} cm.s^{-1}, measurements of distance accurate to say one centimeter require timing accuracy of 30 picoseconds with stabilities in the clock of better than one part in 10^8 (around 1 s per three years). While stringent, these accuracies are possible.

To simplify our discussion so far, we have assumed that the transmitter sends out a sharp pulse. Actually, it first generates a sharp pulse that is then sent through a dispersive filter to produce a long pulse that can carry more radio energy. After reception, this long pulse is passed through the inverse of the filter to produce the original sharp pulse shape. This technique of pulse compression (Rihaczek, 1969:51) does not alter the concept of an altimeter, but it does make it easier to build the transmitter.

Estimating the arrival of a pulse with a precision of a few hundredths of a

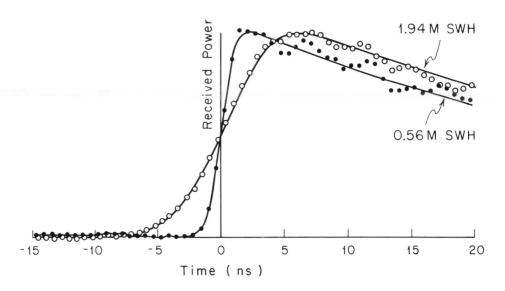

Figure 14.2 In order to determine the arrival time of the altimeter pulses to within a fraction of a nanosecond, and to determine significant wave height (SWH), a smooth curve is fitted through the averaged shape of many altimeter pulses (from Walsh, Uliana and Yaplee, 1978).

nanosecond requires pulses of wide bandwidth, further complicating the radar. From Fourier transform theory (Bracewell, 1965:160), a pulse of duration τ_p has an approximate bandwidth in frequency of τ_p^{-1}; thus a 30 ps pulse has a bandwidth of around 30 GHz. This is unacceptably wide, and by international agreement is not allowed in the frequency bands used by satellites. Pulse compression does not relax the requirement: high time resolution always requires wide bandwidth. Instead, a relatively narrowband 0.3 GHz (3 ns) pulse is used, and the reflected pulse is fitted to a curve to determine its arrival time to 150 ps or better (fig. 14.2). This technique requires a large signal-to-noise ratio in the received pulse. This in turn requires that as many as 1000 individual pulses be averaged together. Thus the altimeter makes a determination of height every few tenths of a second while transmitting thousands of pulses per second.

To be useful, the reflected altimeter pulse should come from a small area on the surface. Two approaches can be used. The first uses a large antenna to direct a narrow beam toward the surface. This beam-limited geometry requires expensive antennas and very good pointing accuracy. The second technique, the pulse-limited geometry, uses a relatively wide beam and short pulses (fig. 14.3). A spherically expanding pulse of rectangular cross section striking a smooth surface at first illuminates a small spot which grows until the trailing edge of the pulse reaches the surface, during which time the scattered power increases linearly with time. Then the illuminated area becomes a spreading annulus of constant area, and the scattered power remains constant until the annulus reaches the edge of the antenna beam. This is the technique used by satellite altimeters.

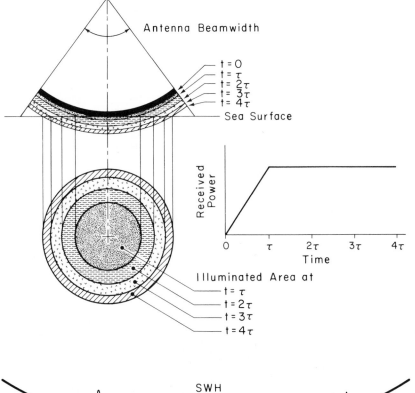

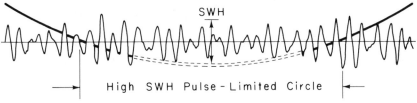

Figure 14.3   The length of the altimeter pulses plus the height of the waves on the sea surface determine the surface area observed by an altimeter. Two extremes are illustrated. The top and middle show a time sequence for a rectangular pulse incident on a sea surface whose wave height is very small compared with the pulse width. The bottom shows a 3 ns pulse incident from an altitude of 843 km on a sea surface with 10 m waves and 300 m dominant wavelength (the vertical scale is magnified by a factor of 100). (From Walsh, Uliana and Yaplee, 1978.)

The diameter of the surface spot illuminated by the radar just before it becomes an annulus is (Walsh, 1977)

$$d = 2\sqrt{hc\tau}$$

$$\tau^2 = \tau_p^2 + \frac{16H^2 \ln 2}{c^2}$$

where h is the satellite height, c is the velocity of light, τ_p is the pulse duration, and H is the standard deviation of wave height. For Seasat, h was

800 km and τ was 3 ns, so r was 900 m for calm seas. If the surface is not smooth but is disturbed by waves, not only is the initial spot increased but the reflected pulse is spread in time, scattering first from crests and then from troughs. This spread is measured by the radar, and allows observations of significant wave height (fig. 14.2). The scattering cross section at vertical incidence is inversely proportional to wind speed; thus the strength of the radar echo can be used to measure surface wind speed. The sensitivity to wind is less than at large incidence angles, however.

Table 14.2 Satellite Altimeters

Satellite (year)	Frequency GHz	Bandwidth MHz	Range resolution/ (averaging time)	Pulse compression	Wave height accuracy
Skylab (1973)	13.9	100	1.0m (0.3s)	13	1—2m
Geos-3 (1975)	13.9	80	0.5m (0.2s)	30	± 25% (4—10m)
Seasat (1978)	13.5	320	0.1m (1.0s)	1000	± 10% (1—20m)
Geosat (1984)	13.5	320	0.1m (1.0s)	30,000	± 10% (1—20m)
ERS-1 (1988)	13.5	400	0.1m (1.0s)	8,000	± 10% (1—20m)
Topex (1988)	13.5	320	0.03m (3.0s)	58,000	± 10% (1—20m)

Three altimeters have been flown in space: on Skylab, Geos-3, and Seasat (table 14.2). Each was an improvement on the previous ones, culminating in the altimeter on Seasat (see McGoogan, 1975 and McGoogan et al., 1974 for further details of these satellite systems). A number of other altimeters are planned for the future. The U.S. Navy plans to fly an altimeter like that of Seasat on a new satellite, Geosat, to measure the oceanic geoid in 1984. NASA plans to fly a very precise altimeter on Topex, a satellite to measure ocean topography and hence surface geostrophic currents and tides; and the European Space Agency plans to fly an altimeter similar to that of Seasat on its ERS-1 satellite in 1986. Finally, the French, Canadians, and Japanese are all considering altimeters for their national space programs.

14.3 Altimeter errors

Accurate altimeter measurements, those with errors of less than 10 cm, are difficult to make, and several sources of error must be considered. Two classes are important: those that influence the measurement of the height of the ocean surface, and those that influence the interpretation of the measurements. We are concerned here with the former, and will discuss the latter in the last section.

The major sources of error include the following (table 14.3):

(a) *Orbit*. The orbit is the reference frame from which the altimeter measurements are made. Any error in the radial component of the orbit directly produces an error in the measurement of the height of the sea surface (fig. 14.4). Any error in the along-track position, when multiplied by the slope of the orbit relative to the surface, also produces a height error. In a similar

Table 14.3 Typical Altimeter Errors

Source of Error	Uncorrected (Seasat) (cm)	Corrected (Seasat) (cm)	Corrected (Topex) (cm)	Wavelength (km)
Geoid	100(m)	100−200	10−50	200−40,000
Orbits	5(km)	100−200	5−10	10,000
Coordinate system	100−200	100−200	10	10,000
Ionosphere	0.2−20	0.2−5	1.3	20−10,000
Mass of air	230	0.7	0.7	1000
Water vapor	6−30	2.0	1.2	50−1000
Electromagnetic bias*	4	2	2.0	100−1000
Altimeter (noise)	5	5	2.0	6−20
Altimeter (tracker)*	10	4	−	100−1000
Altimeter (calibration)	50	5−10	2.0	∞
Altimeter (timing)	−	5	0	20,000

*Assumes 2 m waveheight, and 0.1 wave skewness.

manner, any *clock error* is equivalent to an along-track error. The error depends primarily on the density and height of the satellite, on a knowledge of gravity at satellite heights, and on the accuracy of the satellite tracking (§15.4). The best ephemerides of satellites still have errors of 0.3−0.5 m, and this is the dominant error in altimeter measurements.

There are ways, however, of minimizing the influence of orbit errors. The errors tend to have a wavelength of once per orbit and are relatively small over distances of a few thousand kilometers. Thus measurements made along such short arcs are influenced only by the slope and height error of the short arc. The slope and height can be arbitrarily removed by assuming that the ocean has no long wavelength slope, and the residual altimeter heights are used to study oceanic variability with wavelengths shorter than the arc length. The assumption is not strictly correct, but it is useful in this context.

(b) *Coordinate system.* Closely allied with orbital errors are those introduced by uncertainty in the coordinate systems used for the measurement. A surprisingly large number of coordinate systems are important for altimetry (Mueller, 1981): i) the coordinate system of the network used to track the satellite's position; ii) the coordinate system used to describe the geoid; iii) the inertial reference frame defined by observations of quasars; and iv) the astro-photographic system of star positions used to define the length of day and the position of the pole (§15.1).

The coordinate systems are neither in agreement nor fixed. They all tend to disagree by 1−2 m and are influenced by polar motion (±10 m), the variable rotation rate of Earth (a point on the equator moves 46 cm per millisecond), by Earth tides (±20 cm), by the distortion of the solid earth produced by the loading of ocean tides (±2 cm), and ultimately, over a period of years, by continental drift (±10 cm).

Again, the influence of these errors may be reduced. Geodesists need increasingly accurate coordinate systems; and they expect that, with continued

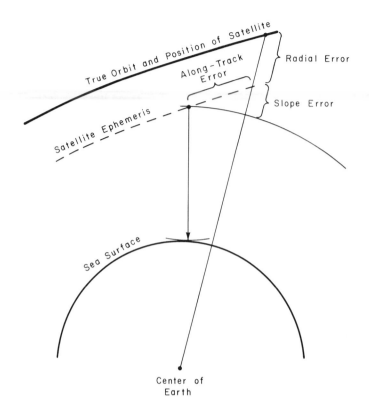

Figure 14.4 Both along-track and radial components of the error in the ephemeris of an altimetric satellite contribute an error to the measurement of surface topography.

work now begun, the various coordinate systems can be brought into agreement with errors of less than 10−20 cm.

(c) *The ionosphere.* The index of refraction n of the ionosphere is (§4.2)

$$n - 1 = N\alpha/(2f^2)$$

where N is the number of free electrons per unit volume, $\alpha = 80.5 \text{ m}^3.\text{s}^{-2}$ is a constant, and f is radio frequency in hertz. The error in range is then

$$\Delta h = \int_0^\infty (n-1)dz = \frac{\alpha}{2f^2} \int_0^\infty N\,dz$$

$$\Delta h = \frac{40.2E}{f^2}$$

where E is the columnar value of free electrons (electrons/m^2) and z is vertical distance. Typically, the electron content varies from day to night (very few free electrons at night), from summer to winter (fewer during the summer), and as a function of the solar cycle (fewer during solar minimum). Depending on the time of the observation, e varies from 10^{16} to 10^{18} electrons/m^2 thus 13 GHz altimeters will have an error of

$$0.2 \text{ cm} < \Delta h < 20 \text{ cm} \quad \text{(ionosphere, 13 GHz)}$$

The magnitude of the ionospheric range error may be estimated from observations of Faraday rotation made at stations on the ground, as was done for Seasat. But the technique has two important limitations. First, Faraday rotation does not measure electron content directly, but the product of electron content and the magnetic field. Secondly, ground-based measurements are sparsely located and extrapolation is necessary, resulting in an inaccuracy as high as 50% of Δh (Johanson, Buonsanto, and Klobuchar, 1978). Alternatively, altimeters could operate at two different radio frequencies, and the observations of range be combined to estimate both electron content and true range. An altimeter operating at 6 and 13.5 GHz could measure range with an inaccuracy of ± 0.5 cm due to free electrons in the ionosphere.

(d) *Atmospheric gases.* Bean and Dutton (1966), following Smith and Weintraub (1953), showed that the refractive index of air may be approximated by

$$(n-1) \times 10^6 = \frac{A}{T} \left(P + \frac{Be}{T} \right)$$

where

$$A = 0.776 \text{ K/Pa}$$

$$B = 4810 \text{ K}$$

where P is barometric pressure of the air in pascals (100 Pa = 1 millibar), T is temperature in K, and e is the partial pressure of water vapor in the air, also in pascals. The approximation requires that the radio frequency be less than 100 GHz; and it is accurate to $\pm 0.5\%$ of $(n-1) \times 10^6$, over the temperature range of $-50°$ to $40°$C.

The range error Δh is

$$\Delta h = \int_0^\infty (n-1) \, dz$$

For an ideal gas, $P/T = R\rho/m$ where R is the universal gas constant (8.317 J.mol^{-1}.K^{-1}), ρ is the density, m_a is the mean molecular weight of air (0.028996 kg.mol^{-1}), and m_w is the molecular weight of water vapor (0.01802 kg.mol^{-1}). Thus

$$\Delta h = \frac{10^{-6}AR}{m_a g} \int_0^\infty g\rho_{\text{air}}(z) dz + \frac{10^{-6}ABR}{m_w} \int_0^\infty \frac{\rho_{\text{water}}(z)}{T} \, dz$$

where g is gravity. Assuming that gravity is constant over the lower atmosphere, the first integral is the surface atmospheric pressure P_s. Assuming further that temperature is constant in the lower regions of the atmosphere where water vapor density is largest

$$\Delta h = 2.27 \times 10^{-5}P_s + 1.723 \, w/T_a$$

where P_s is surface pressure (Pa), w is the columnar value of precipitable water (kg.m^{-2}), and T_a is the average temperature of the lower atmosphere (K).

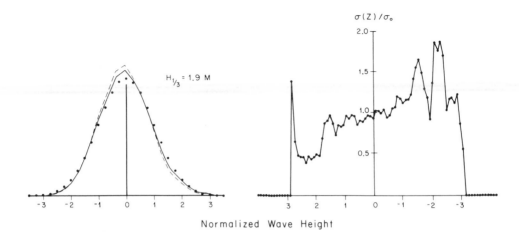

Normalized Wave Height

Figure 14.5 Effect of ocean waves on pulses reflected from the sea. Left: probability density distribution of sea surface elevation (solid curve) and backscattered power of 36 GHz radio signals (dashed curve), together with the normal distribution (dots). The difference in the centroid of the two curves causes the sea level observed by an altimeter to underestimate the height of the sea surface (from Walsh and Kenney, 1982). Right: the scattering cross section at 10 GHz as a function of height of the wave surface relative to mean sea level (from Jackson, 1979). Both figures have been normalized by the standard deviation of wave height. The right curve is also normalized by the scattering cross section per unit area of a large area of sea.

Using typical values of P_s (1.013 × 10^5 Pa), air temperature (290K), and water vapor (1 kg.m^{-2} < w < 5 kg.m^{-2}), then

$$\Delta h = 2.31 \text{ m} \qquad \text{(air mass)}$$

$$6 \text{ cm} < \Delta h < 30 \text{ cm} \qquad \text{(water vapor)}$$

The two terms are often called the *dry* and *wet tropospheric corrections*, but this is not precisely correct. The dry term includes the weight of water molecules; the wet term accounts for their additional influence on the index of refraction.

Of the two sources of error, that due to water vapor is the more important. The influence of dry air is larger, but nearly constant. An uncertainty of ±3 × 10^2 Pa (±3 millibars) in atmospheric pressure results in a range error of ±0.7 cm. Water vapor can be measured using microwave radiometers, and the uncertainty of ±2 kg.m^{-2} in the measurements of precipitable water made by the Seasat SMMR results in a range error of ±1.2 cm. Similar accuracy should be possible using a two-frequency radiometer (Moran and Rosen, 1981), provided the pair of frequencies is chosen to minimize the influence of the temperature profile on the measurements (Wu, 1979).

(e) *Ocean waves.* Troughs of waves tend to reflect altimeter pulses better than do wave crests. Thus the centroid of the distribution of returned power is shifted away from mean sea level toward the troughs of the waves. The shift is often referred to as the *electromagnetic bias*, and it causes the altimeter to underestimate the height of the sea surface. Waves also lengthen the reflected altimeter pulse, thus rendering more difficult the determination of the height of

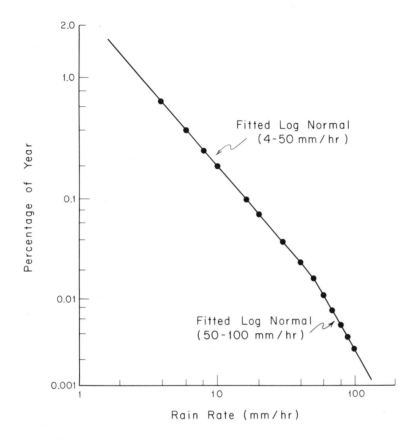

Figure 14.6 The percentage of the time rain rate exceeds a given value at Wallop Island, Virginia. Even very low rain rates of 2–4 mm.hr^{-1} are rare, and altimeter data from rainy areas can usually be ignored in oceanic studies (from Goldhirsh, 1982).

the altimeter above the sea surface. This problem is discussed with instrument error in paragraph (h).

The nature of the electromagnetic bias has been investigated using an airborne surface-contour radar (Kenney et al., 1979) capable of determining the variability of backscattered power per unit area as a function of displacement from mean sea level, for various sea states. Figure 14.5 shows histograms of the distribution of sea surface elevation together with the distribution of vertically backscattered radio power. Each distribution is normalized by the observed standard deviation of the surface height distribution and by its total area in order that the data may be compared with the normal distribution. The data show that the centroid of the distribution of cross section is displaced downward relative to the centroid of surface elevation.

Theoretically, the electromagnetic bias should be a function of wave skewness (Jackson, 1979), but the measurements are inconclusive. Early observations (Yaplee et al., 1971) indicate a bias of around 5% of the significant wave height, albeit with considerable scatter. More recent observations using

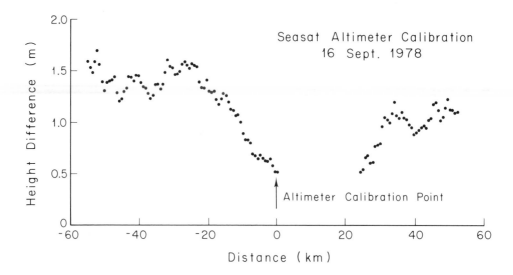

Figure 14.7 Altimeter calibration at Bermuda: the height of the satellite measured by an altimeter on Seasat minus the height measured by a laser was extrapolated out to the oceanic area observed by the altimeter. The shape of the curve is due primarily to errors in estimates of the height of the local geoid, and the gap in the curve is produced by the presence of the island (from Born, Wilkerson, and Lame, 1979).

10 GHz, 36 GHz and laser radiation (Walsh and Kenney, 1982) indicate that the bias decreases with frequency, being $-3.2 \pm 0.3\%$ of significant wave height at 10 GHz, $-1.0 \pm 0.3\%$ at 36 GHz, and $1.4 \pm 0.8\%$ at light wavelengths. The scatter in the observations at 36GHz is a function of wave steepness, wave kurtosis, and wind speed. Thus there is hope that bias may be predictable with an error of less than 0.1% of wave height at 13.9 GHz, using information from the altimeter signal; but this has not yet been done.

(f) *Rain*. Rain attenuates the altimeter pulse, and heavy rain greatly reduces the echo from the sea surface. Light rain tends to produce rapid changes in the strength of the echo as the altimeter crosses rain cells. Both effects degrade the performance of the altimeter; and altimeters usually fail to operate accurately if rain rate exceeds 5 mm/hr. Fortunately, rain is rare (fig. 14.6), and ignoring data from rainy areas is acceptable.

(g) *Calibration*. Determination of the geoid and mean oceanic topography require accurate calibration of the altimeter measurements. This calibration is done by using lasers to measure the range to the satellite while the satellite passes directly over an island, Bermuda in the case of Seasat. The height of the laser above the sea is calculated from level lines to a tide gauge, assuming there is little change in oceanic topography from the tide gauge to the deeper area offshore observed by the altimeter. More difficult is the determination of the local geoid by a combination of gravimetric and astrophotographic methods. The geoid is not smooth near islands, and detailed local geoids must be determined (see figs. 14.7 and 14.9).

Lasers can measure range to reflectors on a satellite with an accuracy of

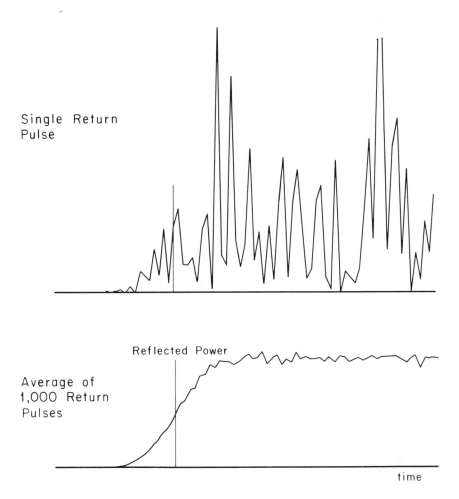

Single Return
Pulse

Reflected Power

Average of
1,000 Return
Pulses

time

Figure 14.8 The received pulse is the superposition of scatter from many wave facets on the sea surface; hence it is a random function. Many pulses must be averaged together to obtain a smooth estimate of the pulse shape. The residual uncertainty in the shape of the averaged pulse contributes noise to the measurement of altimeter height (from Townsend, McGoogan, and Walsh, 1981).

± 5 cm (Vonbun, 1977), and the primary error in calibration results from extrapolating the measurement to nearby oceanic areas. Of course, a laser floating on a barge directly beneath the satellite would yield more straightforward data; but this technique has not been used. The calibration of Seasat using a laser on Bermuda was tedious, but produced a calibration of the altimeter with an accuracy of ± 7 cm (Kolenkiewicz and Martin, 1982).

(h) *Instrument errors.* Because of the limited radio-frequency bandwidth, an altimeter must find the leading edge of a rather broad averaged pulse. The inaccuracies in this calculation result from several processes. First, the shape of the pulse is uncertain due to random variability of the pulses used to calculate the mean echo (fig. 14.8). This variability produces instrument noise. Second, the amplitude of the received signal varies as the cross section of the surface varies. An automatic gain control (AGC) circuit attempts to compensate for this, but rapid changes in echo strength mislead the circuit that tracks the position of the leading edge of the pulse, thus producing *tracker error*. Third, rapidly changing satellite height also produces tracker errors. Fourth, if the subsatellite point is near the edge of the area illuminated by the altimeter, the radar echo is distorted. This is a *pointing error* that is best avoided by accurate control of the satellite's attitude. Fifth, large, steep waves stretch the pulse and distort its shape; thus the leading edge of the stretched pulse is more difficult to determine, leading to greater instrument noise, and the distorted shape of the pulse produced by wave skewness leads to tracker error. Fortunately, waves increase the size of the area on the sea surface illuminated by the radar, decreasing the time required to obtain independent pulses from the sea surface (Walsh, 1982) and allowing more pulses to be averaged together to produce more accurate estimates of the shape of the pulse in areas with long waves. The range error introduced by waves was roughly 2% of significant wave height for Geos-3 and $5 \pm 1\%$ of wave height for Seasat (Hayne and Hancock, 1982; Lipa and Barrick, 1981). This error, plus a 2% electromagnetic bias, caused the Seasat altimeter to read low by $6.4 \pm 0.3\%$ of wave height (Born, Richards, and Rosborough, 1982; Douglas and Agreen, 1983).

14.4 Altimeter observations of the geoid

The largest geophysical signal observed by altimeters is that due to the undulations of the geoid. They vary between -104 m and $+64$ m, while the oceanic topography varies between ± 1.5 m. Thus, with no information about sea surface currents but knowing oceanic tides, the altimeter can map the geoid with an accuracy of ± 0.5 m, an accuracy far better than that provided by other measurements (fig. 14.9). By using information about the internal density field of the oceans, this uncertainty can be further reduced to ± 10 cm (Roemmich and Wunsch, 1982). Maps of the *altimetric geoid*, based on measurements by Geos-3 and Seasat, have a resolution of $100-300$ km (fig. 14.10; plate 14), and are now being used to study the bathymetry of the ocean, the strength of the lithosphere, and the structure of the mantle.

Various wavelengths in the geoid provide various information about the Earth. Over distances of less than a few hundred kilometers, the geoid undulations are due primarily to bathymetry (Cazenave et al., 1982; Dixon et al., 1983). Subsea mountains and seamounts are supported by the lithosphere, which acts as a rigid plate, and their uncompensated weight produces undulations of $1-10$ m in the geoid over distances of tens of kilometers (fig. 14.11). As a result, maps of the geoid can be used to locate bathymetric features in unsurveyed oceanic areas (Dixon and Parke, 1983). Note, however, that the geoid is less sensitive to small features in the bathymetry than is gravity. The geoid is an integral property of the gravitational field and responds to global distributions of mass, while gravity is a more sensitive indicator of local mass distribution (fig. 14.12). The two are related by

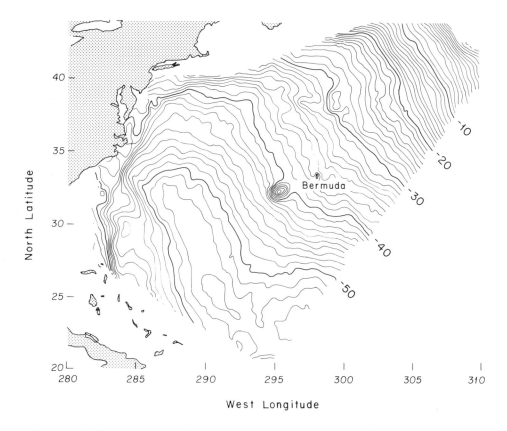

Figure 14.9 The mean sea surface in the Northwest Atlantic produced from satellite altimeter measurements. The surface is determined almost entirely by the local geoid (from J. G. Marsh, NASA Goddard Space Flight Center).

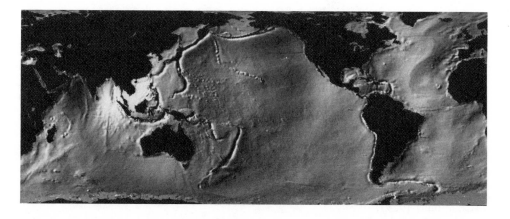

Figure 14.10 The relationship between gravity and the shape of the sea surface. This map of the oceanic surface was produced using altimetry data from Seasat and is designed to illustrate small features on the ocean surface. It shows oceanic trenches, mid-ocean ridges, seamounts, and fracture zones as well as features due to processes deeper in the earth (from M. Parke, Jet Propulsion Laboratory).

[275]

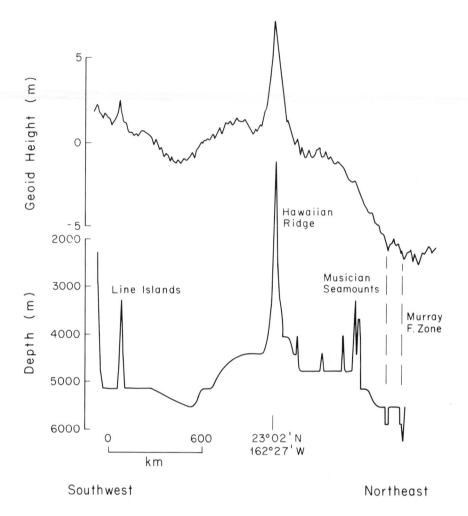

Figure 14.11 Height of the geoid measured by an altimeter on Geos-3 and the corresponding bathymetry along the subsatellite track. The geoid profile has had a regional geoid subtracted from it to emphasize the influence of bathymetry (from Watts, 1979).

$$F_N = \frac{1}{g\,|k|}\, F_g$$

where F_g is the Fourier transform of the vertical component of the gravity field, F_N is the Fourier transform of the geoid height, g is normal gravity, and $k = 2\pi/\lambda$ is the wavenumber associated with undulations of wavelength λ (Chapman, 1979).

Over distances greater than a few hundred kilometers, the weight of bathymetric features is compensated by buoyancy forces under the plate resulting from its sinking. The amount of compensation, and hence the rigidity of the

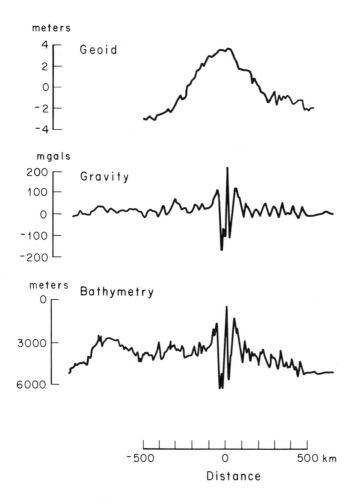

Figure 14.12 The difference between gravity, a force, and the geoid, a potential. Gravity is a more sensitive measure of bathymetry than is the geoid (profile over Southwest Indian Ridge from Chapman et al., 1979).

lithosphere, is observed by calculating the admittance function between the geoid and the bathymetry, where the admittance is the cross spectrum of the geoid and bathymetry divided by the spectrum of bathymetry (Watts, 1979). The admittance is a function of the strength, thickness, and density of the lithospheric plate supporting the bathymetric features, such as seamounts; thus the geodetic information alone cannot be used to determine rigidity uniquely. However, when combined with other (seismic and geochemical) information about the lithosphere, the geoid can give useful information about the strength of the lithosphere and its response to loads (fig. 14.13).

Over distances of several thousand kilometers, the undulations in the geoid

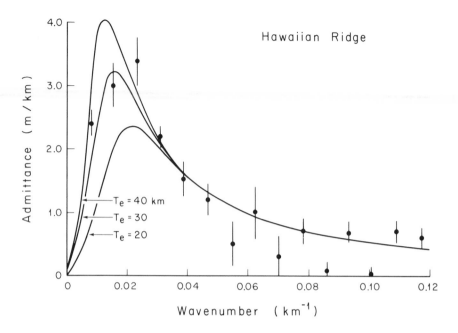

Figure 14.13 The admittance between the geoid observed by Geos-3 and bathymetry in the Pacific, compared with theoretical curves based upon a model of the lithospheric plate, with elastic thickness T_e as a parameter. The wavenumber is 2π/wavelength (from Watts, 1979).

are related to processes deeper in the Earth, particularly those that may produce continental drift. Both laboratory studies and theory suggest that convection may be possible in the mantle, and that it may consist of a series of connected cells of warm, rising material interspersed with regions of cool, sinking material, with typical cell size of 1000 km. Density is controlled primarily by temperature, and the variations in density should produce variations in the height of the geoid (Marsh and Marsh, 1976; McKenzie, 1977).

14.5 Oceanic topography and surface currents

The oceanic topography, the difference between sea level and the geoid, is primarily due to currents and tides, although in shallow water storm surges also contribute. The typical variation in the height of the topography is around 1 m at best (table 14.4), and measurements of the heights require very precise and accurate altimeters. In contrast with other types of satellite measurements, however, observations of oceanic topography are directly related to the variables of interest. Of particular importance is the observation of surface currents (Munk and Wunsch, 1982).

To a good approximation, surface currents are in *geostrophic equilibrium* and hydrostatic balance (Wunsch and Gaposchkin, 1980). In a local Cartesian coordinate system (x,y,z) with x positive to the East, y to the North, and z upward, the horizontal components of velocity (u,v) in the (x,y) directions are related to pressure p through

Table 14.4 Typical Oceanic Topography

Phenomenon	Typical Surface Expression	Period of Variability	Comments
Western boundary currents (Gulf Stream, Kuroshio)	130cm/100km	days to years	Variability in position, and 25% variability in transport
Large gyres	50cm/3000cm	one to many years	25% variability expected
Eastern boundary currents	30cm/100km	days to years	100% variability expected, maybe reversals of direction
Mesoscale eddies	25cm/100km	100 days	100% variability
Rings	100cm/100km	weeks to years	100% variability, growth and decay
Equatorial currents	30cm/5000km	months to years	100% variability
Tides	100cm/5000km	hours to years	aliased to low frequency

$$-fv = -\frac{1}{\rho}\frac{\partial p}{\partial x}$$

$$fu = -\frac{1}{p}\frac{\partial p}{\partial y}$$

$$0 = -\frac{\partial p}{\partial z} - \rho g$$

where

$$f = 2\Omega \sin(\theta)$$

is the Coriolis parameter, $\Omega = 7.272 \times 10^{-5}$ rad.s^{-1} is the rotation rate of Earth, θ is latitude, g is local acceleration of gravity, and ρ is the density of sea water (table 14.5).

Table 14.5 Notation for Describing Geostrophic Currents

D	Ocean depth [m]
f	Coriolis parameter [rad.s^{-1}]
g	Gravity [m.s^{-2}]
L	Characteristic length [m]
p	Pressure [Pa]
R_0	Rossby number
u,v	Horizontal components of geostrophic current [m.s^{-1}]
U	Characteristic velocity [m.s^{-1}]
v_s	Surface geostrophic current [m.s^{-1}]
x,y,z	Cartesian coordinates [m]
z_0	Reference level [m]
ν	Turbulent eddy coefficient [m^2.s^{-1}]
ρ	Density of water [kg.m^{-3}]
ζ	Sea surface topography [m]
Ω	Rotation rate of the Earth [7.3×10^{-5} rad.s^{-1}]

Geostrophic Currents

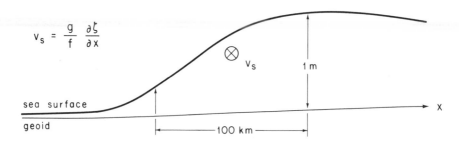

Figure 14.14   The slope of the sea surface relative to the geoid $(\partial\zeta/\partial x)$ is directly related to surface geostrophic current v_s. The slope of 1 m/100 km is typical of western boundary currents.

Combining the equation for hydrostatic equilibrium with, say, the equation for v yields (Fomin, 1964:4ff.)

$$v(x,z) = \frac{g}{\rho f} \int_{z_0}^{z} \frac{\partial \rho}{\partial x} \, dz + v_0(x)$$

where v_0 is an unknown constant of integration depending on the arbitrary level z_0, the reference level.

Traditionally, oceanographers have assumed that at some great depth there is a level of no motion where surfaces of constant pressure are level surfaces. But the ocean is in motion at all depths, and alternatives are required in order to study the mean circulation; hence the attraction of altimetry, for it allows the surface to be the reference level. The altimeter measures the sea surface slopes $(\partial\zeta/\partial x, \partial\zeta/\partial y)$, and these are directly related to geostrophic velocity (u_s, v_s) through (fig. 14.14).

$$v_s = \frac{g}{f} \frac{\partial \zeta}{\partial x}$$

Hence, the velocity at any depth is

$$v(x,z) = \frac{g}{\rho f} \int_{z}^{0} \frac{\partial \rho}{\partial x} \, dz + v_s(x)$$

with similar expressions for $u(y,z)$. This technique requires very accurate measurements of the oceanic topography, however, and accuracies much better than ± 10 cm are required to study the circulation (Roemmich and Wunsch, 1982).

Although we have assumed so far that surface currents are in geostrophic equilibrium, this is not quite true. Strictly applied, geostrophy requires a balance between pressure gradients and the Coriolis force, and the flow is invariant in time. The small deviations from geostrophic balance allow currents to evolve; thus the circulation is really only *quasigeostrophic*.

Acceleration, friction, and nonlinear forces that modify the geostrophic balance tend to be most important at the surface and along coasts. Their influence is roughly approximated by the Rossby number $Ro = U/(fL)$, where U is a characteristic velocity and L is a characteristic length, roughly 100 km (§2.3). Over most of the ocean $Ro \leqslant 10^{-2}$, although it can be 10 or more times larger in strong currents like the Gulf Stream. Wunsch and Gaposchkin (1980), quoting Stommel (1965), point out that the slopes across strong boundary currents give the velocity along the axis of the stream, but that the downstream slope cannot be in geostrophic balance. Rather, the stream flows "downhill."

At the surface, the friction of the wind on the water drives a transient surface current, the local wind drift. The pressure gradients associated with this current are small, however, on order $(\nu/f)^{1/2}/D$ times smaller than the slopes due to geostrophic flow, where ν is a turbulent eddy coefficient (with units of kinematic viscosity, $m^2.s^{-1}$ in SI units) and D is the depth of the ocean. Thus an altimeter does *not* measure surface current, but only the geostrophic component of the surface current. This fact allows the two to be clearly separated, an advantage for some studies.

At the equator, the Coriolis force vanishes and there can be no geostrophic balance. Even here, however, currents are related to sea surface slopes, but the relationship is not as simple as the geostrophic relationship.

The accuracy of altimeter measurements of currents depends on the accuracy with which the oceanic geoid is known. The altimeter measures the height of the sea surface relative to the center of the Earth. This measurement must be referred to the geoid to obtain the slope of the sea surface and hence the surface geostrophic current. Then, knowing the density of the water as a function of depth, the geostrophic and hydrostatic equations can be integrated to obtain the geostrophic currents within the ocean.

The altimetric geoid includes errors due to permanent ocean currents, so it cannot be used for studies of ocean circulation. Instead, independent geoids are required. The long wavelength components of the geoid are known from their influence on satellite orbits, and are measured by accurately tracking satellites (King-Hele, 1976; Lerch et al., 1979). The short wavelength components of the geoid, the *gravimetric geoids*, are known from detailed surveys of gravity. Such surveys are costly and have been made in only a few regions, most notably in the northwest Atlantic (Marsh and Chang, 1978). Taken together, the gravimetric and satellite geoids are uncertain by 10–20 cm when averaged over areas typical of ocean basins; but the errors are much larger in smaller regions, as large as 1–5 m over areas a few hundred kilometers on a side, especially in poorly surveyed areas (Chapman and Talwani, 1979). In well surveyed regions, the errors are much less than 1 m, even for small areas. Thus it is clear that future studies of the general circulation of the ocean will require improvements in our knowledge of the geoid. Continued improvements in the tracking of satellites, and special gravity-measuring satellites, such as the proposed Geomagnetic Research Mission, are expected to contribute.

Variations in Topography

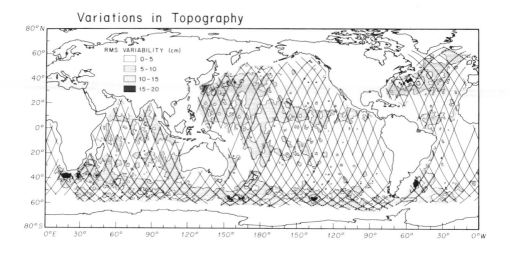

Mean Topography

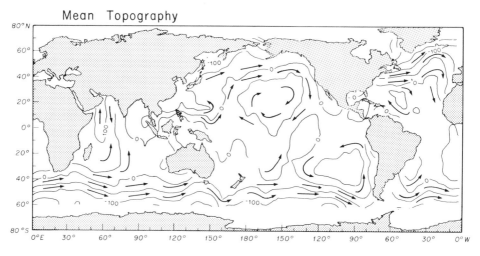

Figure 14.15   Altimeter measurements of surface geostrophic currents. Top: map of mesoscale variability from repeated colinear measurements of topography made by the altimeter on Seasat during September 1978 (from Cheney, Marsh, and Beckley, 1983). Bottom: map of permanent topography, in meters, calculated from a spatially smoothed geoid (PGS-S4) using Seasat altimeter data (from R. E. Cheney and J. G. Marsh, Goddard Space Flight Center).

14.6 Altimeter observations of the sea

Altimeters, particularly that on Seasat, have been used to measure surface currents, tides, winds, and waves. Of these measurements, those of currents are particularly important because the slope of the sea surface can be directly related to surface geostrophic currents. Both orbit errors and uncertainty in the height of the geoid constrain the accuracy of the observations, however; and studies so far have been limited to observations of the temporal variability of currents, the position of strong permanent currents, and the distribution of the larger features in the mean (time-averaged) circulation of the ocean basins.

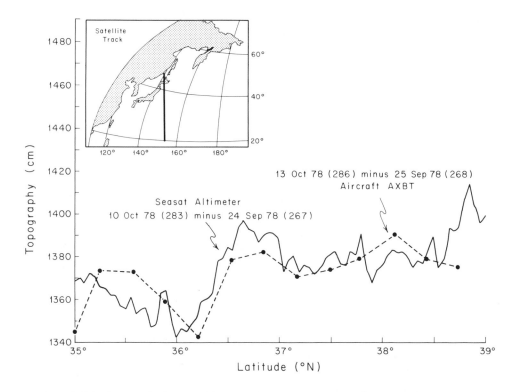

Figure 14.16 Changes in the oceanic topography due to variability of the Kuroshio Current system as observed by the altimeter on Seasat, compared with variability computed from measurements of temperature by air-dropped expendable bathythermographs (AXBT) (from Bernstein, Born, and Whritner, 1982).

The measurements of the variability of the ocean currents are the easiest. Several techniques are used. If the satellite's orbit recurs, any changes in the measured topography of the sea over distances of less than thousands of kilometers must be due to the variability of the surface geostrophic currents. The geoid is constant, provided the subsatellite tracks are colinear within about ±1 km; and orbital errors have much longer wavelengths. If the orbit does not recur, variability is measured at crossings of the subsatellite track. Again, any variability in the measured height at these points must be due to variable currents.

Of course, in either case, heights measured along various passes of the satellite over a region will be offset due to orbital error. Usually, a constant height and slope are removed from the observations. For colinear passes, this is easily done by requiring the mean height and slopes to agree. For a grid of crossing tracks, the height and slope of each pass is adjusted to minimize the differences in height at the crossover points.

Both colinear passes and grids of passes from Geos-3 and Seasat have been used to observe eddies near the Gulf Stream (Huang, Leitao, and Parra, 1978; Mather, Coleman, and Hirsch, 1980; Mather, Rizos, and Coleman, 1979), and globally to map typical variability in oceanic topography (fig. 14.15; plate 15). In addition, the wavenumber spectra of the variability were calculated by Fu

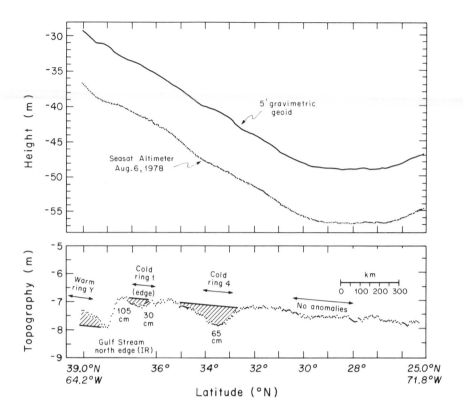

Figure 14.17　Seasat altimeter observations of the Gulf Stream. When the altimeter observations are subtracted from the local geoid they yield the oceanic topography, due primarily to ocean currents in this example (from Cheney and Marsh, 1981a).

(1983) for wavelengths greater than 100 km, using colinear passes of Seasat observations.

The accuracy of altimetric measurements of surface geostrophic currents was tested by Bernstein, Born, and Whritner (1982), who used air-expendable bathythermographs to measure temperature as a function of depth across the Kuroshio along a line observed by the Seasat altimeter. These observations, together with the known relationship between temperature and salinity in that region, allowed the hydrostatic equation to be integrated from some assumed level of no motion. The temporal variability of the topography, and hence the variability of surface currents calculated from hydrography, agreed well with the variability of the topography calculated from altimetry (fig. 14.16).

The measurements of the mean (time-averaged) surface geostrophic currents are more difficult. Both the geoid and the currents contribute to the signal, and the accuracy of the measurements of currents depends on the accuracy of the geoid used in the computations. In a few regions, such as the northwest Atlantic, the short wavelength components of the geoid are fairly well known and, when subtracted from the altimeter measurements, allow the Gulf Stream and its rings and eddies to be observed (fig. 14.17). Over ocean basins, the long-

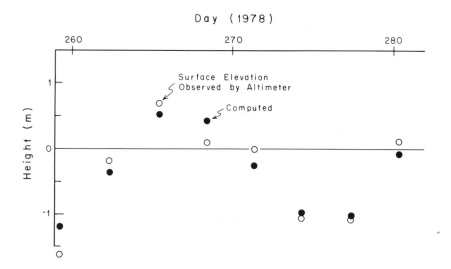

Figure 14.18   The ocean tide at 52.9°N and 343.9°E in the Northeast Atlantic observed by the Seasat altimeter, compared with the tide computed from deep-sea tide gauges (from Cartwright and Alcock, 1981).

wavelength components of the geoid are known well enough to allow larger features of the general circulation to be mapped.

Using the best recent calculations of the geoid, Tai and Wunsch (1983) mapped the circulation in the Pacific, showing that the currents were very similar, on large scales, with those calculated by Wyrtki (1979). At about the same time, Cheney and Marsh (1982) mapped the global distribution of mean currents (fig. 14.15). Both maps agree well with our understanding of ocean circulation, but neither is accurate enough to contribute important new information, which will require measurements of the topography with errors less than 10 cm.

The same techniques used to observe ocean currents have also been used to observe oceanic tides. If the altimeter signal is not corrected for tides, as is done for studies of currents, then the topography observed by the altimeter varies from day to day due to time-variable currents and to tides. But the tides have definite, known frequencies, and that part of the signal coherent with the tidal potential is assumed to be the tide. Note, however, that the altimeter samples the surface at a periodic rate and that this aliases the tidal frequencies. If the signal is not aliased to zero, annual, or semi-annual frequencies, then the assumption is correct. The technique works so long as the altimeter observes many tidal periods, and so long as orbital errors are not too large after adjustment for an offset in height and slope over the oceanic area. Using various implementations of this technique, Cartwright and Alcock (1981) used several months of Seasat data to map the M2 tide in the north Atlantic (fig. 14.18); Brown and Hutchinson (1981) mapped the M2 tide in the north Pacific; and Mazzega (1983) calculated the tides in the Indian Ocean.

The observations of winds and waves by altimeters is considerably easier than

Significant Wave Height

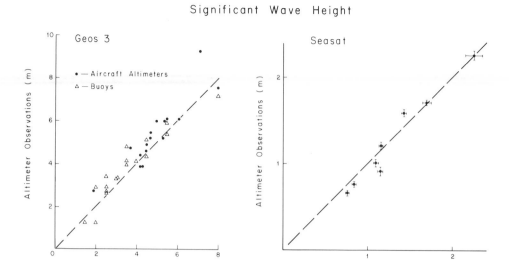

Figure 14.19 Significant wave height measured by aircraft and buoys, compared with satellite altimeter observations of waves in the same area. Left: Geos-3 observations (from Parsons, 1979). Right: Seasat observations of low waves (from Webb, 1981).

Seasat Altimeter Wave Height (m)

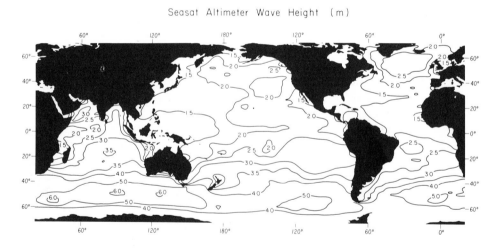

Figure 14.20   Average significant wave height between 7 July and 10 October 1978 observed by the Seasat altimeter (from Chelton, Hussey, and Parke, 1982).

the observations of currents and tides. The height of waves is calculated from the shape of the reflected pulse. An initially short transmitted pulse is lengthened upon reflection, first from wave crests and then from wave troughs, provided the pulse is shorter then the height of the waves. Thus the accuracy of the measurement of low waves depends in part on the length of the transmitted pulse. Seasat, which transmitted a 3.6 ns (1.1 m) pulse could be

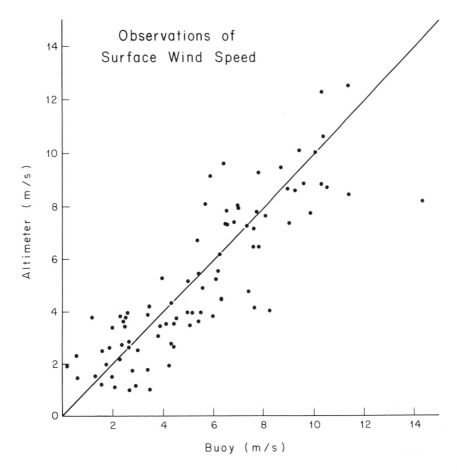

Figure 14.21 Wind speed at 10 m above the surface measured by buoys compared with satellite altimeter observations of wind in the same area (from Fedor and Brown, 1982).

expected to make more accurate measurements than did Geos-3, which transmitted a 12.5 ns (3.8 m) pulse. The expectation was confirmed by comparisons of altimeter observations of wave height with heights measured by buoys, ships, and aircraft in the region observed by the satellite (fig. 14.19). The comparison indicated that Geos-3 could measure waves with an accuracy of ± 0.6 m, but that the satellite tended to underestimate high wave heights (Parsons, 1979). The Seasat altimeter had an accuracy of ± 10% for waves higher than 1 m (Webb, 1981; Fedor and Brown, 1982), an accuracy comparable to that available from well calibrated surface instruments (Stewart, 1980), but it tended to overestimate the height of large waves. Both instruments have been used to map the global distribution of wave heights (fig. 14.20) (Chelton, Hussey, and Parke, 1981; McMillan, 1981).

The altimeter observations of wind speed are based on measurements of the scattering cross section at nadir. The stronger the wind, the lower the cross section and the weaker the reflection from the surface. Comparisons of winds measured by the altimeter with winds measured by ships and buoys (fig. 14.21) indicate that both Geos-3 and Seasat had an accuracy of ± 2 m (Fedor and Brown, 1982; Brown, 1979).

Finally, the Seasat and Geos altimeters could track the height of continental ice sheets, and the data from the former were used to map the height of the ice covering Greenland and portions of Antarctica (Zwally et al., 1983).

Effective strategies for sampling the temporal and spatial variability of the sea surface, and the accuracies required of the observations, have been explored for future satellite systems (NASA, 1981). In order to avoid aliasing high frequency variability, the satellite orbit should recur often enough; and recurring tracks should be close together to avoid aliasing the high wavenumber variability. The two requirements are incompatible. But a useful compromise would be a satellite orbit that recurs every 10—12 days. This is considerably faster than the highest frequencies typical of the open ocean, and produces tracks that are 250 km apart at mid-latitudes.

The sampling strategy must also avoid tidal aliases that compromise the usefulness of the observations. Seasat was nearly synchronous with the P1 tidal constituent (fig. 15.4, chapter 15), and it was aliased to zero frequency. Hence this tidal component is indistinguishable from the time-averaged topography, and errors in estimating the tidal amplitude produce errors in the calculation of the mean circulation. Seasat also aliased the K1 and S2 tidal constituents into a frequency of 2 cycles per year. This not only makes it difficult to use altimeter data to observe tides, but errors in estimating tidal amplitudes cause errors in frequency bands important for studies of the ocean circulation. Future satellites should also avoid orbits that recur at intervals near 14 days, for this would alias the lunar tidal constituents into long periods.

The precision of future observations should be sufficient to resolve the topography of oceanic eddies; ± 2 cm should be adequate. The accuracy should be sufficient to constrain estimates of the time-averaged circulation. The permanent topography due to these currents varies by about ± 1 m, but is already known within ± 10 cm from hydrography. Thus errors in the observation of topography should be less than ± 1 to ± 5 cm. This accuracy is difficult, and will require that the error in the observations of topography be uncorrelated with the topography so that many observations made over a period of months could be averaged together.

15

ORBITS

An understanding of satellite orbits and the factors influencing orbital dynamics is required for many reasons. The orbit of a satellite determines what areas will be viewed by the instruments on the satellite, and how often; it controls the amount of time a satellite will be in sunlight and able to receive solar energy for conversion to electrical power; it is used to predict when a satellite instrument will observe a particular area; and it is required in order to interpret those satellite observations that depend on satellite position, such as signals from satellite altimeters. Carefully tailored satellite orbits enhance the usefulness of satellites by allowing repeated coverage of the same area at the same time each day, or repeated coverage of the same area every few minutes through the day; and a careful selection of satellite altitudes can reduce the complexity of ocean-observing instruments carried into space.

In addition to studying the basic satellite orbit, we often wish to know a particular orbit to great accuracy, to better than 1 m or even 10 cm. How can this be measured, how well can an orbit be specified, and how will the knowledge degrade with time if there are no new measurements of the satellite's orbit? Answers to these questions are particularly important for studies of Earth's geoid and oceanic topography observed by satellite altimeters.

15.1 Coordinate systems

The terms used to describe a satellite's orbit are those of classical celestial mechanics. Three coordinates fix the orientation of a geocentric coordinate system; three more fix the position of the satellite within this system; and a further three — pitch, roll, and yaw — define the orientation of the satellite in orbit.

The origin of the geocentric coordinate system used for satellite orbits is Earth's center of mass C. The coordinate system is further defined by the orientation of two axes, the third axis being perpendicular to these two. The first axis extends from the center of mass through the North Pole. The second axis extends from the center of mass through the Earth's equatorial plane towards Aries; where Aries is a fixed point in the heavens that lies along the

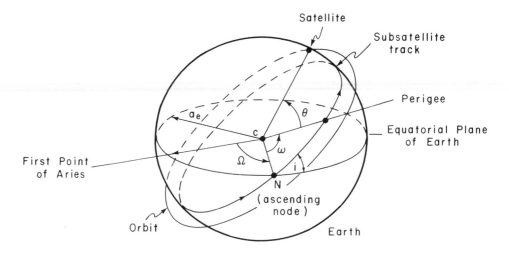

Figure 15.1   Coordinates for describing the orbit of a satellite orbiting around the Earth.

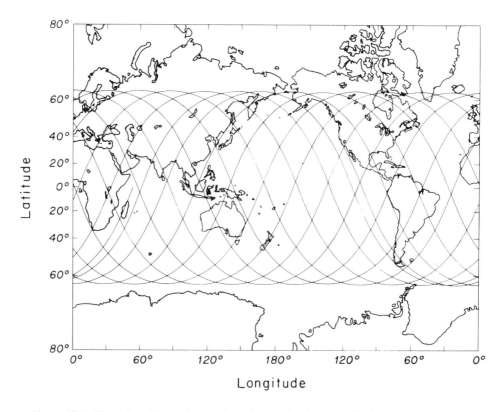

Figure 15.2   The subsatellite track traced out in one day by a satellite in a prograde orbit at a height of 1330 km and an inclination of 63.4°. The maximum latitudinal extent of the track is equal to the inclination of the orbit.

Table 15.1 Notation for Describing Satellite Orbits

a	Semi-major axis of an elliptical orbit [m]
a_e	Equatorial radius of the Earth [6378 km]
A	Cross-sectional area of a satellite [m²]
c	Velocity of light [3×10^8 m.s^{-1}]
d	Integer number of days between recurring orbits
C	Center of mass of the Earth
C_D	Drag coefficient
$\bar{C}_\ell^m, \bar{S}_\ell^m$	Normalized coefficients of the spherical harmonic expansion of a field
e	Eccentricity of an elliptical orbit
F	Force acting on a satellite [N]
GM	Earth's gravitational constant [398,600 km³.s^{-2}]
i	Inclination of orbital plane
i_c	Critical angle [63.4°]
J_2	Second, zonally-averaged, spherical harmonic coefficient [1.086×10^{-3}]
k	Coefficient of reflectivity
k_2	Second Love number
ℓ, m	Integers in spherical harmonic expansion
m	Satellite's mass [kg]
N	Ascending node of an orbit
$p_\ell^m(\sin\phi)$	Normalized associated Legendre polynomials
P_N	Nodal period [s]
P_π	Anomalistic period [s]
r	Radial distance from C to satellite [m]
R	Integer number of revolutions between recurring orbits
S	Solar constant [1350 W.m^{-2}]
v	Satellite velocity [m.s^{-1}]
v_s	Speed of the subsatellite point [m.s^{-1}]
V	Gravitational potential [m².s^{-2}]
α	Incidence angle
γ	Crossing angle between subsatellite tracks at the equator
θ	Angular position of a satellite in its orbital plane (the true anomaly)
ϕ, λ	Latitude, longitude
ρ	Air density [kg.m^{-3}]
ω	Argument of perigee of an elliptical orbit
Ω	Right ascension of orbital plane

line defined by the intersection of the orbital plane of Earth moving around the sun and the equatorial plane of Earth, such that the line from the sun to the Earth points towards Aries at the time of the Vernal Equinox, around March 20.

The orbit of an artificial Earth satellite lies within a plane that rotates a few degrees per day relative to a celestial coordinate system. The terms used to describe this plane and the orbit of the satellite (fig. 15.1 and table 15.1) are:

(a) The *subsatellite point* is that point on the Earth's surface that lies on the line between the satellite and Earth's center of mass.

(b) The *inclination i* of the orbital plane is the angle between the plane of the satellite's orbit and Earth's equatorial plane. Thus the maximum latitudinal extent of the subsatellite point is $\pm i$ (fig. 15.2). If $i < 90°$, the orbit is *pro-*

grade, and the rotation of the satellite, projected on Earth's equatorial plane, is in the same direction as the rotation of the Earth. If $i > 90°$, the rotation is *retrograde*. If $i = 90°$, the satellite is in *polar orbit*; and if $i = 0°$, the orbit is equatorial.

(c) The *ascending node N* is the intersection of the subsatellite track with the equator at the time the satellite crosses the equatorial plane going northward.

(d) The *right ascension* Ω of the ascending node is that angle in the equatorial plane measured eastward from Aries to N.

Thus the two angles i, Ω define the orientation of the plane of the satellite's orbit. A third angle θ gives the position of the satellite within the plane of the orbit. If the orbit is circular, θ is measured relative to N unless $i = 0$. In this case, N is undefined, so θ is measured from Aries. If the orbit is elliptical, θ is measured from *perigee*, the point in the orbit that is closest to C.

Strictly speaking, the coordinates discussed so far define a *coordinate frame* useful for the mathematical description of satellite motion. Other reference frames include the Cartesian and the spherical coordinates. Only when a reference frame is reduced to practice, including means of observing and calculating the coordinates, does it become a *coordinate system*.

Implementation of a coordinate system becomes increasingly difficult as the accuracy of the system is increased. Routine calculations of a satellite's position have an accuracy of a few hundred meters, and these calculations are sufficient to locate data from SHF radiometers and radars having resolutions of $1-50$ km. Images from optical instruments, such as the Thematic Mapper, or from synthetic-aperture radars have resolutions of 25 m, and locating these data requires more careful work. Finally, geodetic satellites and radar altimeters discussed in chapter 14 require positions with an accuracy of a few centimeters. This level of accuracy is not yet possible, and will require coordinate systems more accurate than those now in use.

A number of accurate coordinate systems have been developed over long periods of time for analyzing observations in geodesy, astronomy, and geophysics. These include (Gaposchkin and Kolaczek, 1981):

(a) *Astrophotogrammetric* systems based on photographs of fundamental stars.

(b) *Extragalactic* systems based on observations of quasars by very long baseline interferometers.

(c) *Geocentric* systems based on observations of polar motion and the length of day, using measurements of the times selected stars pass vertically above stations operated by the International Latitude Service and the Bureau International de l'Heure and later supplemented by stations supporting the International Polar Motion Service.

(d) *Geodetic* systems based on measuring the distance between reference points on the surface of the Earth.

(e) *Satellite tracking* systems based on observations of selected satellites, particularly geodetic satellites such as Lageos, Transit, and the Global Positioning System (GPS) satellites.

These accurate coordinate systems are difficult to implement for several fundamental reasons. All observations of the coordinate system must be made from the Earth, yet the Earth moves in an irregular, unpredictable way. The pole moves about due to precision, nutation, and the Chandler wobble, with periods of 24,000 years, 18.6 years, and 13 months respectively (Munk and MacDonald, 1960; Lambeck, 1980). Over a period of a few years, the North Pole moves irregularly about a circle with a diameter of 10–100 m at the surface of the Earth, producing uncertainty in the location of the vertical axis of the geocentric coordinate systems. In addition, the length of day varies irregularly by several milliseconds per day over a period of years, and the cumulative change in time can be a second or more per decade, requiring the leap second to keep solar time in agreement with atomic time. Because a point on the Equator moves 465 m.s^{-1}, the location of the axis pointing toward Aries is uncertain in geocentric coordinates due to uncertainty in the length of day.

The problems introduced by the uncertainty of our knowledge of the rotation of the Earth are compounded by the difficulty of comparing observations made by the different techniques used to implement the coordinate systems. Quasi-stellar objects are difficult to locate visually with telescopes, as are the positions of satellites relative to the fixed stars. As a result, astrophotographic systems can be related to extragalactic systems or to geocentric systems only with limited accuracy.

Typically, the origin of these systems differs in position by a few meters; but the accuracy is improving as more refined techniques and worldwide observations are implemented. For further discussions on the various coordinate systems and their use in geodesy, particularly satellite geodesy, see NASA Special Publication SP 365 (NASA, 1977) on the National Geodetic Satellite Program.

15.2 The motion of a low satellite about Earth

The motion of a small mass about a larger body, the subject matter of celestial mechanics, has been well known for several centuries. But the application of this theory to the motion of artificial satellites is a recent development (King-Hele, 1964; Kaula, 1966). To a first approximation, the satellite moves in an elliptical orbit determined by its launch, but this crude approximation is not of practical use. Earth's equatorial bulge perturbs the orbit; drag of the upper atmosphere causes the satellite to spiral in gradually toward the surface, and solar radiation pushes weakly but persistently.

If we neglect all the smaller forces acting on a satellite, and consider a small mass orbiting in a vacuum about a larger, spherically symmetric mass, then its orbit is an ellipse given by (Moulton, 1914)

$$r = a(1-e^2)/(1+e\cos\theta)$$

where r is the radial distance from C to the satellite, e is the eccentricity of the orbit, and a is the semi-major axis (fig. 15.3). The angular position θ, the *true anomaly*, is measured from perigee, the lowest point in the orbit. The highest point, *apogee*, occurs at $\theta = \pi$. The period of the satellite is

$$P = 2\pi \frac{a^{3/2}}{\sqrt{GM}}$$

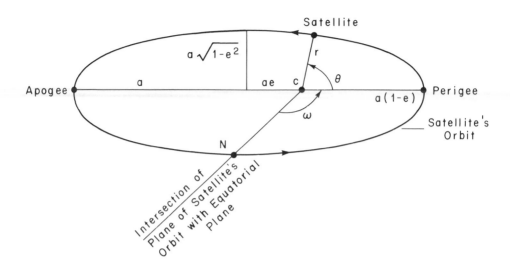

Figure 15.3 Coordinates and notation for describing an elliptical orbit.

where GM is Earth's gravitational constant, the product of the gravitational constant G and Earth's mass M, having an approximate value of 4×10^5 km^3.s^{-2}.

The plane of the ellipse is specified by Ω and i, and the orientation of the ellipse within this plane is specified by the *argument of perigee* ω, the angle between the ascending node N and perigee. Thus the position of a satellite in an elliptical orbit is specified by the six *orbital elements* $\{a, i, \Omega, \omega, e, \theta\}$.

For most purposes, the elliptical orbit is not sufficiently accurate and the forces that perturb the orbit must be considered. Of these, the largest is the gravitational attraction of Earth's equatorial bulge, provided the satellite is not so low that atmospheric drag dominates.

The gravitational field of the Earth is expressed by a spherical harmonic expansion of the gravitational potential V (Stacey, 1977; Jackson, 1962:64):

$$V = -\frac{GM}{r}\left\{ 1 + \sum_{l=2}^{\infty} \left(\frac{a_e}{r}\right)^l \sum_{m=0}^{l} [\bar{C}_l^m \cos m\lambda + \bar{S}_l^m \sin m\lambda]\, p_l^m\, (sin\, \phi) \right\}$$

where $\{r, \phi, \lambda\}$ are the coordinates of range, latitude, and longitude, a_e is the Equatorial radius of Earth, $\bar{C}_l^m$ and $\bar{S}_l^m$ are the coefficients of the expansion, and $p_l^m\, (cos\, \theta)$ are the associated Legendre polynomials (§5.1) of degree l and order m. The expansion is usually normalized so that

$$\frac{1}{4\pi} \int_0^{2\pi} \int_{-1}^{1} p_l^m\, (sin\phi, \sin m\lambda, \cos m\lambda)\cos\phi\, d\phi\, d\lambda = 1$$

although other normalizations are sometimes used.

Some of the constants in the expansion have been determined to great accuracy through tracking of interplanetary spacecraft and Earth satellites and through geodetic measurements. The official values adapted by the International Union of Geodesy and Geophysics in 1980—the Geodetic Reference System 1980—are reported in Moritz (1980). Slightly more accurate values based on more recent observations are:

$$GM = 398600.5 \pm 0.1 \text{ km}^3.\text{s}^{-2} \qquad \text{(Esposito, 1979)}$$
$$a_e = 6378.1349 \text{ km} \qquad \text{(West, 1982)}$$
$$\bar{C}_2^0 = -484.16600 \times 10^{-6} \qquad \text{(Lerch et al., 1982)}$$

The other coefficients $\{\bar{C}_l^m, \bar{S}_l^m\}$ are much smaller, and recent values of some of them are listed in Lerch et al. (1982), Wagner et al. (1977), and Stacey (1977), all with the above normalization. In general, these terms are smaller than one thousandth $\bar{C}_2^0$; the accuracy of the various coefficients have been assessed by Lambeck and Coleman (1983). The spherical harmonic expansion of the gravitational potential is often truncated to

$$V = -\frac{GM}{r} \left[1 - \frac{a_e^2}{r^2} \frac{J_2}{2} (3\sin^2\phi - 1) \right]$$

where $J_2 = -\sqrt{5}\,\bar{C}_2^0 = 1082.63 \times 10^{-6}$ is the harmonic that includes most of the influence of the equatorial bulge. This approximation to the potential is often adequate for many orbit calculations.

The primary influence of the J_2 term in the potential is to produce large secular changes in Ω and ω and smaller short and long period variations in all the orbital elements. While these periodic variations are important for calculating accurate orbits, the secular changes are the most noticeable, and their influence must be considered in estimating orbits for many practical purposes.

The secular change in Ω causes the plane of the satellite's orbit to precess slowly relative to the stars, at a rate (fig. 15.4)

$$\dot{\Omega} = -\frac{3}{2} J_2 \sqrt{\frac{GM}{a}} \frac{a_e^2}{a^3} \frac{\cos i}{(1-e^2)^2}$$

with inclination held nearly constant. If the inclination and height of the satellite are carefully chosen, the rotation of the orbital plane can be made to match the rotation of the Earth about the sun. As a result, the satellite crosses the equatorial plane at the same time of day, and the orbit is *sun synchronous*.

The secular change in ω causes the orientation of the elliptical orbit to rotate slowly within the orbital plane; and the argument of perigee changes at a rate

$$\dot{\omega} = -\frac{3}{4} J_2 \sqrt{\frac{GM}{a}} \frac{a_e^2}{a^3} \frac{(1-5\cos^2 i)}{(1-e^2)^2}$$

This influence is zero at the *critical angle* i_c given by

$$1 - 5\cos^2 i_c = 0$$
$$\cos i_c = 1/\sqrt{5}$$
$$i_c = 63.4°$$

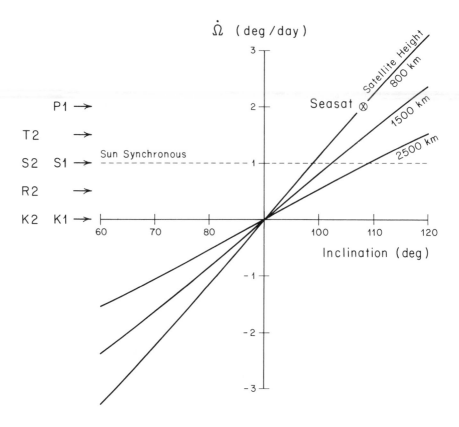

Figure 15.4 Precession rate of the orbital plane as a function of inclination, with height of the satellite as a parameter. The indicated precession rates are those that alias the noted ocean tidal constituents $(P1, S1, \text{etc.})$ to zero frequency when the ocean surface is observed by a satellite altimeter.

Finally, the equatorial bulge influences the period of the orbit. If the Earth were a point mass, the orbital plane would be fixed, and the orbital period would be the time required to return to a fixed point in the orbit. But a satellite moving in a precessing orbit never returns to the same point in geocentric coordinates, so the period is more difficult to define. Several definitions can be used (El'yasberg, 1967:126):

(a) The *nodal period P_N* is the time between successive ascending nodal crossings (fig. 15.5):

$$P_N = \frac{2\pi a^{3/2}}{\sqrt{GM}} \left\{ 1 + \frac{3}{4} J_2 \left[\frac{a_e}{a} \right]^2 \left[(1-3\cos^2 i) + \frac{(1-5\cos^2 i)}{(1-e^2)^2} \right] \right\}$$

(b) The *anomalistic period P_π* is the time between successive perigee passages

$$P_\pi = \frac{2\pi a^{3/2}}{\sqrt{GM}} \left[1 + \frac{3}{4} J_2 \left[\frac{a_e}{a} \right]^2 \frac{(1-5\cos^2 i)}{(1-e^2)^2} \right]$$

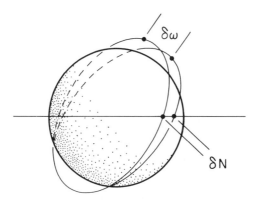

Figure 15.5 Two definitions of the period of a satellite
in a precessing orbit. The time between successive nodal
crossings, the nodal period P_N, is increased by an amount
δN proportional to the rate at which the orbital plane
precesses. Similarly, the times between perigee passages,
the anomalistic period P_π, is increased by an amount $\delta\omega$
due to the precession of perigee.

Two other forces, air drag and solar radiation pressure, also influence satellite
orbits. Their effects are usually calculated numerically, because the force
depends on the orientation of the satellite and hence on the position of the
satellite in the orbit. But they can be estimated by considering the forces acting
on a satellite of constant cross-sectional area A, either in the direction of the
sun or the satellite velocity vector, in a spherical orbit of radius a.

The force due to air drag is

$$\rho A C_D v^2 \qquad \text{(air drag)}$$

where ρ is air density and v is the satellite's velocity. The drag coefficient C_D
is 0.92 for a sphere or 1.1 for a flat plate perpendicular to the flow, provided the
satellite is higher than a few hundred kilometers above the surface of the Earth,
so that the mean free path of an air molecule is greater than the typical dimension of the satellite. If the force is constant, the change in height δa per orbit
can be calculated by integrating eq. (4.85) in King-Hele (1964) to obtain

$$\delta(a) = 4\pi A C_D \rho a^2 / m$$

where m is the mass of the satellite.

Air density in the upper atmosphere can be determined using the empirical
relationships in Jacchia (1971) or those in the NASA Special Publication SP-
8021 (NASA, 1973). In general, density depends on the exospheric temperature, and this, in turn, depends on the absorption of ultraviolet solar radiation
with wavelengths shorter than 0.2 μm. Such radiation cannot be measured at
the surface of the Earth, because the atmosphere is opaque to ultraviolet light;
but fortunately it is well correlated with 10.7 cm radiation from the sun that
does reach the Earth's surface. Hence temperature is calculated using
observations of 10.7 cm radiation monitored by solar observatories, together

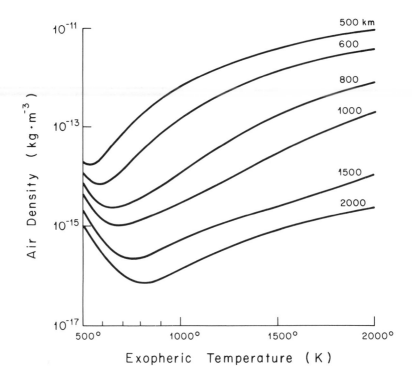

Figure 15.6 Air density as a function of exospheric temperature with height as a parameter. The exospheric temperature varies from day to night, from season to season, and as a function of the solar cycle (from Jacchia, 1971).

with knowledge of the diurnal and seasonal variation of sunlight in the upper atmosphere.

Typically, the exospheric temperature varies from 600K during solar minimum conditions to 900K at solar maximum, with increases up to 1200K or higher during solar storms. Thus air density at 800 km varies between 3×10^{-15} and 8×10^{-15} kg.m^{-3} with increases to 5×10^{-14} kg.m^{-3} (fig. 15.6). As a result of air drag, a satellite with an area-to-mass ratio of 10^{-2} m^2.kg^{-1} (a rather dense satellite) orbiting at a height of 800 km above the Earth, where the air density is 5×10^{-14} kg.m^{-3} will spiral inward at a rate of around to 35 cm per orbit.

Radiation pressure of sunlight, both direct and reflected from the Earth, produces a small but persistent force of magnitude

$$kSA \, (\cos \alpha)/c \qquad \text{(radiation force)}$$

where S is the irradiance, c is the velocity of light, and α is the angle between the incident radiation and the perpendicular to the surface. The coefficient k depends on the orientation, shape, and reflectivity of the satellite's surfaces, and varies from 1.0 for perfect absorbers and reflectors to 1.44 for Lambert surfaces (El'yasberg, 1967:319). For sunlight, the irradiance S is the solar constant, 1350 W.m^{-2}. Thus the radiation pressure of sunlight is about $4.5\,\mu$Pa. Earthlight is about one-third this value for low satellites above the sunlit hemisphere of the Earth.

The maximum excursion δa from a circular orbit due to solar radiation occurs when the Earth—Sun line lies in the orbital plane, and then (El'yasberg, 1967:324)

$$\delta a = \frac{3\pi a^3 ASk}{mcGM}$$

For a satellite orbiting at 800 km, with an area to mass ratio of 0.01 m^2.kg^{-1}, and whose area is primarily perfectly absorbing solar cells oriented perpendicular to the solar radiation, the excursion is 40 cm with a period of once per orbit. Thus solar radiation pressure has an influence comparable to that of air drag at these altitudes.

15.3 Some special orbits

Satellites measure oceanic variables in a particular region only at those times when the region is in view of the satellite. Thus orbits must be carefully chosen to provide a useful combination of temporal and spatial sampling. Several orbits are particularly useful and are widely used. Other orbits, such as that of Seasat, are carefully tailored to meet special requirements (Cutting, Born, and Frautnick, 1978). Particularly useful orbits include:

(a) *Geostationary orbits.* If a satellite is placed in an equatorial orbit, one with $i = 0$, and if the satellite is high enough, then the satellite can have the same angular velocity as a point on the equator. In this case, the satellite will appear to be stationary with respect to an observer on the Earth, and is said to be geostationary. This requires that the satellite have a period of one sidereal day, so $P_N = 86164$ s. This in turn, requires that the semimajor axis be

$$a = \left(\frac{P_N^2 GM}{4\pi^2} \right)^{1/3}, \quad i = 0$$

$$a = 42{,}164 \text{ km}$$

$$a - a_e = 35{,}786 \text{ km}$$

where $(a-a_e)$ is the height of the satellite above the equator. Most communication satellites, the Geostationary Operational Environmental Satellites (Goes), and the Meteosats use this orbit. If the nodal period is a sidereal day but the inclination is not zero, the subsatellite track moves in a "figure-of-eight" centered above a fixed point on the equator (El'yasberg, 1967:10).

A line from a geostationary satellite to the Earth's surface is tangent to the surface at a latitude ϕ of

$$\cos \phi = \frac{a_e}{a}$$

$$\phi = 81.3°$$

However, observations at grazing incidence are not useful for oceanography, and a practical limit to the satellite's field of view is about 60°. At this angular distance, the satellite is 22° above the horizon, and a line from the satellite to the Earth has a local incidence angle of 68°.

(b) *Sun-synchronous orbits.* A second very useful orbit is one whose plane rotates at the same rate as the rotation of the Earth about the sun. In this case, a satellite views points on the Earth's surface at the same local time each day and night. The rotation rate $\dot{\Omega}$ that satisfies this condition is

$$\dot{\Omega} = \frac{2\pi}{365.24} \text{ rad/yr} \approx 1°/\text{day}$$

$$\dot{\Omega} = 1.99 \times 10^{-7} \text{ rad.s}^{-1}$$

Various combinations of the semi-major axis a and inclination i produce such a precession rate. For example, the Noaa-4 satellite viewed the Earth at 09:00 and 21:00 local time each day from a height of 1450 km ($a = 7839$ km). Using these values, the inclination must be

$$\cos i = -\frac{2\dot{\Omega} a^{7/2}}{3 J_2 \sqrt{GM} a_e^2}, \quad e = 0$$

$$\cos i = -0.203$$

$$i = 101.69°$$

This is a retrograde orbit inclined at 78.3°. Note that the inclination for sun-synchronous orbits is only a weak function of satellite height and eccentricity, and that the high inclination allows the satellite to view almost the entire surface of the Earth from pole to pole, a most fortunate happenstance.

(c) *Altimetric satellite orbits.* Altimetric satellites, discussed in the last chapter, measure the height of the sea surface relative to the center of the Earth. Both solid earth and ocean tides contribute to the altimetric signal, and their influence must be subtracted from it. The tides cannot yet be computed with sufficient accuracy and must be measured directly. Thus satellites carrying altimeters require orbits that do not alias the tidal signal into unacceptable frequency bands. For example, a sun-synchronous orbit aliases the $S1$ and $S2$ tidal constituents into zero frequency, and the $P1$, $K1$, $K2$, $T2$, and $R2$ tides into one cycle per year, the same frequency as the large annual variability of the sea surface (fig. 15.4; NASA, 1981). Thus sun-synchronous orbits must be avoided. Orbits with inclinations less than 65° or greater than 120°, for satellites at heights below 1500 km, alias the tides into frequencies greater than two cycles per year, allowing the satellites to measure tides directly. But note that satellites in low orbits cannot sample a particular region often enough, at least four times per day, to avoid aliasing tides; so the problem cannot be avoided, but its consequences can be reduced. In addition to the tidal problem, altimeters should be able to resolve with equal accuracy the two components of surface slope, thus the ascending and descending subsatellite tracks should cross at angles near 90°. At the equator, the crossing angle is (fig. 15.7)

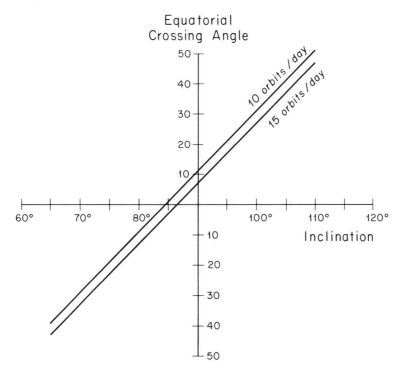

Figure 15.7   Angle between intersections of the subsatellite track as a function of orbital inclination with number or orbits per day as a parameter.

$$\gamma = 2 \tan^{-1} \left[\frac{v_S \cos i - \Omega_e a_e}{v_S \sin i} \right]$$

where v_S is the speed of the subsatellite point and $\Omega_e a_e$ is the velocity of a point on the equator. For a satellite in circular orbit

$$v_S = \left(\frac{a_e GM}{a^3} \right)^{1/2}$$

Away from the equatorial region the crossing angles become larger, being 90° at mid-latitudes, but eventually approaching zero at the latitude equal to the inclination (fig. 15.2). Near polar orbits have very small crossing angles and should be avoided.

(d) *Exactly recurring orbits.* For many purposes, it is useful to make observations from a satellite with an exactly recurring orbit. This requires that (Cutting, Born, and Frautnick, 1978)

$$P_N (\Omega_e - \dot{\Omega}) = \frac{2\pi d}{R}$$

where Ω_e is the rate of rotation of the Earth, d is the integer number of days between exact repeats, and R is the integer number of revolutions between

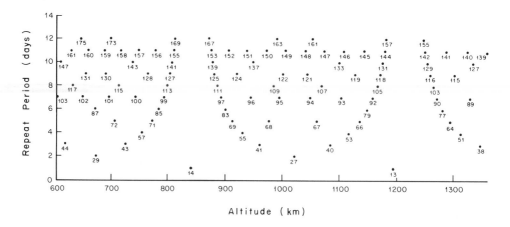

Figure 15.8   Exactly repeating satellite orbits, for orbits with $i = i_c \equiv 63.4°$ and $e = 0$. The integers are the number of satellite revolutions between repetitions of the orbit (from Frautnick and Cutting, 1983).

repeats. Because P_N and $\dot{\Omega}$ are functions of i, e, and a, there are many possible recurring orbits (fig. 15.8), and it is usually possible to find an orbit that recurs in d days for an orbit near any particular altitude. Those with two- and three-day periods are particular common.

15.4 Accurate orbits

In order to interpret the signal from satellite altimeters described in the last chapter, as well as for geodetic studies, some satellite orbits must be known with an accuracy of $10-100$ cm. For these cases, a precessing elliptical orbit is not a sufficiently accurate approximation to the true orbit, and more exact ones must be calculated. This is done by integrating numerically the equations of motion, considering all forces acting on the satellite.

The forces that perturb a satellite's orbit, in relative order of importance, are (fig. 15.9):

(a) *Gravity*. The lack of spherical symmetry in Earth's gravitational field must be accounted for, usually to degree and order 36 in the spherical harmonic expansion of the gravitational potential.

(b) *Atmospheric drag* is calculated using the empirical relationships for air density, together with the known shape and orientation of the satellite.

(c) *Solar and lunar* gravitational attractions, as well as some planetary gravitational attractions, are calculated knowing the mass and location of these bodies.

(d) *Radiation pressure* of both sunlight and earthlight is calculated using the known shape, orientation, and reflectivity of the satellite.

(e) *Oceanic tides* perturb the gravitational potential, and their influence is calculated from a spherical harmonic expansion to degree and order 6 of the tides calculated from hydrodynamic theory.

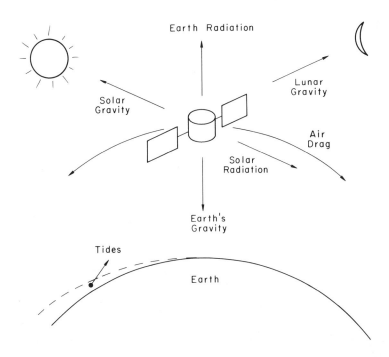

Figure 15.9 Forces influencing a satellite's orbit. Earth's gravity dominates,
but all forces must be considered when calculating very accurate orbits.

(f) *Outgassing* from the satellite must be minimized, for it makes difficult
the calculation of the orbits especially for the first few weeks after launch.

(g) *Gravitational attraction* of most of the planets, and relativistic effects
(non-Newtonian mechanics) are small enough to be neglected.

With this set of vector forces **F**, it is possible, in theory, to integrate
Newton's equation of motion

$$\mathbf{F}(t) = m\dot{\mathbf{v}}(t)$$

where m is the mass of the satellite and $\dot{\mathbf{v}}$ its acceleration, to obtain the
satellite's position in tabular form, the satellite's *ephemeris*. The forces are not
accurately known, however, and the integration must be constrained by measurements of the satellite's position made by a satellite-tracking network. The
more accurate the measurements, and the more frequently they are made, the
more accurate is the satellite's ephemeris.

Depending on the type of tracking data, the equation of motion is usually
integrated along an arc traversed by the satellite in a few revolutions of the
Earth, or in a few days. The solution is compared with observations from the
tracking system, and the initial conditions are adjusted to produce a better fit to
the observations. In this way, satellite dynamics are used to interpolate
between positions determined by the tracking system. Alternatively, the
satellite's position could be tracked continuously and then smoothed to obtain

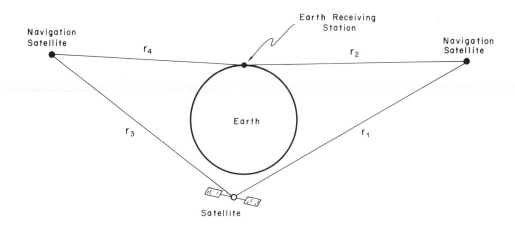

Figure 15.10 Satellite-to-satellite tracking: the doubly-differenced method for tracking a satellite in low Earth orbit. The difference between (r_2-r_1) and (r_3-r_4) can be used to track the position of a satellite without requiring a very accurate clock on the satellite or at the Earth receiving station. The technique is similar to the hyperbolic methods of navigation in oceanography, and should allow satellites to be tracked continuously with an accuracy of around 10 cm.

the satellite's ephemeris. Such tracking is not yet possible, but may be in the near future.

After the ephemeris has been calculated for many arcs, the differences between the calculated and observed satellite positions are analyzed to produce improved calculations of the forces acting on the satellite. From this analysis comes improved knowledge of Earth's gravity field and the density of the upper atmosphere, as well as a better understanding of ocean tides, as described in the last section of this chapter.

The importance of satellite tracking has led to the development of a variety of tracking systems based on radios, which measure range and range rate, and lasers, which measure instantaneous range to satellites when skies are clear. The tracking stations can be at fixed positions on Earth, or on a higher satellite in an orbit less influenced by perturbations in Earth's gravity field or atmosphere. The latter measurement is known as *satellite-to-satellite* tracking.

The accuracy and frequency of the range or range-rate measurements depend on the measurement system. For example, lasers can estimate ranges to satellites with an accuracy of ± 5 cm (Vonbun, 1977), but must operate in cloud-free periods or regions. Proposed modifications to contemporary radiometric tracking systems may result in all-weather measurements of range, but with an accuracy of ± 10 cm, a value that is slightly poorer than that for laser tracking. Future radio ranging systems that use the signal from the Global Positioning Satellite System (GPS) may be able to track satellites with a precision of 1–5 cm relative to ground-based tracking stations (fig. 15.10). When coupled with simultaneous radiometric observations of atmospheric water vapor above the ground-based receivers used by the system, the technique may track satellites with centimeter accuracy even through clouds.

In addition, satellites carrying altimeters can track their own height above the sea surface. Measurements made at the points where the subsatellite tracks cross further constrain the calculation of the orbit, yielding a more accurate ephemeris (Goad, Douglas, and Agreen, 1980).

15.5 Accuracy of satellite orbits

The orbits of a number of geodetic satellites have been accurately calculated (table 15.2). In general, satellites that are dense, that fly above 800 km, and

Table 15.2 Accuracy of geodetic satellite orbits

Satellite	Inclination	Height (km)	Area/Mass (m^2/kg)	Radial Orbit Error (m)
Geos-3	115°	840	.0041	1.0−2.0
Lageos	110°	5900	.0007	0.3−0.4
Starlett	50°	800−1100	.0010	1.5−2.0
Seasat	108°	800	.0114	0.3−0.5

that are observed by laser and radio tracking systems have orbits that can be calculated with an accuracy of a few decimeters in the radial and cross-track direction and a few meters in the along-track direction. The errors in the orbits are predominantly once per orbit (fig. 15.11) and are due primarily to uncertainty in Earth's gravity field. A few satellites, such as Seasat, whose orbits have been analyzed with great care, have orbits known with an accuracy of 30−50 cm in the radial direction. Eventually, high, dense satellites tracked with improved radiometric systems should have orbits whose radial component is known with an accuracy of better than 10 cm.

Two classes of error influence the ephemeris: (a) errors in measurement of the satellite's position, and (b) errors in estimating the forces acting on the satellite. Included within the class of measurement errors are:

(a) *Atmospheric and ionospheric uncertainties*—the inability to know accurately the velocity of radiation in the atmosphere and ionosphere. The mass of the atmosphere, water vapor, and electron density contribute an uncertainty of a few parts per million in the velocity of radiation at the super high radio frequencies used by radio tracking stations.

(b) *Station positions*—the inability to know the precise location of the tracking stations relative to the center of the Earth. At present, these are known with an accuracy of 20−30 cm for the primary laser tracking stations and 50−100 cm for the radio tracking stations. In addition, the coordinate system used to determine station positions is not perfectly known because of polar motion and the variations of the length of the day.

(c) *Station distribution*—the inability to have many stations well distributed around Earth. The laser stations tend to be concentrated in the northern hemisphere and on continents rather than distributed uniformly.

(d) *Satellite mass variations*—the mass of satellites varies because of outgas-

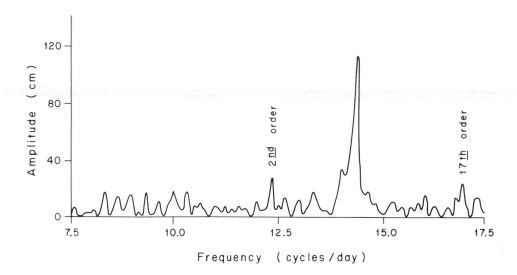

Figure 15.11 An estimate of the error spectrum for the radial component of the Seasat ephemeris. The amplitude spectrum is dominated by a peak at once per orbit, and the smaller peaks are due to errors in the second and seventeenth-order terms in the spherical harmonic expansion of gravity (from Lerch et al., 1982).

sing and because of mass expelled during maneuvers. This change in mass not only changes the position of the satellite's center of mass (which is the point of reference for the orbit computation) but also enters directly into the dynamical computations of very accurate orbits.

The predominant errors in estimating the forces acting on a satellite in near Earth orbit include:

(a) *Gravitational potential*—the uncertainty in the knowledge of the gravitational field, at satellite heights, due to the uneven distribution of Earth's mass. At present, only the lower-order harmonics of the spherical-harmonic expansion of Earth's gravity field are known relatively well.

(b) *Unknown forces*—including atmospheric drag and pressure due to solar radiation and upwelling radiation from Earth. The satellite's acceleration can vary between 10^{-8} and 10^{-10} g over times of minutes to years, depending on the satellite's mass and area, solar activity, sunspot intensity, the distribution of Earth's albedo, and the satellite's reflectivity and orientation.

The two classes of error influence the determination of satellite orbits in different ways. The measurement errors influence estimates of the original position and velocity of the satellite, and the dynamic errors introduce uncertainties into the equations used to predict motion away from the original position. Therefore, the latter contribute errors in predicted position that grow with time, and new observations of position must be continuously introduced into the calculation to limit the growth of error in the predicted orbit. Hence the trend toward satellite-to-satellite tracking, which allows a satellite to be observed almost continuously with a small number of satellites and receiving stations.

From the accurate tracking of satellites has come important new geophysical information. Foremost is a greatly improved understanding of Earth's gravity field (King-Hele, 1976; Lerch et al., 1982) and the density of the upper atmosphere (Jacchia, 1971). As orbits have become yet more accurate, the calculations are yielding better observations of the shape of the Earth (Anderle, 1980; Yoder, et al., 1983) and of pole position and the length of the day (Robertson et al., 1983; Anderle and Oesterwinter, 1980). In the future, the positions of tracking stations should be known accurately enough that continental drift should produce noticeable effects. And, of course, this information will eventually result in the ability to produce still more accurate orbits.

15.6 Tides from satellite orbits

The gravitational attraction of the tidal bulge in the solid earth and in the oceans perturbs satellite orbits. The perturbation is small but observable, and precise observations of satellite orbits can be used to calculate the lower-order coefficients of the spherical harmonic expansion of the tides, primarily $O1$, $P1$, and $S2$ oceanic tides. The dominant semi-diurnal ocean tidal constituent, the lunar $M2$ tide, produces much smaller effects and is more difficult to observe. But it too can be calculated, using the very accurate orbits now available for some geodetic satellites.

The tidal bulge produces small but long-period variations in the orbital elements, especially in the inclination i and ascending node Ω. At the same time, the dominant errors in the elements are short period, being mostly at a period of once per orbit. Thus it is possible to calculate the mean elements with much better accuracy than their instantaneous values at any point in the orbit, and very small tidal influences can be observed.

To obtain the mean elements, the precise ephemeris is calculated once per minute for a day or longer using the coordinate system of the Kepler elements $\{a, i, e, \omega, \Omega, \theta\}$. That is, each one-minute segment of the orbital arc is approximated by an elliptical orbit that is tangent to the true orbit. The elements of these ellipses are the *osculating elements*. From the series of osculating elements are subtracted all known perturbations to the orbit, including those due to gravity, solar radiation pressure, and air drag. This produces a set of elements with greatly reduced variability that can be further smoothed to produce the mean elements. Finally, the mean elements are fitted to a secularly precessing Kepler ellipse to calculate $\bar{a}, \bar{e}, \bar{i}, \bar{\omega}, \dot{\omega}, \bar{\Omega}$, and $\dot{\Omega}$ once per day (Douglas, Marsh, and Mullins, 1973; Felsentreger, Marsh, and Williamson, 1979).

Careful application of this or similar procedures produces mean elements with an accuracy of a few centimeters for $\bar{a}$ and a few hundredths of an arc second for $\bar{i}$ ($5 \times 10^{-2} \mu$rad is 0.01 arc second), at intervals of once per day for periods of hundreds of days. This accuracy must be compared with the small tidal influence due to the $M2$ tide, which perturbs $\bar{i}$ with an amplitude of 0.02–0.07 arc second and a period of 10–20 days for typical satellites. Thus $M2$ tidal perturbations can be observed with an accuracy of better than 10% (fig. 15.12).

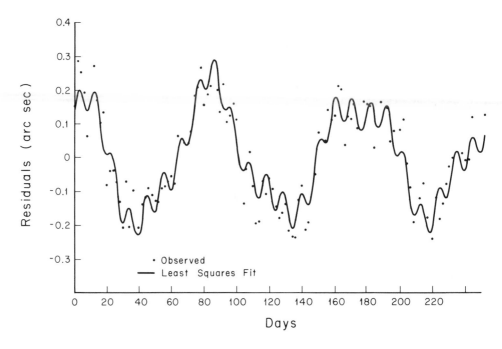

Figure 15.12 The influence of the $M2$ tide on the orbit of the Starlett satellite. The $M2$ tide produces a perturbation with a period of exactly 10.5 days and an amplitude of 0.042 arc seconds. This perturbation is seen in the time series of the inclination, and is calculated with an accuracy of better than 10% from the spectrum of the same time series.

Observations of tidal perturbations of satellite orbits yield the lowest-order coefficients, C_{22}^+ and C_{42}^+, in the spherical harmonic expansion of the tides, where $(C_{\ell m}^+)^2 = (\bar{C}_\ell^m)^2 + (\bar{S}_\ell^m)^2$ and where the $+$ sign refers to the direction of propagation of the tidal bulge. Both solid earth and oceanic tides contribute to the perturbation of the satellite's orbit, and in order to separate the two, it is necessary to have good models for one or the other. Of the two, the earth tides are the better known. Seismic and gravimetric observations indicate that the Earth responds almost exactly as an elastic body, with no phase lag between the tidal response and the tidal potential, and with an amplitude that is linearly related to the tidal potential. Thus the amplitude of the tide is obtained from the second Love number k_2, where $k_2 = 0.30$ is the ratio of the gravitational potential produced by the tidal deformation of the solid earth to the deforming potential. Note that the earth is also deformed by the weight of the ocean tides, the ocean loading; but this influence is considered as part of the oceanic tides.

Subtracting such calculated earth tides from C_{22} and C_{42} yields the contributions of the oceanic tide to the coefficients. The values of the tidal coefficients calculated in this way have an uncertainty of around $\pm 10\%$, and are more accurate than values calculated from hydrodynamic theory. For example, the various hydrodynamic calculations yield C_{22} in the range of 3.6 to 5.4 cm, while the satellite data give 3.42 ± 0.24 cm (Felsentreger, Marsh, and Williamson,

1979). The corresponding phase lag is calculated to be $325.5 \pm 3.9°$.

The calculations of tidal coefficients are particularly useful for estimating tidal dissipation in the oceans because this is closely related to the amplitude and phase lag of the tidal bulge. The dissipation causes a deceleration in the rotation rate of the moon about the Earth, an important geophysical variable. Astronomic values of the deceleration differed from values calculated from estimates of the dissipation of oceanic tidal energy, leading to speculation that the difference must be due to dissipation in the Earth's core or mantle. Now the estimates of the deceleration of the lunar longitude $\dot{N}$ based on satellite data are in good agreement with astronomical values. The satellite data yield $\dot{N} = -25 \pm 3$ arc seconds/(century)^2 and the astronomical values based on ancient eclipses, observations of the transits of Mercury, and lunar ranging give $\dot{N} = -25$ to -22 arc seconds/(century)^2 (Felsentreger, Marsh, and Williamson, 1979; Goad and Douglas, 1978; Cazenave and Daillet, 1981).

16

SATELLITE SYSTEMS

A satellite system, in its broadest sense, includes sensors to view the ocean; a structure called a bus that supports the sensors and provides them with power; memories to store the observations; radios to transmit them to Earth; electrical and mechanical devices and small rockets to control the orientation of the satellite; Earth stations to receive the transmitted data; and a network to process and distribute the information from the sensors. Clearly, the sensors—which have been the focus of our discussion—are the smallest, least costly portion of the system. Often they can generate data at such a prodigious rate that only a complex and expensive network on Earth can process and distribute the information to the many who will ultimately use the information.

Satellite systems can be divided into two broad classes: (a) experimental satellites, and (b) operational satellite systems. The former class consists of one-of-a-kind satellites to test new sensors and a unique temporary network to analyze and distribute data from them. Examples include Geos-3 and Seasat. The latter class consists of a series of similar satellites to measure some phenomena continuously, and a continuing, evolving organization to use and distribute the satellite data. An example is the NOAA series of meteorological satellites operated by the National Environmental Satellite, Data, and Information Service (NOAA/NESDIS) to observe Earth's weather. Within the USSR, the experimental satellites are those in the Cosmos series, while the operational meteorological satellites are included in the Meteor series.

Altogether, thousands of satellites have been launched, and perhaps several hundred have been designed specifically to view Earth. Of these, some have provided data of direct interest to a broad group of oceanographers, while many more have studied the atmosphere, solar radiation, and electric currents in the ionosphere, returning data of interest to a more restricted group. This chapter will describe primarily the major series of Earth-observing satellites developed and launched by NASA, including the NOAA satellites, but a few European and Japanese systems will also be mentioned. Together these provide the bulk of the oceanic observations, and incorporate the most advanced sensors.

[310]

16.1 Names of satellites

Satellites are given names, usually acronyms, that roughly describe their function. For example, ERTS stands for Earth Resources Technology Satellite. Before launch, the name is appended by a letter denoting the satellite's position in a series of similar satellites, ERTS-A. After a successful launch this is changed to a number, ERTS-1. Occasionally, the original name is cumbersome and it is replaced by one more catchy or descriptive; thus ERTS became Landsat. Needless to say, these changing names are confusing, and are sometimes not even applied systematically. The TIROS-N satellite was still known by that name long after launch, rather than as TIROS-11.

16.2 The major U.S. satellite series for viewing the Earth

The first series of experimental, Earth-viewing satellites was TIROS, the Television Infra Red Observational Satellite, with ten successful launches of ever more useful satellites (table 16.1; Widger, 1966). Naturally, this extensive development led to the first operational weather satellites: the TOS (Tiros Operational Satellites), renamed the ESSA series after the Environmental Science Services Administration, but also standing for Environmental Survey Satellite, with nine successful launches (table 16.2). These were nearly identical to the TIROS-9 and provided cloud pictures through March 1976. The operational weather satellites continued, with the Improved Tiros Operational Satellite Series (ITOS) having a new bus and improved sensors (table 16.3; Fortuna and Hambrick, 1974). This series was later renamed the NOAA series after it was taken over by the new National Oceanic and Atmospheric Administration (NOAA).

Concurrently with the civilian satellite program, the Department of Defense operated its own system, details of which were revealed in early 1973. The Defense Meteorological Satellite Program (DMSP) continues to evolve, with some data available to non-military users (table 16.4). The latest satellites in this series, the Block 5D satellites, strongly influenced the development of the latest NOAA satellites. The newest NOAA series of operational satellites, which began operating in 1979, is based on the DSMP Block 5D bus, but with slightly different sensors (table 16.5; Hussey, 1977; Schwalb, 1978). This new series of NOAA weather satellites is particularly useful for oceanography because the satellites carry an improved infrared radiometer to measure sea surface temperature, the Advanced Very High Resolution Radiometer, AVHRR, and a system, ARGOS, to collect data from drifting buoys while determining the buoy's position. In the future, both systems will use an imroved version of this bus, the Advanced Tiros-N (ATN) spacecraft, the first of which is NOAA-E (Schwalb, 1982).

All of these operational satellites, TIROS, ESSA, NOAA, and DMSP, are in low, sun-synchronous, near-polar orbits and provide twice-daily coverage from pole to pole. To provide higher temporal resolution of cloud motion in equatorial and mid-latitude regions, a new series of operational satellites, the Geo-

Table 16.1 TIROS Satellites

TIROS	1	2	3	4	5	6	7	8	9	10
launch	1Apr60	23Nov60	12Jul61	8Feb62	19Jun62	18Sep62	19Jun63	21Dec63	22Jan65	2Jul65
life (d)	77	76	145	120	320	388	1089	1287	1238	730
inclination	48.4°	48.5°	47.9°	48.3°	58.1°	58.3°	58.2°	58.5°	96.3°	98.6°
height (km)	722	675	728	777	781	698	636	727	1643	788
eccentricity	0.003	0.007	0.006	0.009	0.027	0.002	0.002	0.004	0.117	0.007
instruments:										
5-channel radiometer		x	x	x			x			
medium angle radiometer		x	x	x						
omni-directional radiometer			x	x			x			
wide angle visible	x	x	x	x	x	x	x	x		
narrow angle visible	x	x							x	x
medium angle visible				x	x	x				
APT								x		

TIROS = Television Infra Red Observational Satellite
APT = Automatic Picture Transmission

Table 16.2 ESSA/TOS Satellites

ESSA	1	2	3	4	5 (TOS-C)	6 (TOS-D)	7 (TOS-E)	8 (TOS-F)	9
launch	3Feb66	28Feb66	20Oct66	26Jan67	20Apr67	10Nov67	16Aug68	15Dec68	27Feb69
life (d)	861	1692	738	465	1034	763	571	2644	1726
inclination	97.8°	100.9°	100.9°	102.0°	101.8°	102.0°	101.7°	101.8°	101.8°
height (km)	770	1386	1438	1384	1390	1449	1454	1440	1467
eccentricity	0.010	0.004	0.006	0.008	0.004	0.005	0.003	0.003	0.005
instruments	#	APT	AVCS	APT	AVCS	APT	AVCS	APT	AVCS

ESSA = Environmental Science Services Administration — Environmental Survey Satellite
TOS   = Tiros Operational Satellite

\#   = Tiros Wide Angle Vidicon Camera on ESSA-1 only
APT   = Automatic Picture Transmission
AVCS = Advanced Vidicon Camera System

Table 16.3 ITOS/NOAA Satellites

	ITOS-1 (TIROS-M)	NOAA-1 (ITOS-A)	(ITOS-B)	NOAA-2 (ITOS-D)	(ITOS-E)	NOAA-3 (ITOS-F)	NOAA-4 (ITOS-G)	NOAA-5 (ITOS-H)
launch	23Jun70	11Dec70	21Oct71	15Oct72	16Jul73	6Nov73	15Nov74	29Jul76
life (d)	510	252		837		1029	1463	990
inclination	102.0°	101.9°	failed	101.7°	failed	101.7°	101.7°	102.1°
height (km)	1456	1447	on	1450	on	1505	1451	1511
eccentricity	0.003	0.003	launch	0.001	launch	0.001	0.001	0.001
instruments	APT	APT		APT		APT	APT	APT
	AVCS	AVCS		VHRR		VHRR	VHRR	VHRR
	FPR	FPR		VTPR		VTPR	VTPR	VTPR
	SPM	SPM		SPM		SPM	SPM	SPM
	SR	SR		SR		SR	SR	SR

ITOS	=	Improved Tiros Operational Satellite
NOAA	=	U.S. National Oceanic and Atmospheric Administrtion

APT	=	Automatic Picture Transmission
AVCS	=	Advanced Vidicon Camera System
FPR	=	Flat Plate Radiometer
SR	=	Scanning Radiometer
SPM	=	Solar Proton Monitor
VHRR	=	Very High Resolution Radiometer
VTPR	=	Vertical Temperature Profiling Radiometer

Table 16.4 DMSP Satellites: Block 5D Series

DMSP	F1	F2	F3	F4	F5	F6
launch	11Sep76	5Jun77	1May78	6Jun79	14Jul80	20Dec82
inclination	98.7°	99.0°	97.6°	98.7°	failed	99.0°
height (km)	833	833	608	828	on	833
eccentricity	0.0020	0.004	0.0064	0.001	launch	0.001
instruments	OLS	OLS	OLS	OLS		OLS
	SSH	SSH	SSH	SSH		SSJ/4
		SSJ	SSJ	SSM/T		SSI/E
		SSI/E		SSC		SSB/A
		SSI/P		SSD		

DMSP	=	Defense Meteorological Satellite Program

OLS	=	Optical Line Scanner (Visible and Infrared)
SSB/A	=	X-ray Scanner
SSC	=	Snow/Cloud Discriminator
SSD	=	Atmospheric Density Sensor
SSH	=	Infrared Temperature and Humidity Sounder (16 channel, $350-1022$ cm^{-1})
SSI/E	=	Ion and Electron Density Profiler
SSI/P	=	Passive HF Ionospheric Monitor ($1-13$ MHz)
SSJ	=	Electron Spectrometer
SSM/T	=	Microwave Temperature Sounder

[315]

Table 16.5 The NOAA Meteorological Satellites: The TIROS-N Series

	TIROS-11 TIROS-N	NOAA-6 (NOAA-A)	(NOAA-B)	NOAA-7 (NOAA-C)	NOAA-8 (NOAA-E)*
launch	13Oct78	27Jun79	29May80	23Jun81	28Mar83
inclination	98.9°	98.8°	failed	99.0°	98.8°
height (km)	854	820	on	847	815
eccentricity	0.001	0.001	launch	0.001	0.002
instruments	ARGOS	ARGOS		ARGOS	ARGOS
	AVHRR	AVHRR/1		AVHRR/2	AVHRR/1
	SEM	SEM		SEM	SEM
	TED	TED		TED	TED
	MEPD	MEPD		MEPD	MEPD
	HEPAD	HEPAD		HEPAD	
	TOVS	TOVS		TOVS	TOVS
	HIRS/2	HIRS/2		HIRS/2	HIRS/2
	SSU	SSU		SSU	SSU
	MSU	MSU		MSU	MSU
					SAR

ARGOS	=	Data Collection and Platform Location System
AVHRR	=	Advanced Very High Resolution Radiometer
HEPAD	=	High Energy Proton and Alpha-particle Detector (part of SEM)
HIRS	=	High Resolution Infrared Sounder (part of TOVS)
MEPD	=	Medium-Energy Proton/Electron Detector (part of SEM)
MSU	=	Microwave Sounding Unit (part of TOVS)
SAR	=	Search and Rescue
SEM	=	Space Environment Monitor
SSU	=	Stratospheric Sounding Unit
TED	=	Total Energy Detector (part of SEM)
TOVS	=	Tiros Operational Vertical Sounder

*NOAA-D is held in storage.

[316]

Table 16.6 Geosynchronous Meteorological Satellites

	SMS-1 (SMS-A)	SMS-2 (SMS-B)	GOES-1 (SMS-C)	GOES-2	GOES-3 (GOES-C)	GOES-4 (GOES-D)
launch	17May74	6Feb75	16Oct75	16Jun77	16Jun78	9Sep80
inclination	1.9°	1.0°	1.0°	0.9°	0.3°	0.0°
height (km)	35788	35789	35783	35778	35785	35786
eccentricity	0.0002	0.0002	0.0003	0.0120	0.0000	0.0003
instruments	VISSR	VISSR	VISSR	VISSR	VISSR	VAS
	DCS	DCS	DCS	DCS	DCS	DCS
	SEM	SEM	SEM	SEM	SEM	SEM

	GOES-5 (GOES-E)	GOES-6 (GOES-F)	GMS-1	GMS-2	METEOSAT-1	METEOSAT-2
launch	22May81	28Apr83	14Jul77	11Aug81	23Nov77	19Jun81
inclination	0.2°	0.0°	1.0°	0.3°	0.7°	0.1°
height (km)	35786	35786	35786	35788	35787	35788
eccentricity	0.0002	0.0002	0.0003	0.0007	0.009	0.0001
instruments	VAS	VAS	VISSR	VISSR	IR	IR
	DCS	DCS	DCS	DCS	DCS	DCS
	SEM	SEM	SEM	SEM		

GMS	=	Geostationary Meteorological Satellite
GOES	=	Geostationary Operational Environmental Satellite
SMS	=	Synchronous Meteorological Satellite

DCS	=	Data Collection System
IR	=	Imaging Radiometer
SEM	=	Space Environment Monitor
VAS	=	VISSR Atmospheric Sounder
VISSR	=	Visible and Infra Red Spin Scan Radiometer

[317]

stationary Operational Environmental Satellites (GOES) was started in 1975 based on experience with the Synchronous Meteorological Satellites (SMS). These satellites operate in high sun-synchronous orbits and continuously view a 120° sector of Earth centered on the equator, providing coverage between 55°N and 55°S every thirty minutes (table 16.6; NOAA, 1979; Hughes, 1980). The satellites are particularly useful for observing tropical regions seldom observed by ships, and a constellation of four nearly identical satellites was launched for the World Weather Experiment in 1979. Two were operated by the U.S., one by Japan (the GMS), and one by the European Space Agency (the Meteosat).

The sensors on the operational weather satellites have been designed specifically to study the atmosphere. Oceanic observations are a more or less unintentional byproduct, and are limited to observations of patterns in the sun glint and thermal infrared. Specific oceanic sensors have been flown only on experimental spacecraft, with the exception of the AVHRR instrument on the latest NOAA weather satellites. Thus the operational satellites, which provide the bulk of the spaceborne observations, are perhaps less important for many oceanographic studies than are the experimental satellites, which have provided a means to develop and test dozens of special instruments to study the space, atmospheric, and oceanic environments.

The longest series of experimental satellites, the Nimbus series, includes seven satellites launched between 1964 and 1978 carrying many different sensors mounted on essentially the same bus (table 16.7; Horan, 1978). Of particular importance to oceanographers are the microwave radiometers and coastal zone color scanner, instruments designed specifically to study the world's oceans. Of equal importance are many unique satellites. Notable among these are Seasat, the first oceanographic satellite; Geos-3, which carried an altimeter to study the oceanic geoid; Lageos, for study of Earth's gravity; and the Landsats, which provide clear, high resolution pictures of the sea and coastal regions (table 16.8).

The development of sensors to observe the ocean reached a zenith with Seasat, which carried four SHF radio instruments to observe all of the oceans every three days regardless of clouds. The satellite was in a low, near-polar orbit but was not sun-synchronous, so it could observe phenomena at various times of the day. The sensors responded to wind, rain, sea temperature, currents, and waves; and by operating together, their overall accuracy was improved over that of a single instrument. Although the satellite survived for only 100 days between June and October 1978, the large quantities of data it produced are being intensively studied to determine the ability of satellite systems to measure oceanic phenomena.

Details on the currently active satellites are listed in the yearly *Report on Active and Planned Spacecraft and Experiments*, available in libraries and from:

> National Space Sciences Data Center, Code 601.4
> NASA/Goddard Space Flight Center
> Greenbelt, Maryland 20771
> Telephone: (301) 344-6695

By consulting back issues of this publication, it is possible to learn details of satellites active since 1974. I know of no catalog of spacecraft and instruments for earlier periods; but the *RAE Table of Earth Satellites* (King-Hele et al., 1981) at least lists satellites launched from 1957 to 1980, together with their orbital elements.

Table 16.7 Nimbus Satellites

	1 (A)	2 (B)	3 (B2)	4 (D)	5 (E)	6 (F)	7 (G)
launch	24Aug64	16May66	14Apr69	8Apr70	11Dec72	12Jun75	24Oct78
life (d)	24	978	1078	3827	>2920	>1825	>1460
inclination	98.7°e	100.4°	99.9°	99.9°	99.9°	100.0°	99.3
height (km)	683	1139	1105	1100	1102	1100	950
eccentricity	0.032	0.005	0.004	0.001	0.001	0.001	0.001
instruments	APT	APT	HRIR	BUV	ESMR	ERB	CZCS
	AVCS	AVCS	IDCS	FWS	ITPR	ESMR	ERB
	HRIR	HRIR	IRIS	IDCS	NEMS	HIRS	LIMS
			IRLS	IRIS	SCMR	LRIR	SAM II
			MRIR	IRLS	SCR	PMR	SBUV
			MUSE	MUSE	TJOR	SCAMS	SCAMS
			SIRS	SIRS		TJOR	SMMR
				SCR		TWERLE	THIR
				THIR			

APT	=	Automatic Picture Transmission
AVCS	=	Advanced Vidicon Cameras
BUV	=	Backscatter Ultraviolet Spectrometer
CZCS	=	Coastal Zone Color Scanner
ERB	=	Earth Radiation Budget
ESMR	=	Electronically Scanned Microwave Radiometer
FWS	=	Filter Wedge Spectrometer
HIRS	=	High Resolution Infrared Spectrometer
HRIR	=	High Resolution Infrared Radiometer
IDCS	=	Image Disector Camera Subsystem
IRIS	=	Infrared Interferometer Spectrometer
IRLS	=	Interrogation Recording Location System
ITPR	=	Infrared Temperature Profiling Radiometer
LIMS	=	Limb Infrared Monitoring of the Stratosphere
LRIR	=	Limb Radiance and Inversion Radiometer
MRIR	=	Medium Resolution Infrared Radiometer
MUSE	=	Monitor of Solar Ultraviolet Energy
NEMS	=	Nimbus E Microwave Spectrometer
PMR	=	Pressure Modulated Radiometer
SAM II	=	Stratospheric Aerosol Measurement
SAM	=	Stratospheric and Mesospheric Sounder
SBUV	=	Solar Backscatter Ultraviolet and Total Ozone Mapping Spectrometer
SCAMS	=	Scanning Microwave Spectrometer
SCMR	=	Surface Composition Mapping Radiometer
SCR	=	Selective Chopper Radiometer
SIRS	=	Satellite Infrared Spectrometer
SMMR	=	Scanning Multifrequency Microwave Radiometer
THIR	=	Temperature Humidity Infrared Radiometer
TWERLE	=	Tropical Wind Energy Conversion and Reference

Table 16.8 Special Satellites

	Landsat-1 (ERTS-A)	Landsat-2	Landsat-3	Landsat-4	Geos-3	Lageos	HCMM (AEM-A)	Seasat
launch	23Jul72	22Jan75	3May78	16Jul82	9Apr75	4May76	26Apr78	26Jun78
inclination	99.1°	99.1°	99.1°	98.2°	115.0°	109.9°	97.6°	108.0°
height (km)	907	916	906	700	844	5891	601	791
eccentricity	0.001	0.001	0.001	0.001	0.001	0.004	0.006	0.002
instruments	MSS RBV	MSS RBV	MSS/2 RBV/2	MSS TM	ALT/1	Corner Reflectors for laser	HCMR	SMMR SAR SCAT ALT/2 VIRR

AEM = Applications Explorer Mission
ERTS = Earth Resources Technology Satellite
GEOS = Geodynamic Experimental Ocean Satellite (Geodetic Earth Orbiting Satellite)
HCMM = Heat Capacity Mapping Mission

ALT = Altimeter
VHRR = Advanced Very High Resolution Radiometer
HCMR = Heat Capacity Mapping Radiometer
MSS = Multispectral Scanner
RBV = Return Beam Vidicon
SAR = Synthetic Aperture Radar
SCAT = Scatterometer
SMMR = Scanning Multifrequency Microwave Radiometer
TM = Thematic Mapper
VIRR = Visible and Infrared Radiometer

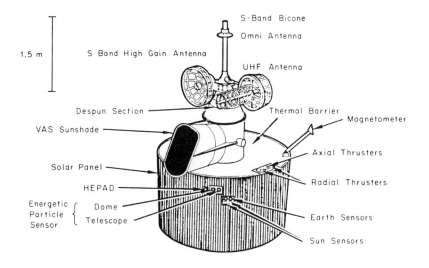

GOES SPACECRAFT

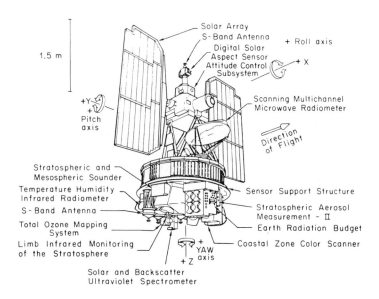

NIMBUS - 7 SPACECRAFT

Figure 16.1 Two examples of recent spacecraft. Top: the Geostationary Environmental Observation Satellite, Goes-4. Bottom: the Nimbus-7 experimental spacecraft.

16.3 The satellite structure

The structure or bus that supports the satellite instruments dominates the sensors and decides to a great extent the types of measurements that can be made. Power ultimately comes from the sun through solar cells; thus the larger the bus, the larger the solar panels, the more power is available, and the more elaborate are the sensors and electronics that can be supported by the bus.

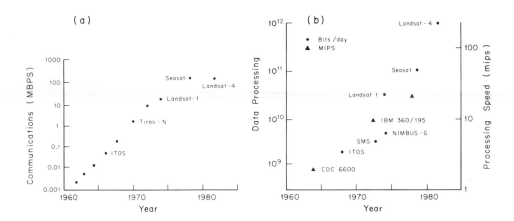

Figure 16.2 (a) Space data transmission requirements in millions of bits per second (MBPS). (b) Space data processing requirements, including processing rates in units of millions of instructions per second (MIPS) (partly after McElroy et al., 1977).

The series of operational satellites, and many of the experimental satellites, are based on just a few different structures: (a) the TIROS series; (b) the TOS/ESSA series; (c) the Nimbus/Landsat series; (d) the ITOS/NOAA series; (e) the TIROS-N/DMSP/NOAA group 5-D bus; (f) the SMS/GOES/GMS-bus; and (g) the Applications Explorer Mission (AEM) bus. Satellites in each series are basically the same, but differ slightly in the ways they transmit data to the ground and in the types and placement of instruments within the structure. However, each series differs very substantially from the others. The most recent satellites tend to be large, expensive structures. For example, the GOES weighs 400 kg, is 4.3 m high, uses 6200 solar cells to generate 350 watts of power, and is designed to last seven years; while the Nimbus-7 spacecraft weighed 907 kg, was 3 m high, and generated 550 watts (fig. 16.1).

Besides supporting instruments, the bus carries radios to transmit data to Earth and to help determine the position of the satellite; sensors to determine the environment within the structure, such as its temperature or the amount the power it produces or its orientation; and small rockets to adjust the satellite's orbit.

Data from the satellites are usually split into two parallel streams. One stream is recorded for one to two orbits and then transmitted at high rates to special data collection sites. Major ones include Fairbanks, Alaska; Goldstone, California; Wallops Island, Virginia; and Kiruna, Sweden. The second stream of data is continuously transmitted for reception at simple stations located anywhere on Earth in sight of the satellite. In the future, satellites may also broadcast to geostationary satellites for retransmission to the Earth, especially as the quantities of data from satellites rise each year, now being as large as 10^8–10^9 bits of information per second for some satellites (fig. 16.2a).

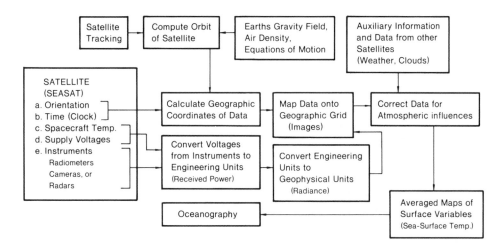

Figure 16.3 Some of the steps necessary to convert satellite data into oceanographic information. A typical example of the information produced at some of these steps is given in parentheses (from Stewart, 1982).

16.4 Data processing and dissemination systems

Processing the prodigious quantity of satellite data is beyond the scope of all but the largest facilities, and storage is impossible except for a small subset of all data collected. Even for this subset, photographic reproduction of satellite images is the usual storage medium, a form that greatly degrades the original information. Some data are now available in digital form, however, especially data from experimental satellites and from Landsats.

Processing of large quantities of data, usually as images generated by computers, is the subject of many textbooks and will not be discussed here. It should be sufficient to state that the problem is complex and solutions are costly. For those who are interested in small or local regions, the data continuously broadcast from some satellites can be received and processed with simpler, less costly equipment (Evans, Kent, and Seidman, 1980); but even this solution is expensive and beyond the scope of smaller institutions.

To understand the complexity of the problem, consider a list of the computations that must be performed (fig. 16.3):

(a) The data must be put into a form suitable for further computation.

(b) Orbit information must be merged with sensor information to determine the coordinates of the area viewed by the satellite.

(c) Data from the sensors must be corrected, converted to scientific units, and mapped onto some coordinate system.

(d) Auxiliary information from other sensors must also be processed and mapped in the same coordinates.

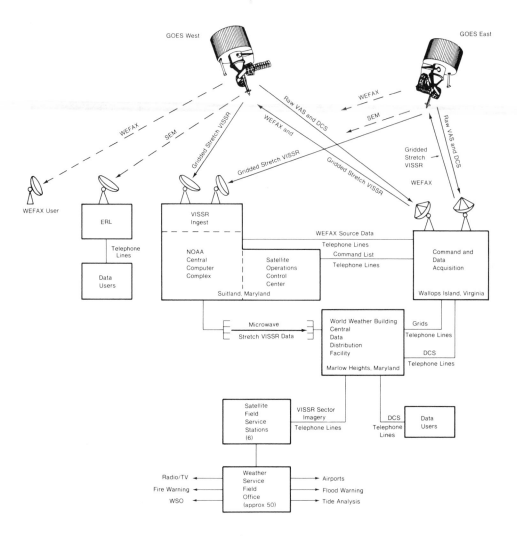

Figure 16.4   A practical implementation of a system to process and distribute satellite data, the system for the Goes meteorological satellites (from Hughes, 1980).

(e) The images must be printed or stored in a form suitable for study by the user.

Remember that these steps must be repeated millions of times; that even simple routines, such as producing a distortionless map, involve many computations; and that instruments can generate $10^6 - 10^8$ bits of information per second. Thus the practical implementation of a system to process and distribute satellite data requires large, fast computers (fig. 16.2b), high-speed data transmissions, and large storage facilities.

To the calculations necessary to process data must be added those necessary to determine the orbit of the satellite, including the reduction of data from tracking networks and the integration of the equations of satellite motion. For some satellites, those with sensors having coarse spatial resolution, the orbit need not be known to an accuracy of better than a few kilometers in order to determine where on Earth the satellite was looking. Such calculations are performed routinely by organizations that keep track of all objects in space. For other satellites, such as the geodetic satellites or Seasat, the orbit must be known to within a fraction of a meter, and many perturbing influences must be included in the computations, making them very complex. This is a job best done by special groups, such as those at the Smithsonian Astrophysical Observatory, the Naval Surface Weapons Laboratory, NASA's Goddard Space Flight Center, or the University of Texas.

The methods of processing data from operational satellites is fairly well described in NOAA publications, particularly those by Bristor (1975) and Fortuna and Hambrick (1974). The practical implementations of systems to handle meteorological satellite data, and to distribute the information to many weather forecasters before the data are too old to be useful, tend to be large, elaborate, and expensive (fig. 16.4). They are efficient for the task, but are not designed to handle requests for specially processed data required by many research problems.

Networks for processing and distributing experimental satellite data are ephemeral, are designed to produce specially processed data, but tend not to be well documented in generally available publications. It is often necessary to contact directly the project managing the satellite or, if the project no longer exists, to refer to contractor's reports and catalogs of data published by various archives. Of course, the scientific results of the analyses are often published in the literature, but those who wish to go back and use old data often face formidable difficulties.

To help reduce the problems of finding and using satellite data, NASA has experimented with several pilot data systems to collect, store, and distribute data from a variety of sources. The one for oceanographic data is the

> Pilot Ocean Data System
> NASA/Jet Propulsion Laboratory
> 4800 Oak Grove Dr.
> Pasadena, California 91109

It contains, in machine readable form, the Seasat and Geos-3 data, the extensive sets of surface observations used to calibrate the instruments on these satellites, and a bibliography of documents, reports, and papers that describe the satellites, their instruments, the calibration of the instruments, and the application of the satellite data to oceanography. The system is particularly useful, not only because it contains data that have been processed to produce maps of oceanic variables in geographic coordinates and in units used by oceanographers but also because the data have formed the basis for a number of oceanographic studies. This last step is important, because such use of the data detects and removes many errors both large and small found in most satellite observations. Thus the system is, in a sense, an information archive, in contrast with conventional archives that either hold data in less well processed forms or have data of less well known accuracy and usefulness.

A similar archive for satellite data has been started by the European Space Research Institute (ESRIN) of the European Space Agency, located at:

> ESRIN/Information Retrieval Service
> Online Services Division
> Via Galileo Galilei
> 00044 Frascati (Rome), Italy

At present the archive contains extensive listings of scientific and engineering publications, but in the future it will contain a catalog of satellite data, and eventually sets of satellite data in machine readable form.

16.5 Sources of data

Selected sets of satellite data are stored in various national archives. The major sources for routine observations are listed below. Much specialized information, particularly that from experimental satellites, and even some observations from operational satellites, are available only from the project managing the satellite or from the scientists supported by these projects. (For more information about sources of oceanographic data, see Cornillon, 1982.)

Data from the NOAA operational satellites, from experimental meteorological satellites, and from Seasat are described in Dismachek (1977), Dismachek, Booth, and Leese (1980), and NOAA (1979), and are available from the National Environmental Satellite, Data, and Information Service (NESDIS) at:

> NOAA/NESDIS
> Satellite Data Services Division
> Room 100, World Weather Building
> Washington, D.C. 20233
> Telephone: (301) 763-8111

Summaries of the NOAA Meteorological satellite data are printed in *Environmental Satellite Imagery* and in the *Oceanographic Monthly Summary*, both published monthly by the NESDIS. More general information is contained in the *Satellite Data Users Bulletin*, published at irregular intervals by the same group.

Data from the space shuttle, including data from the Shuttle Imaging Radar (SIR-A), the Ocean Color Experiment (OCE), and the Shuttle Multispectral Infrared Radiometer (SMIRR), as well as data from the Heat Capacity Mapping Mission, are available from:

> National Space Science Data Center
> World Data Center A for Rockets and Satellites
> NASA/Goddard Space Flight Center Code 601
> Greenbelt, Maryland 20771

Photographs from the manned space flights (Gemini, Apollo, and Skylab) and the Landsat data sets are available from:

NOAA Landsat Customer Services
EROS Data Center
Sioux Falls, South Dakota 57198
Telephone: (605) 594-6515

Much valuable information about the Landsats and other land-observing systems is included in the monthly *Landsat Data Users Notes*, published by the EROS data center and available from them.

Data from the Defense Meteorological Satellite Program is stored at NOAA/NESDIS, and at the Cooperative Institute for Research in Environmental Sciences (CIRES) through:

World Data Center A for Glaciology
CIRES
Campus Box 449
University of Colorado
Boulder, Colorado 80309

Within Europe, data from the NASA Landsat, Nimbus-7, HCMM, and Seasat satellites are distributed through the Earthnet Program Office of the European Space Research Institute (ESRIN) at:

ESRIN/Earthnet Program Office
Via Galileo Galilei
00044 Frascatti (Rome), Italy
Telephone: (06)9422401

Meteosat data are distributed by the European Space Operations Center (ESOC) at:

Meteosat Data Services
Meteosat Data Management Department
ESOC
Robert-Bosch-Strasse 5
D-6100 Darmstadt, West Germany

16.6 Data collection systems

In addition to observing the oceans from space, many satellites now carry equipment to receive data from buoys on the ocean and, in some instances, to determine the buoy's position. While a description of drifting buoys, their instruments, and their role in oceanography is beyond the scope of this book, the space segment of the data collection network does deserve mention.

Two systems are widely used: the Data Collection System on the GOES satellites, and the Argos system on the NOAA series of meteorological satellites.

The Data Collection System (DCS) routinely collects data many times a day from many surface platforms within the large field of view of the GOES satellites. The system is designed to handle 10,000 transmissions per hour, each of thirty seconds' duration. Three modes of operation are used: (a) the data from the surface may be transmitted at regularly timed intervals; (b) the transmissions may be in response to a command from the satellite; or (c) the transmissions may be emergency messages in response to unusual surface conditions, such as high winds or floods on rivers. To handle the large number of messages, the system uses both time-division and frequency division multiplexing, and a total of 231 radio channels are available for the various modes of operation and users of the system (NOAA, 1979; Nelson, 1980).

While the GOES system is used primarily to collect data from meteorological platforms in remote areas of land, it also collects data from ships and large fixed buoys. Moreover, the system broadcasts a very accurate time-of-day code, and the signal from the satellite can be used to synchronize clocks on the surface with an accuracy of $\pm 100 \, \mu$s and a precision of $\pm 20 \, \mu$s (Easton et al., 1976).

The French Argos system carried on the NOAA series of meteorological satellites collects small amounts of data from inexpensive platforms, and can determine the positions of the platforms several times a day. To discriminate among various platforms within its field of view, the system relies on time-division multiplexing. Each surface platform may transmit only for a duration of 360–920 ms, at irregular intervals of 40–200 s. Thus the probability of two platforms transmitting simultaneously is very small. But even during that unlikely event, the system can track four simultaneous transmissions because the platforms will have slightly different Doppler shifts (Bessis, 1981).

The position of each platform is calculated from the Doppler shift of the received signal, using several transmissions spanning a period of 500 s or longer. The accuracy of the position depends on the stability and accuracy of the radio frequency transmitted by the platform. If the frequency stability is better than 10^{-9} over a period of 1200 s, then the positions can be calculated with a typical accuracy of ± 1 km. For this calculation, stability is defined to be $\Delta f / f$ where f is the frequency of the transmission and Δf the variation in frequency over a 1200 s interval. The positioning technique is also used in reverse, to track the position of the satellite by using radio transmissions from precisely known stations having accurate clocks synchronized with the satellite's clock.

Surface platforms reporting to the Argos system need carry only cheap, low-power transmitters that can be powered by batteries, so the system is widely used to track buoys drifting at the sea surface. Suitably instrumented and drogued buoys can relay back air and sea temperatures and atmospheric pressure, and successive buoy positions give surface currents. The system was particularly helpful in obtaining weather and currents in the Southern Ocean during the Global Weather Experiment (fig. 1.3), and is widely used to study surface currents.

Further information about Argos can be obtained from:

> Service Argos
> Centre National d'Etudes Spatiales
> 18, av. Edouard Belin
> 31055 Toulouse Cedex, France

BIBLIOGRAPHY

Allison, L. J., Rodgers, E. D., Wilheit, T. T., and Fett, R. W. 1974. Tropical cyclone rainfall as measured by the Nimbus-5 electrically scanning microwave radiometer. *Bull. Am. Meteorol. Soc.* **55** (9): 1074–1089.

Alpers, W. R., and Hasselmann, K. 1978. The two-frequency microwave technique for measuring ocean-wave spectra from an airplane or satellite. *Boundary-Layer Meteorology* **13**: 215–230.

Alpers, W. R., Ross, D. B., and Rufenach, C. L. 1981. On the detectability of ocean surface waves by real and synthetic aperture radar. *J. Geophys. Research* **86** (C7): 6481–6498.

Alpers, W. R., and Rufenach, C. L. 1979. The effects of orbital motions on synthetic aperture imagery of ocean waves. *IEEE Trans. Antennas and Propag.* **AP-27**: 685–690.

Alpers, W. R., Schroter, J., Schlude, F., Muller, H. J., and Koltermann, K. P. 1981. Ocean surface current measurements by an L band two-frequency microwave scatterometer. *Radio Science* **16** (1): 93–100.

Altshuler, E. E., and Telford, L. E. 1980. Frequency dependence of slant path rain attenuation at 15 and 35 GHz. *Radio Science* **15** (4): 781–796.

Anderle, R. J. 1980. Accuracy of mean earth ellipsoid based on Doppler, laser, and altimetric observations. *Bull. Géodésique* **54** (4): 521–527.

Anderle, R. J., and Oesterwinter, C. 1980. Determination of high frequency variations in Earth's rotation from Doppler satellite observations. *Bull. Géodésique* **54** (4): 544–522.

Apel, J. R., Byrne, M. H., Proni, J. R., and Channell, R. L. 1975. Observations of oceanic internal and surface waves from the Earth Resources Technology Satellite. *J. Geophys. Research* **80** (6): 865–881.

Aumann, H. H., and Chahine, M. T. 1976. Infrared multidetector spectrometer for remote sensing of temperature profiles in the presence of clouds. *Appl. Optics* **15** (9): 2091–2094.

Austin, R. W. 1979. Coastal zone color scanner radiometry. *Proc. Soc. Photo-optical Instrumentation Engineers* **208**: 170–177.

Baird, K. M. 1983. Frequency measurements of optical radiation. *Physics Today* **36** (1): 52–57.

Barber, N. F., and Tucker, M. J. 1962. Wind waves. Chap. 19 in: *The sea*, Vol. 1, edited by M. N. Hill, 664–699. New York: John Wiley & Sons.

Barnett, E. C. 1974. *Climatology from satellites*. London: Methuen, 418 pp.

Barnett, T. P., Patzert, S. C., Webb, W. C., and Bean, B. R. 1979. Climatological use-fulness of satellite determined sea-surface temperature in the tropical Pacific. *Bull. Am. Meteorol. Soc.* **60** (3): 197–205.

Barnum, J. R., Maresca, J. W., and Serebreny, S. M. 1977. High-resolution mapping of oceanic wind fields with skywave radar. *IEEE Trans. Antennas and Propag.* **AP-25**: 128–132.

Barrick, D. E. 1968. Rough surface scattering based on the specular point theory. *IEEE Trans. Antennas and Propag.* **AP-16**: 449–454.

––––––. 1972. First-order theory and analysis of MF/HF/UHF scatter from the sea. *IEEE Trans. Antennas and Propag.* **AP-20**: 2–10.

––––––. 1973. The use of skywave radar for remote sensing of sea states. *Marine Technical Society Journal* 7 (1): 29–33.

Barrick, D. E., Evans, M. W., and Weber, B. L. 1977. Ocean surface currents mapped by radar. *Science* **198** (4313): 138–144.

Barrick, D. E., Headrick, J. M., Bogle, R. W., and Crombie, D. D. 1974. Sea back-scatter at HF: Interpretation and utilization of the echo. *Proc. IEEE* **62** (6): 673–680.

Barrick, D. E., and Snider, J. B. 1977. The statistics of HF sea-echo Doppler spectra. *IEEE Trans. Antennas and Propag.* **AP-25**: 19–28.

Basharinov, A. E., and Shutko, A. M. 1980. Research into the measurement of sea state, sea temperature and salinity by means of microwave radiometry. *Boundary-Layer Meteorology* **18**: 55–64.

Bass, F. G., and Fuks, I. M. 1979. *Wave scattering from statistically rough surfaces*, translated by C. B. and J. F. Vesecky. Oxford: Pergamon Press, 525 pp.

Battan, L. J. 1973. *Radar Observations of the Atmosphere.* Revised Edition. Chicago: The University of Chicago Press, 324 pp.

Beal, R. C. 1980. Spaceborne imaging radar: Monitoring of ocean waves. *Science* **208** (4450): 1373–1375.

––––––. 1981. Spatial evolution of ocean wave spectra. In: *Spaceborne synthetic aperture radar for oceanography*, edited by R. C. Beal, P. S. De Leonibus, and I. Katz, 110–127. Baltimore: The Johns Hopkins University Press.

Beal, R. C., De Leonibus, P. S., and Katz, I., eds. 1981. *Spaceborne synthetic aperture radar for oceanography.* Baltimore: The Johns Hopkins University Press, 215 pp.

Bean, B. R., and Dutton, E. J. 1966. *Radio meteorology.* Washington: U.S. Government Printing Office, 435 pp.

Beckmann, P., and Spizzichino, A. 1963. *The scattering of electromagnetic waves from rough surfaces.* New York: Macmillan.

Bernstein, R. L. 1982. Sea surface temperature estimation using the NOAA 6 satellite advanced very high resolution radiometer. *J. Geophys. Research* **87** (C12): 9455–9465.

Bernstein, R. L., Born, G. H., and Whritner, R. H. 1982. Seasat altimeter determination of ocean current variability. *J. Geophys. Research* **87** (C5): 3261–3268.

Bernstein, R. L., Breaker, L., and Whritner, R. 1977. California current eddy formation: ship, air, and satellite results. *Science* **195**: 353–359.

Bernstein, R. L., and White, W. B. 1977. Zonal variability of eddy energy in the mid-latitude north Pacific Ocean. *J. Phys. Oceanog.* **7** (1): 123–126.

Bessis, J. L. 1981. Operational data collection and platform location by satellite. *Remote Sensing of Environment* **11** (2): 93–111.

Blume, H. C., Kendall, B. M., and Fedors, J. C. 1978. Measurement of ocean temperature and salinity via microwave radiometry. *Boundary-Layer Meteorology* **13**: 295–308.

Born, G. H., Wilkerson, J., and Lame, D. 1979. *Seasat Gulf of Alaska workshop report:*

Vol 1, Panel Reports. Pasadena: Jet Propulsion Laboratory, Internal Document 622–101.

Born, G. H., Richards, M. A., and Rosborough, G. W. 1982. An empirical determination of the effects of sea state bias on Seasat altimetry. *J. Geophys. Research* **87** (C5): 3221–3226.

Born, M. and Wolf, E. 1970. *Principles of Optics.* Fourth Edition. Oxford: Pergamon Press, 808 pp.

Bowers, R. J., and Frey, J. 1972. Technology assessment and microwave diodes. *Scientific American* **226** (2): 13–21.

Bracalente, E. M., Boggs, D. H., Grantham, W. L., and Sweet, J. L. 1980. The SASS scattering coefficient $\sigma°$ algorithm. *IEEE J. Oceanic Engineering* **OE-5** (2): 144–154.

Bracewell, R. 1965. *The Fourier transform and its applications.* New York: McGraw-Hill, 381 pp.

Bristor, C. L., ed. 1975. *Central processing and analysis of geostationary satellite data.* Washington: NOAA Tech. Memo. NESS 64, 155 pp.

Brower, R. L., Gohrband, H. S., Pichel, W. G., Signore, T. L., and Walton, C. C. 1976. *Satellite derived sea-surface temperatures from NOAA spacecraft.* Washington: NOAA Tech. Memo. NESS 78, 74 pp.

Brown, G. S. 1978. Backscattering from a Gaussian-distributed perfectly conducting rough surface. *IEEE Trans. Antennas and Propag.* **AP-26**: 472–482.

––––––. 1979. Estimation of surface wind speeds using satellite-borne radar measurements at normal incidence. *J. Geophys. Research* **84** (B8): 3974–3978.

Brown, G. S., Stanley, H. R., and Roy, N. A. 1981. The wind-speed measurement capability of spaceborne radar altimeters. *IEEE Trans. Oceanic Engineering* **OE-6** (2): 59–63.

Brown, R. A., and Liu, W. T. 1982. An operational large-scale marine planetary boundary layer model. *J. Appl. Meteorol.* **21** (3): 261–269.

Brown, R. D., and Hutchinson, J. K. 1981. Ocean tide determination from satellite altimetry. In: *Oceanography from space,* edited by J. F. R. Gower, 897–906. New York: Plenum Press.

Brown, W. L., Elachi, C., and Thompson, T. W. 1976. Radar imaging of ocean surface patterns. *J. Geophys. Research* **81** (15): 2657–2667.

Brown, W. M., and Porcello, L. J. 1969. An introduction to synthetic-aperture radar. *IEEE Spectrum* **6**: 52–62.

Bulban, E. J. 1979. Airborne sensor detects fish at night. *Aviation Week and Space Technology* **110** (9): 60–63.

Bullrich, K. 1964. Scattered radiation in the atmosphere and the natural aerosol. *Advances in Geophysics,* edited by H. E. Landsberg and J. Mieghem, 99–260. New York: Academic Press.

Busch, N. E. 1973. The surface boundary layer. *Boundary-Layer Meteorology* **4**: 213–240.

Businger, J. A., Wyngaard, J. C., Izumi, Y., and Bradley, E. F. 1971. Flux-profile relationships in the atmospheric surface layer. *J. Atmos. Sci.* **28** (2): 181–189.

Butman, S., and Lipes, R. G. 1975. *The effect of noise and diversity on synthetic array radar imagery.* Pasadena: Jet Propulsion Laboratory, Deep Space Network Progress Report 42-29, 46–53.

Campbell, W. J., Wayenberg, J., Ramseyer, J. B., Ramseier, R. O., Vant, M. R., Weaver, R., Redmond, A., Arsenault, L., Gloersen, P., Zwally, H. J., Wilheit, T. T., Chang, A. T. C., Hall, D., Gray, L., Meeks, D. C., Bryan, M. L., Barath, F. T., Elachi, C., Leberl, F., and Farr, T. 1978. Microwave remote sending of sea ice in the

AIDJEX main experiment. *Boundary-Layer Meteorology* **13**: 309–336.

Carlson, H., Richter, K., and Walden, H. 1967. Messungen der statistichen verteilung der auslenkung der meeresoberfläsche im seegang. *Deutsche Hydrographische Zeitschrift* **20** (2): 59–64.

Cartwright, D. E., and Alcock, G. A. 1981. On the precision of sea surface elevations and slopes from SEASAT altimetry of the northeast Atlantic Ocean. In: *Oceanography from space*, edited by J. F. R. Gower, 885–895. New York: Plenum Press.

Cavalieri, D. C., Gloersen, P., and Campbell, W. J. 1981. Observations of sea ice properties with the Nimbus-7 SMMR. In: *Proceedings of the 1981 International Geoscience and Remote Sensing Symposium*. New York: IEEE Press.

Cazenave, A., and Daillet, S. 1981. Lunar tidal acceleration from earth satellite orbit analyses. *J. Geophys. Research* **86** (B3): 1659–1663.

Cazenave, A., Lago, B., and Dominh, K. 1982. Geoid anomalies over the north-east Pacific fracture zones from satellite altimeter data. *Geophys. J. Roy. astr. Soc.* **69**, 15–31.

Chahine, M.T. 1974. Remote sensing of cloudy atmospheres. 1. The single cloud layer. *J. Atmos. Sci.* **31** (1): 233–243.

_____. 1981. Remote sensing of sea surface temperature in the 3.7μm CO_2 band. In: *Oceanography from space*, edited by J. F. R. Gower, 87–95. New York: Plenum Press.

Chahine, M.T. Aumann, H. H., and Taylor, F. W. 1977. Remote sounding of cloudy atmospheres. III. Experimental verification. *J. Atmos. Sci.* **34** (5): 758–765.

Chandrasekhar, S. 1960. *Radiative transfer*. New York: Dover, 393 pp.

Chang, A. T. C., Gloersen, P., Schmugge, T., Wilheit, T. T., and Zwally, H. J. 1976. Microwave emission from snow and glacier ice. *J. Glaciology* **16** (74): 23–39.

Chang, A. T. C., and Wilheit, T. T. 1979. Remote sensing of atmospheric water vapor, liquid water, and wind speed at the ocean surface by passive microwave techniques from the Nimbus 5 satellite. *Radio Science* **14** (5): 793–802.

Chapman, M. E. 1979. Techniques for interpretation of geoid anomalies. *J. Geophys. Research* **84** (B8): 3793–3801.

Chapman, M. E., and Talwani, M. 1979. Comparisons of gravimetric geoids with Geos-3 altimetric geoids. *J. Geophys. Research* **84** (B8): 3803–3816.

Chapman, M. E., Talwani, M., Kahle, H., and Bodine, J. H. 1979. *Shape of the ocean surface and implications for the Earth's interior*. NASA Contractor Report 156859.

Charnock, H. 1955. Wind stress on a water surface. *Quart. J. Roy. Meteorol. Soc.* **81** (350): 639–640.

Chelton, D. B., Hussey, K. J., and Parke, M. E. 1981. Global satellite measurements of water vapor, wind speed and wave height. *Nature* **294** (5841): 529–532.

Cheney, R. E., and Marsh, J. G. 1981a. Seasat altimeter observations of dynamic topography in the Gulf Stream region. *J. Geophys. Research* **86** (C1): 473–483.

_____. 1981b. Global mesoscale variability from Seasat collinear altimeter data. *EOS* **62** (17): 298 (abstract).

_____. 1982. Large-scale circulation from satellite altimetry. Unpublished manuscript.

Cheney, R. E., Marsh, J. G., and Beckley, B. D. 1983. Global mesoscale variability from collinear tracks of Seasat altimeter data. *J. Geophys. Research* **88** (C7): 4343–4354.

Chevrel, M., Courtois, M., and Weil, G. 1981. The Spot satellite remote sensing mission. *Photogrammetric Engineering and Remote Sensing* **47** (8): 1163–1171.

Chou, S. H., and Atlas, D. 1982. Satellite estimates of ocean-air heat fluxes during cold air outbreaks. *Monthly Weather Review* **110** (10): 1434–1450.

Choudhury, B. J., and Chang, A. T. C. 1979. The solar reflectance of a snow field. *Cold Regions Science and Technology* **1**: 121–128.

Claassen, J. P., Fung, A. K., Moore, R. K., and Pierson, W. J. 1972. Radar sea return and the RADSCAT satellite anemometer. In: *IEEE Int. Conf. Engineering in the Ocean Environment*, New York: IEEE Press.

Clarke, G. L., Ewing, G. C., and Lorenzen, C. J. 1970. Spectra of backscattered light from the sea obtained from aircraft as a measure of chlorophyll concentration. *Science* **167** (3921): 1119–1121.

Coakley, J. A., and Bretherton, F. P. 1982. Cloud cover from high resolution scanner data: Detecting and allowing for partially filled fields of view. *J. Geophys. Research* **87** (C7): 4917–4932.

Cornillon, P. 1982. *A guide to environmental satellite data.* Narragansett: University of Rhode Island, Marine Tech. Rept. 79, 469 pp.

Cox, C. S., and Munk, W. H. 1954. Measurements of the roughness of the sea surface from photographs of the sun's glitter. *J. Opt. Soc. Am.* **44** (11): 838–850.

Crombie, D. D. 1955. Doppler spectrum of sea echo at 13.56 Mc/s. *Nature* **175** (4449): 681–682.

Curran, R. J. 1972. Ocean color determination through a scattering atmosphere. *Appl. Optics* **11** (8): 1857–1866.

Cutrona, L. J., Leith, E. N., Porcello, L. J., and Vivian, W. E. 1966. On the application of coherent optical processing techniques to synthetic-aperture radar. *Proc. IEEE* **54** (8): 1026–1032.

Cutting, E., Born, G. H., and Frautnick, J. C. 1978. Orbit analysis for Seasat-A. *J. Astronaut. Sci.* **26** (4): 315–342.

Daley, J. C. 1973. Wind dependence of radar sea return. *J. Geophys. Research* **78** (33): 7823–7833.

———. 1974. Reply. *J. Geophys. Research* **79** (18): 2756.

Davies, J. T., and Vose, R. W. 1965. On the damping of capillary waves by surface films. *Proc. Roy. Soc. London* **286** (1405): 218–234.

Davies, K. 1966. *Ionospheric radio propagation.* New York: Dover Publications, 470 pp.

———. 1980. Recent progress in satellite radio beacon studies with particular emphasis on the ATS-6 radio beacon experiment. *Space Science Reviews* **25** (3): 356–430.

Davies, K., Hartmann, G. K., and Leitinger, R. 1977. A comparison of several methods of estimating the columnar electron content of the plasmasphere. *J. Atmos. and Terrest. Phys.* **39** (5): 571–580.

Davis, R. E. 1977. Techniques for statistical analysis and prediction of geophysical fluid systems. *Geophys. Astrophys. Fluid Dynamics* **8**: 245–277.

Deirmendjian, D. 1964. Scattering and polarization properties of water clouds and hazes in the visible and infrared. *Appl. Optics* **3** (2): 187.

de Loor, G. P., Hoogeboom, P., and Spanhoff, R. 1981. A discrepancy between ground-based and airborne radar backscatter measurements. In: *Oceanography from space*, edited by J. F. R. Gower, 651–655. New York: Plenum Press.

Derr, V. E., ed. 1972. *Remote sensing of the troposphere.* Washington: U.S. Government Printing Office.

DeRyke, R. J. 1973. Sea ice motions off Antarctica in the vicinity of the eastern Ross Sea as observed by satellite. *J. Geophys. Research* **78** (36): 8873–8879.

Deschamps, P. Y., and Phulpin, T. 1980. Atmospheric correction of infrared measurements of sea surface temperature using channels at 3.7, 11, and 12 μm. *Boundary-Layer Meteorology* **18**: 131–144.

Develet, J. A. 1964. Performance of a synthetic-aperture mapping radar system. *IEEE Trans. Aerospace and Navigational Electronics* **ANE-11**: 173–179.

Dicke, R. H. 1946. The measurement of thermal radiation at microwave frequencies. *Review of Scientific Instruments* **17** (7): 268–275.

Dismachek, D. C. 1977. *National Environmental Satellite Service catalog of products.* Washington: NOAA Tech. Memo. NESS 88: 102 pp.

Dismachek, D. C., Booth, A. L., and Leese, J. A. 1980. *National Environmental Satellite Service catalog of products.* Third Edition. Washington: NOAA Tech. Memo. NESS 109, 120 pp.

Dixon, T. H., Naraghi, M., McNutt, M. K., and Smith, S. M. 1983. Bathymetric prediction from Seasat altimeter data. *J. Geophys. Research* **88** (C3): 1563–1571.

Dixon, T. H., and Parke, M. E. 1983. Bathymetry estimates in the southern ocean from Seasat altimetry. *Nature* **304** (5925): 406–411.

Douglas, B. C., and Agreen, R. W. 1983. The sea state correction for Geos 3 and Seasat satellite altimeter data. *J. Geophys. Research* **88** (C3): 1655–1661.

Douglas, B. C., Marsh, J. G., and Mullins, N. E. 1973. Mean elements of Geos 1 and Geos 2. *Celestial Mechanics* **7**: 195–204.

Dowd, D. L., and Tapley, B. D. 1979. Density models for the upper atmosphere. *Celestial Mechanics* **13**: 271–295.

Downing, H. D., and Williams, D. 1975. Optical constants of water in the infrared. *J. Geophys. Research* **80** (12): 1656–1661.

Dozier, J., Schneider, S. R., and McGinnis, D. F. 1981. Effect of grain size and snowpack water equivalence on visible and near-infrared satellite observations of snow. *Water Resources Research* **17** (4): 1213–1221.

Driscoll, W. G., and Vaughan, W. 1978. *Handbook of optics.* New York: McGraw-Hill.

Duing, W. 1978. Spatial and temporal variability of major ocean currents and mesoscale eddies. *Boundary-Layer Meteorology* **13**: 7–22.

Duntley, S. Q., Roswell, W., Wilson, W. H., Edgerton, C. F., and Moran, S. E. 1974. *1974 ocean color analysis.* La Jolla: Scripps Institution of Oceanography, Technical Report 74-10.

Easton, R. L., Fisher, L. C., Hanson, D. W., Hellwig, H. W., and Rueger, L. J. 1976. Dissemination of time and frequency by satellite. *Proc. IEEE* **64** (10): 1482–1493.

Eckart, C. 1960. *Hydrodynamics of oceans and atmospheres.* Oxford: Pergamon Press, 290 pp.

Ekman, V. W. 1905. On the influence of the Earth's rotation on ocean currents. *Arkiv for Matematik, Astronomi, och Fysik* **2** (11): 1–53.

Elachi, C. 1978. Radar imaging of the ocean surface. *Boundary-Layer Meteorology* **13**: 165–179.

Elachi, C., and Brown, W. E. 1977. Models of radar imaging of ocean surface waves. *IEEE Trans. Antennas and Propag.* **AP-25**: 84–95.

El'yasberg, P. E. 1967. *Introduction to the theory of flight of artificial Earth satellites,* translated by Z. Lerman. Jerusalem: Israel Program for Scientific Translations, 345 pp.

Escobol, P. R. 1976. *Methods of orbit determination.* New York: John Wiley & Sons.

Esposito, P. B. 1979. Present status and future prospects for the evaluation of the geocentric gravitational constant. *EOS* **60** (18): 234 (abstract).

Etkins, R., and Epstein, E. S. 1982. The rise in global mean sea level as an indicator of climate change. *Science* **215** (4530): 287–289.

Evans, R. H., Kent, S. S., and Seidman, J. B. 1980. *Satellite remote sensing facility for oceanographic applications.* Pasadena: Jet Propulsion Laboratory, Publication 80-40.

Falcone, V. J., Abreu, L. W., and Shettle, E. P. 1979. *Atmospheric attenuation of millimeter and submillimeter waves: Models and computer code.* Cambridge: Air Force Geophysics Laboratory Report 79-0253, 76 pp.

Farmer, C. B. 1974. Infrared measurements of stratospheric composition. *Canadian J. Chemistry* **52** (8): 1544–1559.

Fedor, L. S., and Brown, G. S. 1982. Waveheight and wind speed measurements from the Seasat altimeter. *J. Geophys. Research* **87** (C5): 3254–3260.

Felsentreger, T. L., Marsh, J. G., and Williamson, R. G. 1979. M_2 ocean tide parameters and the deceleration of the moon's mean longitude from satellite orbit data. *J. Geophys. Research* **84** (B9): 4675–4679.

Flatté, S. M., Dashen, R. D., Munk, W. H., Watson, K. M., and Zachariasen, F. 1979. *Sound transmission through a fluctuating ocean.* Cambridge: Cambridge University Press, 299 pp.

Fomin, L. F. 1964. *The dynamic method in oceanography.* New York: Elsevier, 212 pp.

Foote, G. B., and du Toit, P. S. 1969. Terminal velocity of raindrops aloft. *J. Appl. Meteorol.* **8** (2): 249–253.

Fortuna, J., and Hambrick, L. N. 1974. *The operation of the NOAA polar satellite system.* Washington: NOAA Tech. Memo. NESS 60, 127 pp.

Frank, L. A., Craven, J. D., Ackerson, K. L., English, M. R., Eather, R. H., and Carovillano, R. C. 1981. Global auroral imaging instrumentation for the Dynamics Explorer mission. *Space Science Instrumentation* **5** (4): 369–393.

Fraser, R. S. 1964. *Theoretical investigation: The scattering of light by a planetary atmosphere.* Redondo Beach: TRW, Inc.

Frautnick, J. C., and Cutting, E. 1983. *Flight path design issues for the Topex mission.* American Institute of Aeronautics and Astronautics 21st Aerospace Sciences Meeting, Paper AIAA-83-0197, 9 pp.

Friedman, D. 1969. Infrared characteristics of ocean water $(1.5 - 15 \mu)$. *Appl. Optics* **8** (10): 2073–2078.

Fu, L. L. 1983. On the wavenumber of oceanic mesoscale variability observed by the Seasat altimeter. *J. Geophys. Research* **88** (C7): 4331–4341.

Fu, L. L., and Holt, B. 1982. *Seasat views oceans and sea ice with synthetic-aperture radar.* Pasadena: NASA Jet Propulsion Laboratory Publication 81-120: 200 pp.

———. 1983. Some examples of detection of oceanic mesoscale eddies by the Seasat Synthetic-Aperture Radar. *J. Geophys. Research* **88** (C3): 1844–1852.

Gaposchkin, E. M., and Kolaczek, B. 1981. *Reference coordinate systems for Earth dynamics.* Dordrecht: D. Reidel: 396 pp.

Garrett, W. D. 1967. Damping of capillary waves at the air-sea interface by organic surface-active material. *J. Marine Res.* **25** (3): 279–291.

Gautier, C. 1981. Daily shortwave energy budget over the ocean from geostationary satellite measurements. In: *Oceanography from space*, edited by J. F. R. Gower, 201–206. New York: Plenum Press.

Gautier, C., Diak, G., and Masse, S. 1980. A simple physical model to estimate incident solar radiation at the surface from Goes satellite data. *J. Appl. Meteorol.* **19** (8): 1005–1012.

Gloersen, P., Zwally, H. J., Chang, A. T. C., Hall, D. K., Campbell, W. J., and Ramseier, R. O. 1978. Time-dependence of sea-ice concentration and multiyear ice fraction in the Arctic basin. *Boundary-Layer Meteorology* **13**: 339–359.

Goad, C. C., and Douglas, B. C. 1978. Lunar tidal acceleration obtained from satellite-derived ocean tide parameters. *J. Geophys. Research* **83** (B5): 2306–2310.

Goad, C. C., Douglas, B. C., and Agreen, R. W. 1980. On the use of satellite altimeter data for radial ephemeris improvement. *J. Astronaut. Sci.* **28** (4): 419–428.

Goldhirsh, J. 1982. Slant path fade and rain-rate statistics associated with the COM-STAR beacon at 28.56 GHz for Wallops Island, Virgina over a three-year period. *IEEE Trans. Antennas and Propag.* **AP-30** (2): 191–198.

Gonzales, F. E., Shuchman, R. A., Ross, D. B., Rufenach, C. L., and Gower, J. F. R. 1981. Synthetic aperture radar wave observations during Goasex. In: *Oceanography from space*, edited by J. F. R. Gower, 459–468. New York: Plenum Press.

Goody, R. M. 1964. *Atmospheric radiation*. Oxford: Oxford University Press, 436 pp.

Gordon, A. L. 1981. Seasonality of Southern Ocean sea ice. *J. Geophys. Research* **86** (C5): 4195–4197.

Gordon, H. R. 1978. Removal of atmospheric effects from satellite imagery of the oceans. *Appl. Optics* **17** (10): 1631–1636.

Gordon, H. R., and Clark, D. K. 1980. Atmospheric effects in the remote sensing of phytoplankton pigments. *Boundary-Layer Meteorology* **18**: 299–314.

_____. 1981. Clear water radiances for atmospheric correction of Coastal Zone Color Scanner imagery. *Appl. Optics* **20** (24): 4175–4180.

Gordon, H. R., and Clark, D. K., Brown, J. W., Brown, O. B., Evans, R. H., and Broenkow, W. W. 1983. Phytoplankton pigment concentrations in the Middle Atlantic Bight: comparison of ship determinations and CZCS estimates. *Appl. Optics* **22** (1): 20–36.

Gordon, H. R., Clark, D. K., Mueller, J. L., and Hovis, W. A. 1980. Phytoplankton pigments from the Nimbus-7 Coastal Zone Color Scanner: Comparisons with surface measurements. *Science* **210** (4465): 63–66.

Griggs, M. 1975. Measurements of atmospheric aerosol optical thickness over water using ERTS-1 data. *J. Air Pollution Control Assoc.* **25** (6): 622–626.

_____. 1977. Comment on 'Relative atmospheric aerosol content from ERTS observations' by Y. Mekler, H. Quenzel, G. Ohring, and I. Marcus. *J. Geophys. Research* **82** (31): 4972.

Grody, N. C. 1976. Remote sensing of atmospheric water content from satellites using microwave radiometery. *IEEE Trans. Antennas and Propag.* **AP-24** (2): 155–162.

Grody, N. C., Gruber, A., and Shen, W. C. 1980. Atmospheric water content over the tropical Pacific derived from the Nimbus-6 Scanning Microwave Spectrometer. *J. Appl. Meteorol.* **19** (8): 986–996.

Gruber, A., and Winston, J. S. 1978. Earth-atmosphere radiative heating based on the Noaa scanning radiometer measurements. *Bull. Am. Meteorol. Soc.* **59** (12): 1570–1573.

Guinard, N., and Daley, J. C. 1970. An experimental study of a sea clutter model. *Proc. IEEE* **58** (4): 543–550.

Gunn, K. L. S., and East, T. W. R. 1954. The microwave properties of precipitation particles. *Quart. J. Roy. Meteorol. Soc.* **80** (346): 522–544.

Guymer, T. H., Businger, J. A., Jones, W. L., and Stewart, R. H. 1981. Anomalous wind estimates from Seasat scatterometer. *Nature* **294** (5843): 735–737.

Halliwell, G. R., and Mooers, C. N. K. 1979. The space-time structure and variability of the shelf water-slope water and Gulf Stream surface temperature fronts and associated warm core eddies. *J. Geophys. Research* **84** (C12): 7707–7725.

Halpern, D., and Knox, R. A. 1983. Coherence between low-level cloud motion vectors and surface wind measurements near 0°, 152°W from April 1979 to February 1980. *Atmosphere-Ocean* **21** (1): 82–93.

Hanel, G. 1976. The properties of atmospheric aerosol particles as functions of the relative humidity at thermodynamic equilibrium with the surrounding moist air. In: *Advances in Geophysics*, edited by K. E. Landsberg and J. Van Miegham, 73–188. New York: Academic Press.

Harger, R. O. 1970. *Synthetic aperture radar systems*. New York: Academic Press, 240 pp.

_____. 1981. SAR ocean imaging mechanisms. In: *Synthetic aperture radar for oceanography*, edited by R. C. Beal, P. De Leonibus, and I. Katz, 41–52. Baltimore: The Johns Hopkins University Press.

Harvey, R. R., and Patzert, W. C. 1976. Deep current measurements suggest long waves in the eastern Equatorial Pacific. *Science* 193 (4256): 883–885.

Hasselmann, D. E., Dunckel, M., and Ewing, J. A. 1980. Directional wave spectra observed during JONSWAP 1973. *J. Phys. Oceanog.* 10 (8): 1264–1280.

Hasselmann, K. 1966. Feynman diagrams and interaction rules of wave-wave scattering processes. *Rev. Geophys.* 4 (1): 1–32.

———, 1980. A simple algorithm for the direct extraction of the two-dimensional surface image spectrum from the return signal of a synthetic aperture radar. *International J. Remote Sensing* 1: 219–240.

Hasselmann, K., Barnett, T. P., Bouws, E., Carlson, H., Cartwright, D. E., Enke, K., Ewing, J. A., Gienapp, H., Hasselmann, D. E., Kruseman, P., Meerburg, A., Müller, P., Olbers, D. J., Richter, K., Sell, W., and Walden, H. 1973. Measurements of wind-wave growth and swell decay during the Joint North Sea Wave Project (JONSWAP). *Ergänzungsheft zur Deutschen Hydrographischen Zeitachrift, Reiche A(8°)*, Nr. 12, 95 pp.

Hasselmann, K., Ross, D. B., Muller, P., and Sell, W. 1976. A parametric wave prediction model. *J. Phys. Oceanog.* 6 (2): 200–228.

Hastings, N. A. J., and Peacock, J. B. 1975. *Statistical distributions.* New York: John Wiley & Sons, 130 pp.

Hayne, G. S., and Hancock, D. W. 1982. Sea-state-related altitude errors in the Seasat radar altimeter. *J. Geophys. Research* 87 (C5): 3227–3231.

Henderson-Sellers, A. 1982. De-fogging cloud determination algorithms. *Nature* 298 (5873): 419–420.

Hickey, J. R., Stowe, L. L., Jacobowitz, H., Pellegrino, P., Maschoff, R. H., House, F., and Vonder Haar, T. H. 1980. Initial solar irradiance determinations from NIMBUS 7 cavity radiometer measurements. *Science* 208: 281–283.

Hilsenrath, E., Heath, D. F., and Schlesinger, B. M. 1979. Seasonal and interannual variations in total ozone revealed by the NIMBUS 4 Backscattered Ultraviolet Experiment. *J. Geophys. Research* 84 (C11): 6969–6979.

Hinze, J. O. 1975. *Turbulence.* Second Edition. New York: McGraw-Hill, 790 pp.

Hobson, D. E., Jr., and Williams, D. 1971. Infrared spectral reflectance of sea water. *Appl. Optics* 10 (10): 2372–2373.

Hofer, R., and Njoku, E. G. 1981. Regression techniques for oceanographic parameter retrieval using space-borne microwave radiometry. *IEEE Trans. Geosci. and Remote Sensing* GE-19: 178–189.

Hofer, R., Njoku, E. G., and Waters, J. W. 1981. Microwave radiometric measurements of sea surface temperature from the Seasat satellite: First results. *Science* 212 (4501): 1385–1387.

Hoffman, D., and Karst, O. J. 1975. The theory of the Rayleigh distribution and some of its applications. *J. Ship Research* 19 (3): 172–191.

Hollinger, J. P. 1971. Passive microwave measurements of sea surface roughness. *IEEE Trans. Geoscience Electronics* 9: 165–169.

Horan, J. J. 1978. NIMBUS: The vanguard of remote sensing. *IEEE Spectrum* 15 (9): 36–43.

Houghton, J. T. 1977. *The physics of atmospheres.* Cambridge: Cambridge University Press, 203 pp.

Hovis, W. A., Clark, D. K., Anderson, F., Austin, R. W., Wilson, W. H., Baker, E. T., Ball, D., Gordon, H. R., Mueller, J. L., El-Sayed, S. Z., Sturm, B., Wrigley, R. C., and Yentsch, C. S. 1980. Nimbus 7 Coastal Zone Color Scanner: System description and initial imagery. *Science* 210 (4465): 60–63.

Huang, N. E., Leitao, C. D., and Parra, C. G. 1978. Large-scale Gulf Stream frontal study using Geos-3 radar altimeter data. *J. Geophys. Research* **83** (C9): 4673–4682.

Hughes Aircraft Company. 1980. *Geostationary Operational Environmental Satellite (GOES): GOES D, E, F data book.* Los Angeles: Hughes Aircraft Company, Space and Communications Group, 96 pp.

Hughes, B. A., Grant, H. S., and Chappell, R. W. 1977. A fast response surface-wave slope meter and measured wind-wave moments. *Deep-Sea Res.* **24** (12): 1211–1223.

Huhnerfuss, H., Alpers, W., and Jones, W. L. 1978. Measurements at 13.9 GHz of the radar backscattering cross section of the North Sea covered with an artificial surface film. *Radio Science* **13** (6): 979–983.

Huhnerfuss, H., Alpers, W., Jones, W. L., Lange, P. A., and Richter, K. 1981. The damping of ocean surface waves by a monomolecular film measured by wave staffs and microwave radars. *J. Geophys. Research* **86** (C1): 429–438.

Huhnerfuss, H., Walter, W., and Kruspe, G. 1977. On the variability of surface tension with mean wind speed. *J. Phys. Oceanog.* **7** (4): 567–571.

Hussey, J. W. 1977. *The Tiros-N polar orbiting environmental satellite system.* Washington: NOAA/National Environmental Satellite Service Report, 33 pp.

Institute of Radio Engineers (IRE). 1942. *Standards on radio propagation. Proc. IRE* **30**: suppl.

Irvine, W. J., and Pollack, J. B. 1968. Infrared optical properties of water and ice spheres. *Icarus* **8**: 324–360.

Ishimaru, A. 1978. *Wave propagation and scattering in random media.* 2 vols. New York: Academic Press, 272+336 pp.

Jacchia, L. G. 1971. *Revised static models of the thermosphere and exosphere with empirical temperature profiles.* Cambridge: Smithsonian Institution Astrophysical Observatory Special Report 332, 115 pp.

Jackson, F. C. 1979. The reflection of impulses from a nonlinear random sea. *J. Geophys. Research* **84** (C8): 4939–4934.

Jackson, F. J. 1981. An analysis of short pulse and dual frequency radar techniques for measuring ocean wave spectra from satellites. *Radio Science* **16** (6): 1385–1400.

Jackson, J. D. 1962. *Classical electrodynamics.* New York: John Wiley & Sons, 641 pp.

Jacobowitz, H., and Coulson, K. L. 1973. *Effects of aerosols on the determination of the temperature of Earth's surface from radiance measurements at* 11.2 μm. Washington: NOAA Technical Report NESS 66, 18 pp.

Jacobowitz, H., Smith, W. L., Howell, H. B., Nagle, F. W., and Hickey, J. R. 1979. The first 18 months of planetary radiation budget measurements from the Nimbus 6 ERB experiment. *J. Atmos. Sci.* **36** (3): 501–507.

Jain, A. 1977. Determination of ocean wave heights from synthetic aperture radar imagery. *Appl. Physics* **13**: 371–382.

————. 1978. Focusing effects in synthetic aperture radar imaging of ocean waves. *Appl. Physics* **15**: 323–333.

————. 1981. SAR imaging of ocean waves: theory. *IEEE J. Oceanic Engineering* **OE-6** (4): 130–139.

Jain, A., Medlin, G., and Wu, C. 1982. Ocean wave height measurement with Seasat SAR using speckle diversity. *IEEE J. Oceanic Engineering* **OE-7** (2): 103–107.

Jarvis, N. L., and Kagarise, R. E. 1962. Determination of the surface temperature of water during evaporation studies — a comparison of thermistor with infrared measurements. *J. Colloid. Sci.* **17**: 501–511.

Johanson, J. M., Buonsanto, M. J., and Klobuchar, J. A. 1978. The variability of the ionospheric time delay. In: *Effect of the ionosphere on space and terrestrial systems,* edited by J. M. Johnson, 479–485. Washington: U.S. Navy.

Johnson, J. W., Williams, L. A., Bracalente, E. M., Beck, F. B., and Grantham, W. L.

1980. Seasat-A satellite scatterometer instrument evaluation. *IEEE J. Oceanic Engineering* **OE-5** (2): 138–144.

Jones, W. L., Boggs, D. H., Bracalente, E. M., Brown, R. A., Guymer, T. H., Chelton, D., and Schroeder, L. C. 1981. Evaluation of the Seasat wind scatterometer. *Nature* **294** (5843): 704–707.

Jones, W. L., Delnore, V. E., and Bracalente, E. M. 1981. The study of mesoscale ocean winds. In: *Spaceborne synthetic aperture radar for oceanography*, edited by R. C. Beal, P. S. DeLeonibus, and I. Katz, 87–94. Baltimore: The Johns Hopkins University Press.

Jones, W. L., and Schroeder, L. C. 1978. Radar backscatter from the ocean: dependence on surface friction velocity. *Boundary-Layer Meteorology* **13**: 133–149.

Jones, W. L., Schroeder, L. C., Boggs, D. H., Bracalente, E. M., Brown, R. A., Dome, G. J., Pierson, W. J., and Wentz, F. J. 1982. The Seasat-A satellite scatterometer: The geophysical evaluation of remotely sensed wind vectors over the ocean. *J. Geophys. Research* **87** (C5): 3297–3317.

Jones, W. L., Schroeder, L. C., and Mitchell, J. L. 1977. Aircraft measurements of the microwave scattering signature of the ocean. *IEEE Trans. Antennas and Propagation* **AP-25** (1): 52–61.

Jones, W. L., Wentz, F. J., and Schroeder, L. C. 1978. Algorithm for inferring wind stress from SEASAT-A. *J. Spacecraft and Rockets* **15** (6): 368–374.

Jordan, A. K., and Lang, R. H. 1979. Electromagnetic scattering patterns from sinusoidal surfaces. *Radio Science* **14** (6): 1077–1088.

Jordan, R. L. 1980. The Seasat-A synthetic aperture radar system. *IEEE J. Oceanic Engineering* **OE-5** (2): 154–163.

Katsaros, K. B. 1980. The aqueous thermal boundary layer. *Boundary-Layer Meteorology* **18**: 107–127.

Katsaros, K. B., Taylor, P. K., Alishouse, J. C., and Lipes, R. B. 1981. Quality of Seasat SMMR (Scanning Multichannel Microwave Radiometer) atmospheric water determinations. In: *Oceanography from space*, edited by J. F. R. Gower, 691–706. New York: Plenum Press.

Kaula, W. M. 1966. *Theory of satellite geodesy*. Waltham: Blaisdell, 124 pp.

Keller, W. C., and Wright, J. W. 1975. Microwave scattering and the straining of wind generated waves. *Radio Science* **10** (2): 139–147.

Kelly, K. A. 1983. Swirls and plumes or application of statistical methods to satellite-derived sea surface temperatures. Ph.D. Thesis, University of California, San Diego.

Kenney, J. E., Uliana, E. A., and Walsh, E. J. 1979. The surface contour radar, a unique remote sensing instrument. *IEEE Trans. Microwave Theory and Techniques*, **MTT-27** (12): 1080–1092.

Kerker, M. 1969. *The scattering of light*. New York: Academic Press, 666 pp.

King-Hele, D. G. 1964. *Theory of satellite orbits in an atmosphere*. London: Butterworth's, 165 pp.

———. 1976. The shape of the Earth. *Science* **192** (4246): 1293–1300.

King-Hele, D. G., Pilkington, J. A., Hiller, H., and Walker, D. M. C. 1981. *The RAE table of earth satellites, 1957–1980*. London: The Macmillan Press Ltd., 656 pp.

Kinsman, B. 1965. *Wind waves*. Englewood Cliffs, N. J.: Prentice-Hall, 676 pp.

Kitaigorodskii, S. A. 1973. *The physics of air-sea interaction*. Jerusalem: Israel Program for Scientific Translations, 237 pp.

Klein, L. A., and Swift, C. T. 1977. An improved model for the dielectric constant of sea water at microwave frequencies. *IEEE Trans. Antennas and Propag.* **AP-25**: 104–111.

Kneizys, F. X., Shettle, E. P., Gallery, W. O., Chetwynd, J. H., Abreu, L. W., Selby,

J. E. A., Fenn, R. W., and McClatchey, R. A. 1980. *Atmospheric transmittance radiance: Computer code LOWTRAN 5.* Cambridge: Air Force Geophysics Laboratory Report TR-80-0067, 233 pp.

Kolenkiewicz, R., and Martin, C. F. 1982. Seasat altimeter height calibration. *J. Geophys. Research* **87** (C5): 3189–3198.

Kondo, J. 1975. Air-sea bulk transfer coefficients in diabatic conditions. *Boundary-Layer Meteorology* **9** (1): 91–112.

Kondo, J., Fujinawa, Y., and Naito, G. 1973. High-frequency components of ocean waves and their relation to the aerodynamic roughness. *J. Phys. Oceanog.* **3** (2): 197–202.

Kraus, E. B. 1972. *Atmospheric-ocean interaction.* Oxford: Oxford University Press, 275 pp.

Kraus, J. D. 1966. *Radio astronomy.* New York: McGraw-Hill, 481 pp.

Krishen, K. 1971. Correlation of radar backscattering cross sections with ocean wave height and wind velocity. *J. Geophys. Research* **76** (20): 6528–6539.

Kropotkin, M. A., Kozyrev, B. P., and Zaitsev, V. A. 1966. Infrared reflection spectra of sea and fresh water of several aqueous solutions. *Izvestiya Atmos. and Oceanic Phys.* **2**: 434–435, (pp. 259–260 in translation).

Kropotkin, M. A., and Sheveleva, T. Y. 1975. A study of the reflectance of ocean water and certain aqueous solutions at the wavelength 10.6μm. *Izvestiya Atmos. and Oceanic Phys.* **11**: 211–214, (pp. 124–126 in translation).

Lambeck, K. 1980. *The Earth's variable rotation.* New York: Cambridge University Press, 449 pp.

Lambeck, K., and Coleman, R. 1983. The Earth's shape and gravity data: A report of progress from 1958 to 1982. *Geophys. J. R. Astr. Soc.* **74**: 25–54.

Lane, J., and Saxton, J. 1952. Dielectric dispersion in pure polar liquids at very high radio frequencies. III. The effect of electrolytes in solutions. *Proc. Roy. Soc. London* **214A**: 531–545.

Lawrence, R. S., Little, C. G., and Chivers, H. J. A. 1964. A survey of ionospheric effects upon Earth-space radio propagation. *Proc. IEEE* **52** (1): 4–27.

Laws, S. O., and Parson, D. A. 1943. The relation of raindrop size to intensity. *Trans. Am. Geophys. Union* **24**: 432–460.

Leavitt, E., and Paulson, C. A. 1975. Statistics of surface layer turbulence over the tropical ocean. *J. Phys. Oceanog.* **5** (1): 143–156.

Leberl, F., Raggam, J., Elachi, C., and Campbell, W. J. 1983. Sea ice motion measurements from Seasat SAR images. *J. Geophys. Research* **88** (C3): 1915–1928.

Legeckis, R. V. 1977. Long waves in the Eastern Equatorial Pacific Ocean: a view from a geostationary satellite. *Science* **197** (4309): 1179–1181.

———. 1978. A survey of worldwide sea surface temperature fronts detected by environmental satellites. *J. Geophys. Research* **83** (C9): 4501–4522.

———. 1979. Satellite observations of the influence of bottom topography in the seaward deflection of the Gulf Stream off Charleston, South Carolina. *J. Phys. Oceanog.* **9** (3): 483–497.

Legeckis, R V., and Gordon, A. L. 1982. Satellite observations of the Brazil and Falkland currents — 1975 to 1976 and 1978. *Deep-Sea Research* **29** (3A): 375–401.

Legeckis, R. V., Pichel, W., and Nesterczuk, G. 1983. Equatorial long waves in geostationary satellite observations and in a multichannel sea surface temperature analysis. *Bull. Am. Meteorol. Soc.* **64** (2): 133–139.

Lerch, F. J., Marsh, J. G., Klosko, S. M., and Williamson, R. G. 1982. Gravity model improvement for Seasat. *J. Geophys. Research* **87** (C5): 3281–3296.

Lerch, F. J., Klosko, S. M., Laubscher, R. E., and Wagner, C. A. 1979. Gravity model improvement using Geos 3 (GEM 9 and 10). *J. Geophys. Research* **84** (B8): 3897–3916.

Lewis, B. L., and Olin, I. D. 1980. Experimental study and theoretical model of high resolution radar backscatter from the sea. *Radio Science* **15** (4): 815–828.

Lichy, D. E., Mattie, M. G., and Mancini, L. J. 1981. Tracking a warm water ring. In: *Spaceborne synthetic aperture radar for oceanography*, edited by R. C. Beal, P. S. De Leonibus, and I. Katz, 171–184. Baltimore: The Johns Hopkins University Press.

Liou, K. N. 1980. *An introduction to atmospheric radiation*. New York: Academic Press, 392 pp.

Lipa, B. J. 1977. Derivation of directional ocean-wave spectra by integral inversion of second-order radar echoes. *Radio Science* **12** (3): 425–434.

———. 1978. Inversion of second-order echoes from the sea. *J. Geophys. Research* **83** (C2): 959–962.

Lipa, B. J., and Barrick, D. E. 1980. Methods for the extraction of long-period ocean wave parameters from narrow beam HF radar sea echo. *Radio Science* **15** (4): 843–854.

———. 1981. Ocean surface height-slope probability density function from Seasat altimeter echo. *J. Geophys. Research* **86** (C11): 10921–10930.

Lipa, B. J., Barrick, D. E., and Maresca, J.W. 1981. HF radar measurements of long ocean waves. *J. Geophys. Research* **86** (C5): 4089–4102.

Lipes, R. G. 1982. Description of the Seasat radiometer status and results. *J. Geophys. Research* **87** (C5): 3385–3395.

Litman, V., and Nicholas, J. 1982. *Guidelines for spaceborne microwave remote sensors*. Washington: NASA Reference Publication 1086, 83 pp.

Liu, T. 1984. The effects of the variations in sea surface temperature and atmospheric stability in the estimation of average wind speed by Seasat SASS. *J. Phys. Oceanog.* **14** (2): 392–401.

Liu, W. T., and Large, W. G. 1981. Determination of surface stress by Seasat-SASS: A case study with JASIN data. *J. Phys. Oceanog.* **11** (12): 1603–1611.

Lleonart, G. T., and Blackman, D. R. 1980. The spectral characteristics of wind-generated capillary waves. *J. Fluid Mech.* **97** (3): 455–479.

Long, A., and Trizna, D. B. 1973. Mapping the North Atlantic winds by HF radar sea backscatter interpretation. *IEEE Trans. Antennas and Propag.* **AP-21**: 680–685.

Longuet-Higgins, M. S. 1963. The effect of non-linearities on statistical distributions in the theory of sea waves. *J. Fluid Mech.* **17** (3): 459–480.

Longuet-Higgins, M. S., Cartwright, D. E., and Smith, N. D. 1963. Observations of the directional spectrum of sea waves using the motions of a floating buoy. In: *Ocean wave spectra*. New York: Prentice-Hall: 111–136.

Longuet-Higgins, M. S. and Stewart, R. W. 1964. Radiation stresses in water waves: A physical discussion with applications. *Deep Sea Res.* **11**: 529–562.

Maresca, J. W. and Barnum, J. R. 1977. Measurement of oceanic wind speed from HF sea scatter by skywave radar. *IEEE Trans. Antennas and Propag.* **AP-25** (1): 132–135.

Mariner's Weather Log. 1977. Smooth log, north Pacific weather. **21** (4): 259–269.

Marsh, B. D., and Marsh, J. G. 1976. On global gravity anomalies and two scale mantle convection. *J. Geophys. Research* **81** (29): 5267–5280.

Marsh, J. G., and Chang, E. S. 1978. 5′ detailed gravimetric geoid in the northwestern Atlantic ocean. *J. Marine Geodesy* **1** (3): 253–261.

Marshall, J. S., and Palmer, W. M. K. 1948. The distribution of raindrops with size. *J. Meteorology* **5** (4): 165–166.

Martin, P. J. 1981. Direct determination of the two-dimensional image spectrum from raw synthetic aperture radar data. *IEEE Trans. Geoscience and Remote Sensing* **GE19** (4): 194–203.

Mather, R. S. 1974. On the solution of the Geodetic boundary value problem for the definition of sea surface topography. *Geophys. J. R. Astr. Soc.* **39** (1): 87–109.

———. 1978. On the realization of a system of reference in four dimensions for ocean dynamics. *Boundary-Layer Meteorology* **13**: 231–244.

Mather, R. S., Coleman, R., and Hirsch, B. 1980. Temporal variations in regional models of the Sargasso Sea from Geos-3 altimetry. *J. Phys. Oceanog.* **10** (2): 171–185.

Mather, R. S., Rizos, C., and Coleman, R. 1979. Remote sensing of surface ocean circulation with satellite altimetery. *Science* **205** (4401): 11–17.

Maul, G. A. 1981. Application of Goes visible-infrared data to quantifying mesoscale ocean surface temperatures. *J. Geophys. Research* **86** (C9): 8007–8021.

Maul, G. A., deWitt, P. W., Yanaway, A., and Baig, S. R. 1978. Geostationary satellite observations of Gulf Stream meanders: infrared measurements and time series analysis. *J. Geophys. Research* **83** (C12): 6123–6135.

Mazzega, P. 1983. The M2 oceanic tide recovered from Seasat altimetry in the Indian Ocean. *Nature* **302** (5908): 514–516.

McCalister, E. D. 1964. Infrared-optical techniques applied to oceanography 1. Measurements of total heat flow from the sea surface. *Appl. Optics* **3** (5): 609–612.

McCartney, E. J. 1976. *Optics of the atmosphere: scattering by molecules and particles.* New York: John Wiley & Sons, 408 pp.

McClain, E. P. 1974. Environmental earth satellites for oceanographic-meteorological studies of the Bering Sea. In: *Oceanography of the Bering Sea*, edited by D. W. Hood and E. J. Kelley. Fairbanks: University of Alaska.

———. 1978. Eleven year chronicle of one of the world's most gigantic icebergs. *Mariner's Weather Log* **22** (5): 328–333.

———. 1981. Multiple atmospheric-window techniques for satellite-derived sea surface temperatures, In: *Oceanography from space*, edited by J. F. R. Gower, 73–85. New York: Plenum Press.

McClatchey, R. A., Benedict, W. S., Clogh, S. A., Burch, D. E., Calfee, R. F., Fox, K., Rothman, L. S., and Garing, J. S. 1973. *AFCRL Atmospheric absorption line parameters compilation.* Bedford, Mass.: Air Force Cambridge Research Laboratory Report TR-73-0096, 78 pp.

McClatchey, R. A., Fenn, R. W., Selby, J. E., Volz, F. E., and Garing, J. S. 1978. Optical properties of the atmosphere. In: *Handbook of optics*, edited by W. G. Driscoll and W. Vaughan. New York: McGraw-Hill.

McElroy, J. H., McAvoy, N., Johnson, E. H., Degnan, J. J., Goodwin, F. E., Henderson, D. M., Nussmeier, T. A., Stokes, L. S., Peyton, B. J., and Flattau, T. 1977. CO_2 laser communication systems for near-Earth space application. *Proc. IEEE* **65** (2): 221–223.

McGoogan, J. T. 1975. Satellite altimetry applications. *IEEE Trans. Microwave Theory and Technique* **MTT-23** (12): 970–978.

McGoogan, J. T., Miller, L. S., Brown, G. S., and Hayne, G. S. 1974. The S-193 Radar altimeter experiment. *Proc. IEEE* **62** (6): 793–803.

McKenzie, D. P. 1977. Surface deformation, gravity anomalies and convection. *Geophys. J. R. Astr. Soc.* **48** (2): 211–238.

McMillan, J. D. 1981. A global atlas of Geos-3 significant waveheight data and comparison of data with national buoy data. Ph.D. Thesis, University of Texas, Austin.

Meeks, M. L., and Lilly, A. E. 1963. The microwave spectrum of oxygen in the earth's atmosphere. *J. Geophys. Research* **68** (6): 1683–1703.

Mie, G. 1908. Contributions to the optics of suspended media, specifically colloidal metal suspensions. *Annelen d. Physik* **25** (4): 377–455.

Miles, J. W. 1967. Surface-wave damping in closed basins. *Proc. Roy. Soc. London* **A297**: 459–475.

Moon, P. 1940. Proposed standard solar radiation curves for engineering use. *J. Franklin Institute* **230** (5): 583–617.

Moore, R. K. 1979. Tradeoffs between picture element dimensions and noncoherent averaging inside-looking airborne radar. *IEEE Trans. Aerospace and Electronic Systems* **AES-15** (5): 697–708.

Moore, R. K., Birrer, I. J., Bracalente, E. M., Dome, G. J., and Wentz, F. J. 1982. Evaluation of atmospheric attenuation from SMMR brightness temperature for the Seasat satellite scatterometer. *J. Geophys. Research* **87** (C5): 3337–3354.

Moran, J. M., and Rosen, B. R. 1981. Estimation of the propagation delay through the troposphere from microwave radiometer data. *Radio Science* **16** (2): 235–244.

Morel, A., and Prieur, L. 1977. Analysis of variations on ocean color. *Limnology and Oceanography* **22** (4): 709–722.

Morel, A., and Smith, R. C. 1982. Terminology and units in optical oceanography, *Marine Geodesy* **5** (4): 335–349.

Moritz, H. 1980. Geodetic Reference System 1980. *Bull. Géodésique* **54** (3): 395–405.

Moulton, F. R. 1914. *An introduction to celestial mechanics.* Second Edition. New York: Macmillan, 437 pp.

Mueller, I. I. 1981. Reference coordinate systems for Earth dynamics: a preview. In: *Reference coordinate systems for Earth dynamics*, edited by E. M. Gaposchkin and B. Kolaczyk, 1–22. Holland: D. Reidel.

Muench, R. D., and Ahlnas, K. 1976. Ice movement and distribution in the Bering Sea from March to June 1976. *J. Geophys. Research* **81** (24): 4467–4476.

Munk, W. H., and MacDonald, J. F. 1960. *The rotation of the Earth.* New York: Cambridge University Press, 323 pp.

Munk, W. H., and Nierenberg, W. A. 1969. High frequency radar sea return and Phillips saturation constant. *Nature* **224** (5226): 1285.

Munk, W. H. and Wunsch, C. 1982. Observing the ocean in the 1990s. *Phil. Trans. Roy. Soc. London* **A307**: 439–464.

Namais, J. 1978. Multiple causes of the North American abnormal winter 1976-77. *Monthly Weather Review* **106** (3): 279–295.

NASA. 1973. *Models of Earth's atmosphere* (90–2500 km). Washington: NASA Special Publication SP-8021, 54 pp.

———. 1977. *The national geodetic satellite program.* Washington: NASA Special Publication SP-365, 2 Vol., 1030 pp.

———. 1981. *Satellite altimetric measurements of the ocean: Report of the TOPEX Science Working Group.* Pasadena: Jet Propulsion Laboratory Report 400-111, 78 pp.

Neckel, H., and Labs, D. 1981. Improved data of solar spectral irradiance from 0.33 to 1.25 μ. *Solar Physics* **74**: 231–249.

Nelson, M. 1980. *GOES data collection program.* Washington: NOAA Tech. Memo. 110.

Nichols, D. A. 1975. *Block 5D compilation.* Los Angeles: U.S. Air Force Space and Missile Systems Organization, DMSP Program Office Report, 599 pp.

Nilsson, B. 1979. Meteorological influence on aerosol extinction in the 0.2–40 μm wavelength range. *Appl. Optics* **18** (20): 3457–3473.

Njoku, E. G. 1980. Antenna pattern correction procedures for the Scanning Multichannel Microwave Radiometer (SMMR). *Boundary-Layer Meteorology* **18**: 79–98.

Njoku, E. G., Christiansen, E. J., and Cofield, R. E. 1980. The Seasat Scanning Multichannel Microwave Radiometer (SMMR): Antenna pattern corrections — development and implementation. *IEEE Journal of Oceanic Engineering* **OE-5** (2): 125–137.

Njoku, E. G., and Hofer, R. 1981. Seasat SMMR observations of ocean surface temperature and wind speed in the North Pacific. In: *Oceanography from space*, edited by J. F. R. Gower, 673–681. New York: Plenum Press.

NOAA. 1979. *Geostationary Operational Environmental Satellite/Data Collection System.* Washington: NOAA Tech. Rept. NESS 78, 80 pp.

Nyquist, H. 1928. Thermal agitation of electric charge in conductors. *Phys. Rev.* **32**: 110–113.

Pandey, P. C., and Kniffen, S. 1982. *Evaluation of the potential of one to three Seasat-SMMR channels in retrieving sea surface temperature.* Pasadena: Jet Propulsion Laboratory Publication 82-89, 30 pp.

Parker, R. L. 1977. Understanding inverse theory. *Annual Reviews of Earth and Planetary Science* **5**: 35–64.

Parsons, C. L. 1979. Geos 3 wave height measurements: an assessment during high sea state conditions in the North Atlantic. *J. Geophys. Research* **84** (B8): 4011–4020.

Peck, E. R., and Reeder, K. 1972. Dispersion of air. *J. Opt. Soc. Am.* **62** (8): 958–962.

Pedlosky, J. 1979. *Geophysical fluid dynamics.* New York: Springer-Verlag, 624 pp.

Pelaez-Hudlet, J. 1984. Phytoplankton pigment concentrations and patterns in the California current as determined by satellite. Ph.D. Thesis, University of California, San Diego.

Penndorf, R. 1957. Tables of the refractive index for standard air and Rayleigh scattering coefficient for the spectral region between 0.2 and 20.0 μ and their application to atmospheric optics. *J. Opt. Soc. Am.* **47** (2): 176–182.

———. 1962. Angular Mie scattering. *J. Opt. Soc. Am.* **52** (4): 402–408.

Phillips, O. M. 1977. *The dynamics of the upper ocean.* Second Edition. Cambridge: Cambridge University Press, 336 pp.

———. 1981. The structure of short gravity waves on the ocean surface. In: *Spaceborne synthetic aperture radar for oceanography*, edited by R. C. Beal, P. S. De Leonibus, and I. Katz, 24–31. Baltimore: The Johns Hopkins University Press.

Pierson, W. J. 1981. Winds over the ocean as measured by the scatterometer on Seasat. In: *Oceanography from space*, edited by J. F. R. Gower, 563–571. New York: Plenum Press.

Plant, W. J. 1977. Studies of backscattered sea return with a CW dual-frequency, X-band radar. *IEEE Trans. Antennas and Propag.* **AP-25** (1): 28–36.

Plant, W. J., Keller, W. C., and Cross, A. 1983. Parametric dependence of ocean wave-radar modulation transfer functions. *J. Geophys. Research* **88** (C14): 9747–9756.

Plant, W. J., Keller, W. C., and Wright, J. W. 1978. Modulation of coherent microwave backscatter by shoaling waves. *J. Geophys. Research* **83** (C3): 1347–1352.

Plant, W. J. and Schuler, D. L. 1980. Remote sensing of the sea surface using one- and two-frequency microwave techniques. *Radio Science* **15** (3): 605–615.

Pond, S., Phelps, G. T., Paquin, J. E., McBean, G., and Stewart, R. W. 1971. Measurements of the turbulent fluxes of momentum, moisture and sensible heat over the ocean. *J. Atmos. Sci.* **28** (6): 901–917.

Prabhakara, C., Dalu, G., and Kunde, V. G. 1974. Estimation of sea surface temperature from remote sensing in the 11- to 13-μm window region. *J. Geophys. Research* **79** (33): 5039–5044.

Pravdo, S. H., Huneycutt, B., Holt, B. M., and Held, D. N. 1982. *Seasat synthetic-aperture radar: Data users manual.* Pasadena: Jet Propulsion Laboratory Publication 82-90.

Price, R. M. 1976. Radiometers. In: *Methods of experimental physics*, Vol.12B, *Astrophysics*, edited by M. L. Meeks, 201–224. New York: Academic Press.

Rabinove, C. J., Chavey, P. S., Gehring, D., and Holmgren, D. 1981. Arid land monitoring using Landsat albedo difference images. *Remote Sensing of Environment* 11 (2): 133–156.

Ramseir, R. O., and Lapp, D. J. 1981. *Proceedings of the final Sursat ice workshop on active and passive microwave measurements of sea ice and icebergs.* Toronto: Canadian Atmospheric Environment Service Report.

Rao, M. S. V., Abbott, W. V., and Theon, J. S. 1976. *Satellite-derived global oceanic rainfall atlas (1973 and 1974).* Washington: NASA Special Publication SP-410.

Rapp, R. H. 1974. The geoid: definition and determination. *EOS* 55 (3): 118–126.

Ray, P. S. 1972. Broadband complex refractive indices of ice and water. *Appl. Optics* 11 (8): 1836–1844.

Reeves, R. G., ed. 1975. *Manual of remote sensing.* 2 Vols. Falls Church: American Society of Photogrammetry, 2144 pp.

Reference Data for Radio Engineers. Fifth Edition. 1973. Indianapolis: Howard W. Sams and Company.

Rice, S. O. 1951. Reflection of electromagnetic waves from slightly rough surfaces. *Comm. Pure and Applied Math.* 4: 351–378.

Ridenour, L. N., ed. 1947. *Radar system engineering.* New York: McGraw-Hill, 748 pp.

Rihaczek, A. W. 1969. *Principles of high resolution radar.* New York: McGraw-Hill, 498 pp.

Robertson, D. S., Carter, W. E., Eanes, R. J., Schutz, B. E., Tapley, B. D., King, R. W., Langley, R. B., Morgan, P. J., and Shapiro, I. I. 1983. Comparison of Earth rotation as inferred from radio interferometric, laser ranging and astrometric observations. *Nature* 302 (5908): 509–511.

Roemmich, D., and Wunsch, C. 1982. On combining satellite altimetry with hydrographic data. *J. Marine Res.* 40 Supplement: 605–619.

Ross, D., and Jones, L. W. 1978. On the relationship of radar backscatter to wind speed and fetch. *Boundary-Layer Meteorology* 13: 151–163.

Rummel, R., and Rapp, R. H. 1976. The influence of the atmosphere on geoid and potential coefficient determinations from gravity data. *J. Geophys. Research* 81 (32): 5639–5642.

Salomonson, V. V., Smith, L. P., Park, A. B., Webb, U. C., and Lynch, T. J. 1980. An overview of progress in the design and implementation of Landsat-D systems. *IEEE Trans. Geoscience and Remote Sensing* GE-18 (2): 137–146.

Saunders, P. M. 1967. The temperature of the ocean-air interface. *J. Atmos. Sci.* 24 (3): 269–273.

Sawyer, C., and Apel, J. R. 1976. *Satellite images of ocean internal-wave signatures.* Miami: NOAA Atlantic Oceanographic and Meteorological Laboratories, S/T 2401.

Schanda, E., ed. 1976. *Remote sensing for environmental sciences.* Berlin: Springer-Verlag, 367 pp.

Schlichting, H. 1968. *Boundary-layer theory.* New York: McGraw-Hill, 747 pp.

———. 1979. *Boundary-layer theory.* Seventh Edition. New York: McGraw-Hill, 817 pp.

Schroeder, L. C., Boggs, D. H., Dome, G., Halberstam, I. M., Jones, W. L., Pierson, W. J., and Wentz, F. J. 1982. The relationship between wind vector and normalized radar cross section used to derive Seasat-A satellite scatterometer winds. *J. Geophys. Research* **87** (C5): 3318–3336.

Schwalb, A. 1978. *The Tiros-N/Noaa-A-G satellite series.* Washington: NOAA Tech. Memo. NESS-95, 75 pp.

———. 1982. *Modified versions of the Tiros-N/Noaa A-G satellite series (NOAA E-J) - Advanced Tiros-N (ATN).* Washington: NOAA Tech. Memo. NESS-116, 23 pp.

Schwiderski, E. W. 1980. On charting global ocean tides. *Reviews of Geophysics and Space Physics* **18** (1): 243–268.

Scott, J. C. 1972. The influence of surface-active contamination on the initiation of wind waves. *J. Fluid Mech.* **56** (3): 591–606.

———. 1975. The preparation of water for surface-clean fluid mechanics. *J. Fluid Mech.* **69** (2): 339–351.

Selby, J. E. A., and McClatchey, R. A. 1975. *Atmospheric transmittance from 0.25 to 28.5 μm: Computer code LOWTRAN 3.* Cambridge: Air Force Cambridge Research Laboratories, Optical Physics Laboratory Report.

Sellers, W. D. 1965. *Physical climatology.* Chicago: University of Chicago Press, 272 pp.

Shuchman, R. A. 1981. Processing synthetic aperture radar data of ocean waves. In: *Oceanography from space*, edited by J. F. R. Gower, 447–496. New York: Plenum Press.

Shuchman, R. A., and Kasischke, E. S. 1981. Refraction of coastal ocean waves. In: *Spaceborne synthetic aperture aadar for oceanography*, edited by R. C. Beal, P. S. De Leonibus, and I. Katz, 110–127. Baltimore: The Johns Hopkins University Press.

Shuchman, R. A., and Zelenka, J. S. 1978. Processing of ocean wave data from a synthetic aperture radar. *Boundary-Layer Meteorology* **13**: 181–191.

Simpson, J. J., and Paulson, C. A. 1980. Small-scale sea surface temperature structure. *J. Phys. Oceanog.* **10** (3): 399–410.

Slater, P. N. 1980. *Remote sensing, optics and optical systems.* Reading, Mass: Addison-Wesley Pub. Co., 575pp.

SMIC 1971. *Report of the study of man's impact on climate.* Cambridge: The MIT Press, 308 pp.

Smith, E. K., and Weintraub, S. 1953. The constants in the equation for atmospheric refractive index at radio frequencies. *Proc. IRE* **41**: 1035–1037.

Smith, R. C. 1981. Remote sensing and depth distribution of ocean chlorophyll. *Marine Ecology - Progress Series* **5**: 359–361.

Smith, R. C., and Baker, K. S. 1978a. The bio-optical state of ocean waters and remote sensing. *Limnology and Oceanography* **23** (2): 247–259.

———. 1978b. Optical classification of natural waters. *Limnology and Oceanography* **23** (2): 260–267.

———. 1982. Oceanic chlorophyll concentrations as determined by satellite (Nimbus 7 coastal zone color scanner). *Marine Biology* **66** (3): 269–279.

Smith, R. C., Eppley, R. W., and Baker, K. S. 1982. Correlation of primary production as measured aboard ship in Southern California coastal waters and as estimated from satellite chlorophyll images. *Marine Biology* **66** (3): 281–288.

Smith, R. C., and Wilson, W. H. 1981. Ship and satellite bio-optical research in the California bight. In: *Oceanography from space*, edited by J. F. R. Gower, 281–294. New York: Plenum Press.

Smith, S. D. 1980. Wind stress and heat flux over the ocean in gale force winds. *J. Phys. Oceanog.* **10** (5): 709–726.

Smith, W. L., Hickey, J., Howell, H. B., Jocobowitz, H., Hilleary, D. T., and Drummond, A. J. 1977. Nimbus 6 earth radiation budget experiment. *Appl. Optics* **16** (2): 306–318.

Smith, W. L., Hilleary, D. T., Fischer, J. C., Howell, H. B., and Woolf, H. M. 1974.

1974 NIMBUS-5 ITPR experiment. *Appl. Optics* **13**(3): 499–506.

Smith, W. L., Rao, P. K., Koffler, R., and Curtis, W. R. 1970. The determination of sea-surface temperature from satellite high resolution infrared window radiation measurements. *Monthly Weather Review* **98**(8): 604–611.

Smith, W. L., Woolf, H. M., Hayden, C. M., Wark, D. Q., and McMillan, L. M. 1979. The Tiros-N operational vertical sounder. *Bull. Am. Meteorol. Soc.* **60**(10): 1177–1187.

Sobczak, L. W. 1977. Ice movements in the Beufort Sea, 1973-1975: Determination by ERTS imagery. *J. Geophys. Research* **82**(9): 1413–1418.

Sorensen, B. M. 1979. *Recommendations of the international workshop on atmospheric correction of satellite observation of sea water color.* Ispra: Joint Research Center — Ispra Establishment, 53 pp.

Stacey, F. D. 1977. *Physics of the earth.* Second Edition. New York: John Wiley & Sons, 414 pp.

Staelin, D. H. 1966. Measurements and interpretation of the microwave spectrum of the terrestrial atmosphere near 1-centimeter wavelength. *J. Geophys. Research* **71**(12): 2875–2881.

Staelin, D. H., Kunzi, K. F., Pettyjohn, R. L., Poon, R. K. L., Wilcox, R. W., and Waters, J. W. 1976. Remote sensing of atmospheric water vapor and liquid water with the Nimbus-5 microwave spectrometer. *J. Appl. Meteorol.* **15**(11): 1204–1214.

Staelin, D. H., Rosenkranz, P. W., Barath, F. T., Johnston, E. J., and Waters, J. W. 1977. Microwave spectroscopic imagery of Earth. *Science* **197**(4307): 991–993.

Stephens, G. L., Campbell, G. G., and Vonder Haar, T. H. 1981. Earth radiation budgets. *J. Geophys. Research* **86**(C10): 9739–9760.

Stewart, R. H. 1975. Synthetic-aperture radars: mechanisms producing ocean wave images. *Proc. Annual Meeting U.S. National Committee of the International Union of Radio Science*, Boulder, Co.: 209.

———. 1980. Ocean wave measurement techniques. In: *Air sea interaction, instruments and methods*, Edited by F. Dobson, L. Hasse, and R. Davis, 447–470. New York: Plenum Press.

———. 1982. Satellite oceanography: The instruments. *Oceanus* **24**(3): 66–74.

Stewart, R. H., and Barnum, J. R. 1975. Radio measurements of oceanic winds at long ranges: an evaluation. *Radio Science* **10**(10): 853–857.

Stewart, R. H., and Joy, J. W. 1974. HF radio measurements of surface currents. *Deep-Sea Research* **21**(12): 1039–1049.

Stewart, R. H., and Teague, C. C. 1980. Dekameter radar observations of ocean wave growth and decay. *J. Phys. Oceanog.* **10**(1): 128–143.

Stogryn, A. 1967. The apparent temperature of the sea at microwave frequencies. *IEEE Trans. Antennas and Propag.* **AP-15**: 278–286.

Stommel, H. 1965. *The Gulf Stream, a physical and dynamical description.* Berkeley and Los Angeles: University of California Press, 248 pp.

Strong, A. E., and McClain, E. P. 1984. Improved ocean surface temperatures from space-comparisons with drifting buoys. *Bull. Am. Meteorol. Soc.* **65**(2): 138–142.

Strong, A. E., and Pritchard, J. A. 1980. Regular monthly mean temperatures of Earth's oceans from satellites. *Bull. Am. Meteorol. Soc.* **61**(6): 553–559.

Svendsen, E., Kloster, K., Farrelly, B., Johannessen, O. M., Johannessen, J. A., Campbell, W. J., Gloersen, P., Cavalieri, D., and Matzler, C. 1983. Norwegian remote sensing experiment: Evaluation of the Nimbus-7 Scanning Multichannel Microwave Radiometer for sea ice research. *J. Geophys. Research* **88**(C3): 2781–2791.

Swithinbank, C., McClain, P., and Little, P. 1977. Drift tracks of Antarctic icebergs. *Polar Record* **18**(116): 495–501.

Tai, C. K., and Wunsch, C. 1983. Absolute measurement of the dynamic topography of the Pacific Ocean by satellite altimetry. *Nature* **301** (5899): 408–410.

Teague, C. C., Tyler, G. L., and Stewart, R. H. 1975. The radar cross section of the sea at 1.95 MHz: Comparison of in-situ and radar determinations. *Radio Science* **10** (10): 847–852.

Thekakara, M. P. 1972. Evaluating the light from the sun. *Optical Spectra* **6** (3): 32–35.

————. 1974. Extra terrestrial solar spectrum, 3000Å–6100Å at 1-Å intervals. *Appl. Optics* **13** (3): 518–522.

Thompson, T. W., Weissman, D. E., and Gonzalez, F. I. 1983. L band radar back-scatter dependence upon surface wind stress: A summary of new Seasat-1 and aircraft observations. *J. Geophys. Research* **88** (C3): 1727–1735.

Thrane, L. 1978. Evaluation of multi-frequency-microwave-radiometer-system perfor-mance for oceanography. *Boundary-Layer Meteorology* **13**: 373–392.

Tomiyasu, K. 1981. Conceptual performance of a satellite borne, wide swath synthetic aperture radar. *IEEE Trans. Geoscience and Remote Sensing* **GE19** (2): 108–116.

Townsend, W. F., McGoogan, J. T., and Walsh, E. J. 1981. Satellite radar altimeters — present and future oceanographic capabilities. In: *Oceanography from space*, edited by J. F. R. Gower, 625–636. New York: Plenum Press.

Troy, B. E., and Hollinger, J. P. 1977. *The measurement of oil spill volume by a passive microwave imager.* Washington: Naval Research Laboratory Memorandum 3515.

Troy, B. E., Hollinger, J. P., Lerner, R. M., and Wisler, M. M. 1981. Measurements of the microwave properties of sea ice at 90 GHz and lower frequencies. *J. Geophys. Research* **86** (C5): 4283–4289.

Tyler, G. L., Faulkerson, W. E., Peterson, A. M., and Teague, C. C. 1972. Second-order scattering from the sea: ten meter radar observations of the Doppler contin-uum. *Science* **177** (4046): 349–351.

Tyler, G. L., Teague, C. C., Stewart, R. H., Peterson, A. M., Munk, W. H., and Joy, J. W. 1974. Wave directional spectra from synthetic aperture observations of radio scatter. *Deep-Sea Research* **21** (12): 989–1016.

Ulbrich, C. W., and Atlas, D. 1977. A method for measuring precipitation parameters and raindrop size distributions using radar reflectivity and optical extinction. In: *Proc. URSI Symposium on Propagation in Non-Ionizing Media*, La Baule, France.

————. 1978. The rain parameter diagram: methods and applications. *J. Geophys. Research* **83** (C3): 1319–1325.

U.S. Navy Marine Climatic Atlas of the World. Washington: U.S. Government Printing Office.

Valenzuela, G. R. 1978. Theories for the interaction of electromagnetic and oceanic waves — a review. *Boundary-Layer Meteorology* **13**: 61–85.

————. 1980. An asymptotic formulation for SAR images of the dynamical ocean sur-face. *Radio Science* **15**: 105–114.

Valenzuela, G. R., Laing, M. B., and Daley, J. C. 1971. Ocean spectra for the high-frequency waves as determined from airborne radar measurements. *J. Marine Research* **29** (2): 69–84.

Van de Hulst, J. C. 1957. *Light scattering by small particles.* New York: John Wiley & Sons.

Van Kuilenburg, J. 1975. Radar observations of controlled oil spills. In: *Proc. 10th Int. Symp. of Remote Sensing of Environment*, 243–250. Ann Arbor: Environmental Research Institute of Michigan.

Van Vleck, J. H. 1947a. The absorption of microwaves by oxygen. *Phys. Rev.* **71**: 413–424.

———. 1947b. The absorption of microwaves by uncondensed water vapor. *Phys. Rev.* **71**: 425–433.

Vesecky, J. F., Assal, H. M., and Stewart, R. H. 1981. Remote sensing of ocean waveheight spectrum using synthetic-aperture-radar images. In: *Oceanography from space,* edited by J. F. R. Gower, 449–457. New York: Plenum Press.

Vesecky, J. F., and Stewart, R. H. 1982. The observation of ocean surface phenomena using imagery from the Seasat synthetic aperture radar: An assessment. *J. Geophys. Research* **87** (C3): 3397–3430.

Vesecky, J. F., Stewart, R. H., Schuchman, R. A., Assal, H. M., Kasischke, E. S., and Lyden, J. D. 1983. On the ability of synthetic aperture radar to measure ocean waves. In: *Proc. IUCRM Symposium on Wave Dynamics and Radio Probing of the Ocean Surface,* edited by O. M. Phillips.

Villier, M., Tanre, D., and Deschamps, P. Y. 1980. An algorithm for remote sensing of water color from space. *Boundary-Layer Meteorology* **18**: 247–267.

Vonbun, F. O. 1977. Goddard laser systems and their accuracies. *Phil. Trans. Roy. Soc. London* **A284**: 443–450.

Vonder Haar, T. H., and Oort, A. H. 1973. New estimate of annual poleward energy transport by northern hemisphere oceans. *J. Phys. Oceanog.* **3** (2): 169–172.

Wagner, C. A. 1979. The geoid spectrum from altimetry. *J. Geophys. Research* **84** (B8): 3861–3871.

Wagner, C. A., Lerch, F. J., Brownd, J. F., and Richardson, J. A. 1977. Improvement in the geopotential derived from satellite and surface data (GEM 7 and 8). *J. Geophys. Research* **82** (5): 901–914.

Walsh, E. J. 1977. Problems inherent in using aircraft for radio oceanographic studies. *IEEE Trans. Antennas and Propagation* **AP-25** (1): 145–149.

———. 1982. Pulse-to-pulse correlation in satellite radar altimeters (unpublished manuscript).

Walsh, E. J., and Kenney, J. E. 1982. Surface contour radar measurement of altimeter mean sea level EM bias at 36 GHz (unpublished manuscript).

Walsh, E. J., Uliana, E. A., and Yaplee, B. S. 1978. Ocean waves heights measured by a high resolution pulse-limited radar. *Boundary-Layer Meteorology* **13**: 263–276.

Warnecke, G., Allison, L. J., McMillin, L. M., and Szekields, K. H. 1971. Remote sensing of ocean currents and sea surface temperature changes derived from the Nimbus II satellite. *J. Phys. Oceanog.* **1** (1): 45–60.

Warren, B., and Wunsch, C., eds. 1981. *Evolution of physical oceanography.* Cambridge: The MIT Press, 623 pp.

Waters, J. W. 1976. Absorption and emission by atmospheric gases. In: *Methods of experimental physics,* Vol. 12B, *Astrophysics,* edited by M. L. Meeks, 142–176. New York: Academic Press.

Watts, A. B. 1979. On geoid heights derived from Geos-3 altimeter data along the Hawaiian-Emperor Seamount chain. *J. Geophys. Research* **84** (B8): 3817–3826.

Webb, D. J. 1981. A comparison of Seasat 1 altimeter measurements of wave height with measurements made by a pitch-roll buoy. *J. Geophys. Research* **86** (C7): 6394–6398.

Webster, W. J., Wilheit, T. T., Ross, D. B., and Gloersen, P. 1976. Spectral characteristics of the microwave emission from a wind-driven foam-covered sea. *J. Geophys. Research* **81** (18): 3095–3099.

Weissman, D. E. 1973. Two frequency radar interferometry applied to the measurement of ocean wave height. *IEEE Trans. Antennas and Propag.* **AP-21** (5): 649–656.

Wentz, F. J. 1975. A two-scale model for foam-free sea microwave brightness temperatures. *J. Geophys. Research* **80** (24): 3441–3446.

Wentz, F. J., Cardone, V. S., and Fedor, L. S. 1982. Intercomparisons of wind speeds inferred by the SASS, Altimeter, and SMAR. *J. Geophys. Research* **87** (C5): 3378–3384.

Wentz, F. J., Christensen, E. J., and Richardson, K. A. 1981. Dependence of sea-surface microwave emissivity on friction velocity as derived from SMMR/SASS. In: *Oceanography from Space*, edited by J. F. R. Gower, New York: Plenum Press.

Werbowetzki, A., ed. 1981. *Atmospheric sounding user's guide*. Washington: NOAA Technical Report NESS 83, 82 pp.

West, G. B. 1982. Mean Earth ellipsoid determined from Seasat 1 altimetric observations. *J. Geophys. Research* **87** (B7): 5538–5540.

Widger, W. K. 1966. *Meteorological satellites*. New York: Holt, Rinehart and Winston.

Wilheit, T. T. 1972. The electrically scanning microwave radiometer (ESMR) experiment. In: *The Nimbus-5 user's guide*, edited by R. R. Sabatini, 59–105. Greenbelt: NASA Goddard Space Flight Center.

———. 1978. A review of applications of microwave radiometry to oceanography. *Boundary-Layer Meteorology* **13**: 277–293.

———. 1979a. The effect of wind on the microwave emission from the ocean's surface at 37GHz. *J. Geophys. Research* **84** (18): 4921–4926.

———. 1979b. A model for the microwave emissivity of the ocean's surface as a function of wind speed. *IEEE Trans. Geoscience Electronics* **OE-17** (4): 244–249.

Wilheit, T. T., Chang, A. T. C., and Milman, A. S. 1980. Atmospheric corrections to passive microwave observations of the ocean. *Boundary-Layer Meteorology* **18**: 65–77.

Wilheit, T. T., Chang, A. T. C., Rao, M. S. V., Rodgers, E. B., and Theon, J. S. 1978. A satellite technique for quantitatively mapping rainfall rates over the oceans. *J. Appl. Meteorol.* **16** (5): 551–560.

Wilkerson, J. C., Brown, R. A., Cardone, V. J., Coons, R. E., Loomis, A. A., Overland, J. E., Peteherych, S., Pierson, W. J., Woiceshyn, P. M., and Wurtele, M. G. 1979. Surface observations for the evaluation of geophysical measurements from Seasat. *Science* **209** (4400): 1408–1410.

Williams, G. F. 1971. Microwave emissivity measurements of bubbles and foam. *IEEE Trans. Geoscience Electronics* **GE-9**: 221–224.

Wilson, W. H., and Austin, R. W. 1978. Remote sensing of ocean color. *Proc. Soc. Photo-Optical Instrumentation Engineers* **160**: 23–30.

Wilson, W. H., Austin, R. W., and Smith, R. S. 1978. Optical remote sensing of chlorophyll in ocean waters. *Proc. 12th International Symposium on Remote Sensing of Environment*. Ann Arbor: Environmental Research Institute of Michigan, 1103–1113.

Wolfe, W. L. 1965. *Handbook of military infrared technology*. Washington: U.S. Government Printing Office.

Wright, J. W. 1966. Backscattering from capillary waves with application to sea clutter. *IEEE Trans. Antennas and Propag.* **AP-14** (3): 749–754.

———. 1968. A new model for sea clutter. *IEEE Trans. Antennas and Propag.* **AP-16**: 217–223.

Wright, J. W., and Keller, W. C. 1971. Doppler spectra in microwave scattering from wind waves. *Phys. Fluids* **14** (3): 466–474.

Wright, J. W., Plant, W. J., Keller, W. C., and Jones, W. L. 1980. Ocean wave-radar modulation transfer functions from the west coast experiment. *J. Geophys. Research* **85** (C9): 4957–4966.

Wu, C., Barkar, B., Huneycutt, B., Leang, C., and Pang, S. 1981. *An introduction to the* interim digital SAR processor and the characteristics of the associated Seasat SAR imagery. Pasadena: Jet Propulsion Laboratory Publication 81-26, 82 + 32 pp.

Wu, J. 1980. Wind-stress coefficients over sea surface near neutral conditions — a revisit. *J. Phys. Oceanog.* **10** (5): 727–740.

Wu, S. C. 1979. Optimum frequencies of a passive microwave radiometer for tropospheric path-length correction. *IEEE Trans. Antennas and Propag.* **AP-27** (2): 233–239.

Wunsch, C., and Gaposchkin, E. M. 1980. On using satellite altimetry to determine the general circulation of the oceans with application to geoid improvement. *Reviews of Geophysics and Space Physics* **18** (4): 725–745.

Wurtele, M. G., Woiceshyn, P. M., Peteherych, S., Borowski, M., and Appleby, W. S. 1982. Wind direction alias removal studies of Seasat scatterometer-derived winds. *J. Geophys. Research* **87** (C5): 3365–3377.

Wylie, D. P., and Hinton, B. B. 1982a. A comparison of cloud motion and ship wind observations over the Indian Ocean for the year of FGGE. *Boundary-Layer Meteorology* **23** (2): 197–208.

———. 1982b. The wind stress patterns over the Indian Ocean during the summer monsoon of 1979. *J. Phys. Oceanog.* **12** (2): 186–199.

Wyrtki, K. 1979. Sea level variations: monitoring the breath of the Pacific. *EOS Trans. Am. Geophys. Union* **60**: 25–27.

Wyrtki, K., Magaard, L., and Hager, J. 1976. Eddy energy in the oceans. *J. Geophys. Research* **81**: 2641–2646.

Yaplee, B. S., Shapiro, A., Hammond, D. L., Au, B. D., and Uliana, E. A. 1971. Nanosecond radar observations of the ocean surface from a stable platform. *IEEE Trans. Geoscience Electronics* **GE-9**: 171–174.

Yoder, C. F., Williams, J. G., Dickey, J. O., Schutz, B. E., Eanes, R. J., and Tapley, B. D. 1983. Secular variation of Earth's gravitational harmonic J_2 coefficient from Lageos and nontidal acceleration of Earth's rotation. *Nature* **303** (5920): 757–762.

Young, A. T. 1981. On the Rayleigh-scattering optical depth of the atmosphere. *J. Appl. Meteorol.* **20** (3): 328–330.

Young, J. D., and Moore, R. K. 1977. Active microwave measurements from space of sea-surface winds. *IEEE J. Oceanic Engineering* **OE-2** (4): 309–317.

Zuev, V. E. 1970. *Atmospheric transparency in the visible and the infrared.* Moscow: Izdatel "Sovetskoe Radio," translated by Z. Lerman. Jerusalem: Israel Program for Scientific Translations.

———. 1982. *Laser beams in the atmosphere*, translated by J. S. Wood. New York: Consultants Bureau, 504 pp.

Zwally, H. J. 1977. Microwave emissivity and accumulation rate of polar firn. *J. Glaciology* **18** (79): 195–215.

Zwally, H. J., Bindschadler, R. A., Brenner, A. C., Martin, T. V., and Thomas, R. H. 1983. Surface elevation contours of Greenland and Antarctic ice sheets. *J. Geophys. Research* **88** (C3): 1589–1596.

Zwally, H. J., and Gloersen, P. 1977. Passive microwave images of the polar regions and research applications. *Polar Record* **18** (116): 431–450.

Zwally, H. J., Wilheit, T. T., Gloersen, P., and Mueller, J. L. 1976. Characteristics of Antarctic sea ice as determined by satellite-borne microwave imagers. In: *Proc. Symp. Meteorological Observations from Space; Their Contributions to the First GARP Experiment*, 94–97. Boulder: National Center for Atmospheric Research.

INDEX

[353]

Antenna (*continued*)
 stick, 219
 synthesized, 225–226
 synthetic-aperture radar, 228–230
 with tapered illumination, 156
 temperature, 159, 160
 uniformly illuminated aperture, 156
ARGOS, 311
Aries, 289, 293
Atmospheric
 aerosols, 140
 correction problem, 121, 125
 density at satellite height, 297
 errors influencing infrared radiometers, 134–135
 fronts, 254
 stability, 12–13, 16, 214, 222, 253
 water vapor, 61, 171–173, 269–270
 windows, 60, 63, 128, 133, 135
Attenuance. *See* Attenuation
Attenuation, 30
 atmospheric, 76–79, 120–126
 coefficient, 30, 53, 73, 77
 coefficient for irradiance, 114
 cross section, 66
 depth, 114
 of radar signal due to clouds and rain, 80, 221
 due to particles, 66–71
 due to water vapor in atmosphere, 59
Azimuth travelling waves, 239, 242–244

Bathymetry
 observed by altimeters, 274, 276
 observed by synthetic-aperture radar, 256
Bathythermographs, 27, 145–146, 283–284
Bioluminescence, 103
Bio-optical algorithm, 120
Blackbodies, 40–42, 86, 94, 132, 158
Boltzmann's constant, 41
Bouguet's Law, 56
Boundary layer
 atmospheric, 10–17
  logarithmic profile, 13, 16
  velocity profile, 13
 oceanic, 10, 24–25, 142–143
 stability of, 12–13, 24, 214, 222, 253
Brightness temperature, 42, 168
 of the atmosphere, 166–167
 and attenuation of radar signals, 221
 of rain, 177, 221
 of the sea surface, 98–99, 168
 of sea water, 91
 of the solar corona, 44
 of the sun, 166
  at 10.7cm radio wavelengths, 297
Brunt-Väisälä frequency, 26
Bulk coefficients, 12
Buoyancy frequency, 26
Bureau International de l'Heure, 292

Centimeter-atmospheres, 50
Chlorophyll
 errors in measurements of, 124
 images of, 111, 125, 149
 measurements from space, 114–127
 measurements from ships, 127
 relationship to temperature, 151
Climate studies, 2, 150
Cloud detection algorithms, 136–142
Clouds
 cumulus, 134
 high cirrus, 134
 images of, 109–113
 influence on infrared observations, 134–142
 stratus, 134
Coastal Zone Color Scanner, 101–102, 125–127, 318
Collision frequency, 51
Columnar value
 of free electrons in the ionosphere, 268
 of water vapor in the troposphere, 171–173, 269
Conductivity, 87
 sea water, 87
Contrast, 78, 135
 transmittance, 79, 81
Coordinate frame, 292
Coordinate systems, 261, 267–268, 289–293
 accuracy of 293
 astrophotogrammetric, 267, 292
 celestial, 291
 extragalactic, 292
 geocentric, 289, 292
 geodetic, 261, 267, 292
 inertial, 267
 for tracking satellites, 292
Coriolis force, 25, 279, 281
Cosmic background, 96, 166
Currents
 boundary, 281
 boundaries of, 253–255
 Ekman, 25
 geostrophic, 25, 260, 262, 278–280
  measured by altimetry, 278–285
 ocean surface, 24–27
  measured by dekameter radar, 206
 permanent (time-averaged), 281–282, 284
 quasigeostrophic, 26, 281
 variable, 282–284

Data collection systems, 327–328
 ARGOS, 327–328
 Data Collection System on Goes, 327–328
Debye equation, 87, 90–92
Detected signal, 197
Dielectric constant, 29, 87–95
 perfect conductor, 186
 pure water, 89
 sea water, 87–92
 static, 87
Diffraction, 31
Digital data
 from infrared radiometers, 136
 from synthetic-aperture radars, 233